Elementary Linear Algebra

Elementary Linear Algebra

SECOND EDITION

HOWARD ANTON

DREXEL UNIVERSITY

JOHN WILEY & SONS New York Santa Barbara London Sydney Toronto

The book was printed and bound by Quinn & Boden.
It was set in Times New Roman by Syntax International.
The drawings were designed and executed by John Balbalis with the assistance of the Wiley Illustration Department.
Patricia Lawson supervised production.

Library of Congress Cataloging in Publication Data:

Anton, Howard.
Elementary linear algebra.

Includes index.
1. Algebras, Linear. I. Title.
QA184.A57 1977 512'.5 76-25492
ISBN 0-471-03244-1

Printed in the United States of America

10 9 8 7 6 5 4 3 2 1

To Pat and my children: Brian, Davy and Lauren

Preface

This textbook provides an elementary treatment of linear algebra that is suitable for students in their freshman or sophomore year. Calculus is *not* a prerequisite. I have, however, included a number of exercises for students with a calculus background; these are clearly marked: "For students who have studied calculus."

My aim in writing this book is to present the fundamentals of linear algebra in the clearest possible way. Pedagogy is the main consideration. Formalism is secondary. Where possible, basic ideas are studied by means of computational examples (over 200 of them) and geometrical interpretation.

My treatment of proofs varies. Those proofs that are elementary and have significant pedagogical content are presented precisely, in a style tailored for beginners. A few proofs that are more difficult, but pedagogically valuable, are placed at the end of the section and marked "Optional." Still other proofs are omitted completely, with emphasis placed on applying the theorem. Whenever a proof is omitted, I try to motivate the result, often with a discussion about its interpretation in 2-space or 3-space.

It is my experience that Σ-notation is more of a hindrance than a help for beginners in linear algebra. Therefore, I have generally avoided its use.

It is a pedagogical axiom that a teacher should proceed from the familiar to the unfamiliar and from the concrete to the abstract. The ordering of the chapters reflects my adherence to this tenet.

Chapter 1 deals with systems of linear equations, how to solve them, and some of their properties. It also contains the basic material on matrices and their arithmetic properties.

Chapter 2 deals with determinants. I have used the classical permutation approach. In my opinion, it is less abstract than the approach through *n*-linear alternating forms and gives the student a better intuitive grasp of the subject than does an inductive development.

Chapter 3 introduces vectors in 2-space and 3-space as arrows and develops the analytic geometry of lines and planes in 3-space. Depending on the background of the students, this chapter can be omitted without a loss of continuity (see the guide for the instructor that follows this preface).

Chapter 4 and **Chapter 5** develop the basic results about real finite dimensional vector spaces and linear transformations. I begin with a study of R^n and proceed slowly to the general concept of a vector.

Chapter 6 deals with the eigenvalue problem and diagonalization.

Chapter 7 gives some applications of linear algebra to problems of approximation, systems of differential equations, Fourier series, and identifying conic sections and quadric surfaces. Applications to business, biology, engineering, economics, the social sciences, and the physical sciences are included in an optional paperback supplement to this text, *Applications of Linear Algebra*, by Chris Rorres and the author. Included in the supplement are such topics as: curves of best fit to empirical data, population dynamics, Markov processes, optimal harvesting, Leontief models in economics, and engineering applications.

Chapter 8 introduces numerical methods of linear algebra; it does not require access to computing facilities since the exercises can be solved by hand computation or with the use of a pocket calculator. This chapter gives the student a basic understanding of how certain linear algebra problems are solved practically. Too many students complete their linear algebra studies with the naive belief that eigenvalues are found in practice by solving the characteristic equation. Some instructors may wish to use this section in conjunction with a programming course.

I have included a large number of exercises. Each exercise set begins with routine "drill" problems and progresses toward more theoretical problems. Answers to all the computational problems are given at the end of the text.

Since there is more material in this book than can be covered in a one-semester or one-quarter course, the instructor will have to make a selection of topics. To help in this selection, I have provided a guide to the instructor, which follows this preface.

WHAT'S NEW IN THE SECOND EDITION

The wide acceptance of the first edition has been most gratifying and the author is grateful for the many favorable comments and constructive suggestions received from users. In response to these suggestions I have made the following changes:

- The pace of Chapter 1 has been increased slightly (eight sections reduced to seven) to allow added time for more difficult material later in the text.
- A few of the more difficult sections in Chapters 4 and 5 have been rewritten and reorganized to make the ideas clearer.
- A chapter on applications has been added, and an optional paperback supplement with hosts of additional applications has been prepared.
- Some new problems have been added to selected exercise sets.

A Guide for the Instructor

Chapter 3 can be omitted without loss of continuity if the students have previously studied about lines, planes, and geometric vectors in 2-space and 3-space. Depending on the available time and the background of the students, the instructor may wish to add all or part of this chapter to the following suggested core material:

Chapter 1	7 lectures
Chapter 2	5 lectures
Chapter 4	14 lectures
Chapter 5	6 lectures
Chapter 6	5 lectures

This schedule is rather liberal; it allows a fair amount of classroom time for discussion of homework problems, but assumes that little classroom time is devoted to the material marked "optional." The instructor can build on this core as time permits by including lectures on optional material, Chapter 3, Chapter 7, and Chapter 8.

For instructors who want added time to discuss applications or numerical methods, it is possible to delete Sections 5.3 and 5.4 from the core material. If this is done the instructor should omit the optional material at the end of 6.1, begin Section 6.2 with the matrix forms of problems 1 and 2, and delete Example 9 in that section.

Acknowledgments

I express my appreciation for the helpful guidance provided by the reviewers: Professor Joseph Buckley of Western Michigan University, Professor Harold S. Engelsohn of Kingsborough Community College, Professor Lawrence D. Kugler of the University of Michigan, Professor Robert W. Negus of Rio Hondo Junior College, Professor Hal G. Moore of Brigham Young University, Professor William A. Brown of the University of Maine at Portland-Gorham, and Professor Ralph P. Grimaldi of Rose-Hulman Institute of Technology.

I am also grateful to Professor William F. Trench of Drexel University, who read the entire manuscript; his suggestions immeasurably improved both the style and content of the text. For their accurate and skillful typing, I thank Susan R. Gershuni and Kathleen R. McCabe; thanks are also due Professor Dale Lick who encouraged my work in its early stages, to Professor Robert Higgins for assistance in solving the problems, and to Frederick C. Corey of John Wiley who helped make the first edition a reality. Finally, I am indebted to the entire production staff of Wiley and especially to my editor Gary Ostedt for his innovative contributions to this new edition.

Howard Anton

Contents

* Additional applications to business, economics and the physical and social sciences are available in a supplement to this text.

8

INTRODUCTION TO NUMERICAL METHODS OF LINEAR ALGEBRA 291

1 Systems of Linear Equations and Matrices

1.1 INTRODUCTION TO SYSTEMS OF LINEAR EQUATIONS

In this section we introduce basic terminology and discuss a method for solving systems of linear equations.

A line in the xy-plane can be represented algebraically by an equation of the form

$$a_1x + a_2y = b$$

An equation of this kind is called a linear equation in the variables x and y. More generally, we define a ***linear equation*** in the n variables $x_1, x_2, \ldots, x_n$ to be one that can be expressed in the form

$$a_1x_1 + a_2x_2 + \cdots + a_nx_n = b$$

where $a_1, a_2, \ldots, a_n$ and b are real constants.

Example 1

The following are linear equations:

$$x + 3y = 7 \qquad x_1 - 2x_2 - 3x_3 + x_4 = 7$$

$$y = \tfrac{1}{2}x + 3z + 1 \qquad x_1 + x_2 + \cdots + x_n = 1$$

Observe that a linear equation does not involve any products or roots of variables. All variables occur only to the first power and do not appear as arguments for trigonometric, logarithmic, or exponential functions. Each of the following *fails* to be a linear equation:

$$x + 3y^2 = 7 \qquad 3x + 2y - z + xz = 4$$

$$y - \sin x = 0 \qquad \sqrt{x_1} + 2x_2 + x_3 = 1$$

A ***solution*** of a linear equation $a_1x_1 + a_2x_2 + \cdots + a_nx_n = b$ is a sequence of n numbers $s_1, s_2, \ldots, s_n$ such that the equation is satisfied when we substitute $x_1 = s_1, x_2 = s_2, \ldots, x_n = s_n$. The set of all solutions of the equation is called its ***solution set***.

Example 2

Find the solution set of each of the following:

(i) $4x - 2y = 1$ (ii) $x_1 - 4x_2 + 7x_3 = 5$

To find solutions of (i), we can assign an arbitrary value to x and solve for y, or choose an arbitrary value for y and solve for x. If we follow the first approach and assign x an arbitrary value t, we obtain

$$x = t, \qquad y = 2t - \tfrac{1}{2}$$

These formulas describe the solution set in terms of the arbitrary parameter t. Particular numerical solutions can be obtained by substituting specific values for t. For example, $t = 3$ yields the solution $x = 3$, $y = 11/2$ and $t = -1/2$ yields the solution $x = -1/2$, $y = -3/2$.

If we follow the second approach and assign y the arbitrary value t, we obtain

$$x = \tfrac{1}{2}t + \tfrac{1}{4}, \qquad y = t$$

Although these formulas are different from those obtained above, they yield the same solution set as t varies over all possible real numbers. For example, the previous formulas gave the solution $x = 3$, $y = 11/2$ when $t = 3$, while these formulas yield this solution when $t = 11/2$.

To find the solution set of (ii) we can assign arbitrary values to any two variables and solve for the third variable. In particular, if we assign arbitrary values s and t to x_2 and x_3 respectively and solve for x_1, we obtain

$$x_1 = 5 + 4s - 7t, \qquad x_2 = s, \qquad x_3 = t$$

A finite set of linear equations in the variables $x_1, x_2, \ldots, x_n$ is called a ***system of linear equations***. A sequence of numbers $s_1, s_2, \ldots, s_n$ is called a ***solution*** of the system if $x_1 = s_1, x_2 = s_2, \ldots, x_n = s_n$ is a solution of every equation in the system. For example, the system of two equations in three unknowns

$$\begin{aligned} 4x_1 - x_2 + 3x_3 &= -1 \\ 3x_1 + x_2 + 9x_3 &= -4 \end{aligned}$$

has the solution $x_1 = 1, x_2 = 2, x_3 = -1$ since these values satisfy both equations. However, $x_1 = 1$, $x_2 = 8$, $x_3 = 1$ is not a solution since these values satisfy only the first of the two equations in the system.

Not all systems of linear equations have solutions. For example, if we multiply the second equation of the system

$$\begin{aligned} x + y &= 4 \\ 2x + 2y &= 6 \end{aligned}$$

by 1/2, it becomes evident that there is no solution, since the two equations in the resulting system

$$\begin{aligned} x + y &= 4 \\ x + y &= 3 \end{aligned}$$

contradict each other.

A system of equations that has no solution is said to be ***inconsistent***. If there is at least one solution, it is called ***consistent***. To illustrate the possibilities that can occur in solving systems of linear equations, consider a general system of two linear equations in the unknowns x and y:

$$\begin{aligned} a_1x + b_1y &= c_1 \qquad (a_1, b_1 \text{ both not zero}) \\ a_2x + b_2y &= c_2 \qquad (a_2, b_2 \text{ both not zero}) \end{aligned}$$

The graphs of these equations are lines; call them l_1 and l_2. Since a point (x,y) lies on a line if and only if the numbers x and y satisfy the equation of the line, the solutions of the system of equations will correspond to points of intersection of l_1 and l_2. There are three possibilities (Figure 1.1).

(a) The lines l_1 and l_2 may be parallel, in which case there is no intersection, and consequently no solution to the system.
(b) The lines l_1 and l_2 may intersect at only one point, in which case the system has exactly one solution.
(c) The lines l_1 and l_2 may coincide, in which case there are infinitely many points of intersection, and consequently infinitely many solutions to the system.

Figure 1.1 (*a*) No solution. (*b*) One solution. (*c*) Infinitely many solutions.

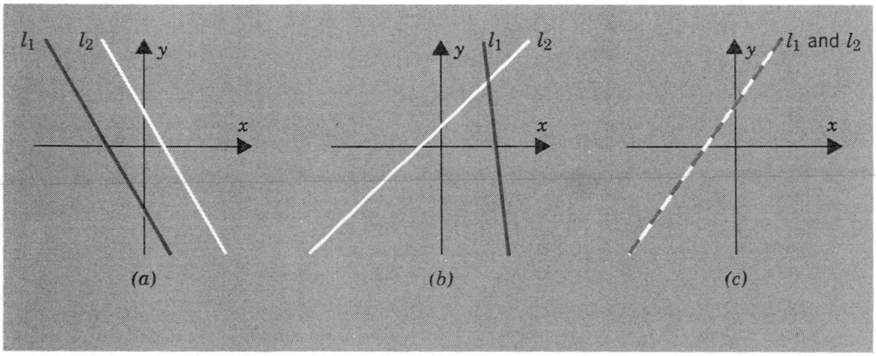

Although we have considered only two equations with two unknowns here, we will show later that this same result holds for arbitrary systems; that is, *every system of linear equations has either no solutions, exactly one solution, or infinitely many solutions.*

An arbitrary system of m linear equations in n unknowns will be written

$$\begin{aligned} a_{11}x_1 + a_{12}x_2 + \cdots + a_{1n}x_n &= b_1 \\ a_{21}x_1 + a_{22}x_2 + \cdots + a_{2n}x_n &= b_2 \\ \vdots \qquad \vdots \qquad\quad\; \vdots \quad &\quad \vdots \\ a_{m1}x_1 + a_{m2}x_2 + \cdots + a_{mn}x_n &= b_m \end{aligned}$$

where $x_1, x_2, \ldots, x_n$ are the unknowns and the subscripted a's and b's denote constants.

For example, a general system of three linear equations in four unknowns will be written

$$\begin{aligned} a_{11}x_1 + a_{12}x_2 + a_{13}x_3 + a_{14}x_4 &= b_1 \\ a_{21}x_1 + a_{22}x_2 + a_{23}x_3 + a_{24}x_4 &= b_2 \\ a_{31}x_1 + a_{32}x_2 + a_{33}x_3 + a_{34}x_4 &= b_3 \end{aligned}$$

The double subscripting on the coefficients of the unknowns is a useful device that we shall employ to establish the location of the coefficient in the system. The first subscript on the coefficient a_{ij} indicates the equation in which the coefficient occurs, and the second subscript indicates which unknown it multiplies. Thus, a_{12} is in the first equation and multiplies unknown x_2.

If we mentally keep track of the location of the +'s, the x's, and the ='s, a system of m linear equations in n unknowns can be abbreviated by writing only the rectangular array of numbers:

$$\begin{bmatrix} a_{11} & a_{12} & \cdots & a_{1n} & b_1 \\ a_{21} & a_{22} & \cdots & a_{2n} & b_2 \\ \vdots & \vdots & & \vdots & \vdots \\ a_{m1} & a_{m2} & \cdots & a_{mn} & b_m \end{bmatrix}$$

This is called the ***augmented matrix*** for the system. (The term *matrix* is used in mathematics to denote a rectangular array of numbers. Matrices arise in many contexts; we shall study them in more detail in later sections.) To illustrate, the augmented matrix for the system of equations

$$\begin{aligned} x_1 + x_2 + 2x_3 &= 9 \\ 2x_1 + 4x_2 - 3x_3 &= 1 \\ 3x_1 + 6x_2 - 5x_3 &= 0 \end{aligned}$$

is

$$\begin{bmatrix} 1 & 1 & 2 & 9 \\ 2 & 4 & -3 & 1 \\ 3 & 6 & -5 & 0 \end{bmatrix}$$

REMARK. When constructing an augmented matrix, the unknowns must be written in the same order in each equation.

The basic method for solving a system of linear equations is to replace the given system by a new system that has the same solution set, but is easier to solve. This new system is generally obtained in a series of steps by applying the following three types of operations to systematically eliminate unknowns.

1. Multiply an equation through by a nonzero constant.
2. Interchange two equations.
3. Add a multiple of one equation to another.

Since the rows (horizontal lines) in an augmented matrix correspond to the equations in the associated system, these three operations correspond to the following operations on the rows of the augmented matrix.

1. Multiply a row through by a nonzero constant.
2. Interchange two rows.
3. Add a multiple of one row to another row.

These are called ***elementary row operations***. The following example illustrates how these operations can be used to solve systems of linear equations. Since a systematic procedure for finding solutions will be derived in the next section, it is not necessary to worry about how the steps in this example were selected. The main effort at this time should be devoted to understanding the computations and the discussion.

Example 3

In the left column below we solve a system of linear equations by operating on the equations in the system, and in the right column we solve the same system by operating on the rows of the augmented matrix.

$$\begin{aligned} x + y + 2z &= 9 \\ 2x + 4y - 3z &= 1 \\ 3x + 6y - 5z &= 0 \end{aligned} \qquad \begin{bmatrix} 1 & 1 & 2 & 9 \\ 2 & 4 & -3 & 1 \\ 3 & 6 & -5 & 0 \end{bmatrix}$$

Add -2 times the first equation to the second to obtain

Add -2 times the first row to the second to obtain

$$\begin{aligned} x + y + 2z &= 9 \\ 2y - 7z &= -17 \\ 3x + 6y - 5z &= 0 \end{aligned} \qquad \begin{bmatrix} 1 & 1 & 2 & 9 \\ 0 & 2 & -7 & -17 \\ 3 & 6 & -5 & 0 \end{bmatrix}$$

Add -3 times the first equation to the third to obtain

$$\begin{aligned} x + y + 2z &= 9 \\ 2y - 7z &= -17 \\ 3y - 11z &= -27 \end{aligned}$$

Add -3 times the first row to the third to obtain

$$\begin{bmatrix} 1 & 1 & 2 & 9 \\ 0 & 2 & -7 & -17 \\ 0 & 3 & -11 & -27 \end{bmatrix}$$

Multiply the second equation by 1/2 to obtain

$$\begin{aligned} x + y + 2z &= 9 \\ y - \tfrac{7}{2}z &= -\tfrac{17}{2} \\ 3y - 11z &= -27 \end{aligned}$$

Multiply the second row by 1/2 to obtain

$$\begin{bmatrix} 1 & 1 & 2 & 9 \\ 0 & 1 & -\frac{7}{2} & -\frac{17}{2} \\ 0 & 3 & -11 & -27 \end{bmatrix}$$

Add -3 times the second equation to the third to obtain

$$\begin{aligned} x + y + 2z &= 9 \\ y - \tfrac{7}{2}z &= -\tfrac{17}{2} \\ -\tfrac{1}{2}z &= -\tfrac{3}{2} \end{aligned}$$

Add -3 times the second row to the third to obtain

$$\begin{bmatrix} 1 & 1 & 2 & 9 \\ 0 & 1 & -\frac{7}{2} & -\frac{17}{2} \\ 0 & 0 & -\frac{1}{2} & -\frac{3}{2} \end{bmatrix}$$

Multiply the third equation by -2 to obtain

$$\begin{aligned} x + y + 2z &= 9 \\ y - \tfrac{7}{2}z &= -\tfrac{17}{2} \\ z &= 3 \end{aligned}$$

Multiply the third row by -2 to obtain

$$\begin{bmatrix} 1 & 1 & 2 & 9 \\ 0 & 1 & -\frac{7}{2} & -\frac{17}{2} \\ 0 & 0 & 1 & 3 \end{bmatrix}$$

Add -1 times the second equation to the first to obtain

$$\begin{aligned} x \qquad + \tfrac{11}{2}z &= \tfrac{35}{2} \\ y - \tfrac{7}{2}z &= -\tfrac{17}{2} \\ z &= 3 \end{aligned}$$

Add -1 times the second row to the first to obtain

$$\begin{bmatrix} 1 & 0 & \frac{11}{2} & \frac{35}{2} \\ 0 & 1 & -\frac{7}{2} & -\frac{17}{2} \\ 0 & 0 & 1 & 3 \end{bmatrix}$$

Add $-\frac{11}{2}$ times the third equation to the first and $\frac{7}{2}$ times the third equation the second to obtain

$$\begin{aligned} x \qquad &= 1 \\ y \quad &= 2 \\ z &= 3 \end{aligned}$$

Add $-\frac{11}{2}$ times the third row to the first and $\frac{7}{2}$ times the third row to the second to obtain

$$\begin{bmatrix} 1 & 0 & 0 & 1 \\ 0 & 1 & 0 & 2 \\ 0 & 0 & 1 & 3 \end{bmatrix}$$

The solution

$$x = 1, \qquad y = 2, \qquad z = 3$$

is now evident.

EXERCISE SET 1.1

1. Which of the following are linear equations in x_1, x_2, and x_3?

(a) $x_1 + 2x_1x_2 + x_3 = 2$ (b) $x_1 + x_2 + x_3 = \sin k$ (k is a constant)
(c) $x_1 - 3x_2 + 2x_3^{1/2} = 4$ (d) $x_1 = \sqrt{2}x_3 - x_2 + 7$
(e) $x_1 + x_2^{-1} - 3x_3 = 5$ (f) $x_1 = x_3$

2. Find the solution set for:

(a) $6x - 7y = 3$ (b) $2x_1 + 4x_2 - 7x_3 = 8$
(c) $-3x_1 + 4x_2 - 7x_3 + 8x_4 = 5$ (d) $2v - w + 3x + y - 4z = 0$

3. Find the augmented matrix for each of the following systems of linear equations.

(a)
$$\begin{aligned} x_1 - 2x_2 &= 0 \\ 3x_1 + 4x_2 &= -1 \\ 2x_1 - x_2 &= 3 \end{aligned}$$

(b)
$$\begin{aligned} x_1 \qquad\quad + x_3 &= 1 \\ -x_1 + 2x_2 - x_3 &= 3 \end{aligned}$$

(c)
$$\begin{aligned} x_1 \qquad + x_3 \qquad\qquad &= 1 \\ 2x_2 - x_3 \qquad + x_5 &= 2 \\ 2x_3 + x_4 \qquad &= 3 \end{aligned}$$

(d)
$$\begin{aligned} x_1 \qquad &= 1 \\ x_2 &= 2 \end{aligned}$$

4. Find a system of linear equations corresponding to each of the following augmented matrices.

(a) $\begin{bmatrix} 1 & 0 & -1 & 2 \\ 2 & 1 & 1 & 3 \\ 0 & 1 & 2 & 4 \end{bmatrix}$ (b) $\begin{bmatrix} 1 & 0 & 0 \\ 0 & 1 & 0 \\ 1 & -1 & 1 \end{bmatrix}$

(c) $\begin{bmatrix} 1 & 2 & 3 & 4 & 5 \\ 5 & 4 & 3 & 2 & 1 \end{bmatrix}$ (d) $\begin{bmatrix} 1 & 0 & 0 & 0 & 1 \\ 0 & 1 & 0 & 0 & 2 \\ 0 & 0 & 1 & 0 & 3 \\ 0 & 0 & 0 & 1 & 4 \end{bmatrix}$

5. For which value(s) of the constant k does the following system of linear equations have no solutions? Exactly one solution? Infinitely many solutions?

$$\begin{aligned} x - y &= 3 \\ 2x - 2y &= k \end{aligned}$$

6. Consider the system of equations

$$\begin{aligned} ax + by &= k \\ cx + dy &= l \\ ex + fy &= m \end{aligned}$$

Discuss the relative positions of the lines $ax + by = k$, $cx + dy = l$, and $ex + fy = m$ when:

(a) the system has no solutions
(b) the system has exactly one solution
(c) the system has infinitely many solutions

7. Show that if the system of equations in Exercise 6 is consistent, then at least one equation can be discarded from the system without altering the solution set.

8. Let $k = l = m = 0$ in Exercise 6; show that the system must be consistent. What can be said about the point of intersection of the three lines if the system has exactly one solution?

9. Consider the system of equations

$$\begin{aligned} x + y + 2z &= a \\ x \qquad + z &= b \\ 2x + y + 3z &= c \end{aligned}$$

Show that in order for this system to be consistent, a, b, and c must satisfy $c = a + b$.

10. Prove: If the linear equations $x_1 + kx_2 = c$ and $x_1 + lx_2 = d$ have the same solution set then the equations are identical.

1.2 GAUSSIAN ELIMINATION

In this section we give a systematic procedure for solving systems of linear equations; it is based on the idea of reducing the augmented matrix to a form that is simple enough so that the system of equations can be solved by inspection.

In the last step of Example 3 we obtained the augmented matrix

$$\begin{bmatrix} 1 & 0 & 0 & 1 \\ 0 & 1 & 0 & 2 \\ 0 & 0 & 1 & 3 \end{bmatrix} \tag{1.1}$$

from which the solution of the system was evident.

Matrix (1.1) is an example of a matrix that is in ***reduced row-echelon form***. To be of this form, a matrix must have the following properties.

1. *If a row does not consist entirely of zeros, then the first nonzero number in the row is a* 1. *(We call this a **leading** 1.)*
2. *If there are any rows that consist entirely of zeros, then they are grouped together at the bottom of the matrix.*
3. *In any two successive rows that do not consist entirely of zeros, the leading* 1 *in the lower row occurs farther to the right than the leading* 1 *in the higher row.*
4. *Each column that contains a leading* 1 *has zeros everywhere else.*

A matrix having properties **1**, **2**, and **3** is said to be in ***row-echelon form***.

Example 4

The following matrices are in reduced row-echelon form.

$$\begin{bmatrix} 1 & 0 & 0 & 4 \\ 0 & 1 & 0 & 7 \\ 0 & 0 & 1 & -1 \end{bmatrix}, \quad \begin{bmatrix} 1 & 0 & 0 \\ 0 & 1 & 0 \\ 0 & 0 & 1 \end{bmatrix}, \quad \begin{bmatrix} 0 & 1 & -2 & 0 & 1 \\ 0 & 0 & 0 & 1 & 3 \\ 0 & 0 & 0 & 0 & 0 \\ 0 & 0 & 0 & 0 & 0 \end{bmatrix}, \quad \begin{bmatrix} 0 & 0 \\ 0 & 0 \end{bmatrix}$$

The following matrices are in row-echelon form.

$$\begin{bmatrix} 1 & 4 & 3 & 7 \\ 0 & 1 & 6 & 2 \\ 0 & 0 & 1 & 5 \end{bmatrix}, \quad \begin{bmatrix} 1 & 1 & 0 \\ 0 & 1 & 0 \\ 0 & 0 & 0 \end{bmatrix}, \quad \begin{bmatrix} 0 & 1 & 2 & 6 & 0 \\ 0 & 0 & 1 & -1 & 0 \\ 0 & 0 & 0 & 0 & 1 \end{bmatrix}$$

The reader should check to see that each of the above matrices satisfies all the necessary requirements.

REMARK. It is not difficult to see that a matrix in row-echelon form must have zeros below each leading 1 (see Example 4). In contrast a matrix in reduced row-echelon form must have zeros above and below each leading 1.

If, by a sequence of elementary row operations, the augmented matrix for a system of linear equations is put in reduced row-echelon form, then the solution set for the system can be obtained by inspection or, at worst, after a few simple steps. The next example will illustrate this point.

Example 5

Suppose that the augmented matrix for a system of linear equations has been reduced by row operations to the given reduced row-echelon form. Solve the system.

(a) $\begin{bmatrix} 1 & 0 & 0 & 5 \\ 0 & 1 & 0 & -2 \\ 0 & 0 & 1 & 4 \end{bmatrix}$ (b) $\begin{bmatrix} 1 & 0 & 0 & 4 & -1 \\ 0 & 1 & 0 & 2 & 6 \\ 0 & 0 & 1 & 3 & 2 \end{bmatrix}$

(c) $\begin{bmatrix} 1 & 6 & 0 & 0 & 4 & -2 \\ 0 & 0 & 1 & 0 & 3 & 1 \\ 0 & 0 & 0 & 1 & 5 & 2 \\ 0 & 0 & 0 & 0 & 0 & 0 \end{bmatrix}$ (d) $\begin{bmatrix} 1 & 0 & 0 & 0 \\ 0 & 1 & 2 & 0 \\ 0 & 0 & 0 & 1 \end{bmatrix}$

Solution to (a). The corresponding system of equations is

$$\begin{aligned} x_1 \qquad\qquad &= 5 \\ x_2 \qquad &= -2 \\ x_3 &= 4 \end{aligned}$$

By inspection, $x_1 = 5, x_2 = -2, x_3 = 4$.

Solution to (*b*). The corresponding system of equations is

$$\begin{aligned} x_1 \qquad + 4x_4 &= -1 \\ x_2 \quad + 2x_4 &= 6 \\ x_3 + 3x_4 &= 2 \end{aligned}$$

Since x_1, x_2, and x_3 correspond to leading 1's in the augmented matrix, we call them ***leading variables.*** Solving for the leading variables in terms of x_4 gives

$$\begin{aligned} x_1 &= -1 - 4x_4 \\ x_2 &= 6 - 2x_4 \\ x_3 &= 2 - 3x_4 \end{aligned}$$

Since x_4 can be assigned an arbitrary value, say t, we have infinitely many solutions. The solution set is given by the formulas

$$x_1 = -1 - 4t, \qquad x_2 = 6 - 2t, \qquad x_3 = 2 - 3t, \qquad x_4 = t$$

Solution to (*c*). The corresponding system of equations is

$$\begin{aligned} x_1 + 6x_2 \qquad\qquad + 4x_5 &= -2 \\ x_3 \quad + 3x_5 &= 1 \\ x_4 + 5x_5 &= 2 \end{aligned}$$

Here the leading variables are x_1, x_3, and x_4. Solving for the leading variables in terms of the remaining variables gives

$$\begin{aligned} x_1 &= -2 - 4x_5 - 6x_2 \\ x_3 &= 1 - 3x_5 \\ x_4 &= 2 - 5x_5 \end{aligned}$$

Since x_5 can be assigned an arbitrary value, t, and x_2 can be assigned an arbitrary value, s, there are infinitely many solutions. The solution set is given by the formulas

$$x_1 = -2 - 4t - 6s, \qquad x_2 = s, \qquad x_3 = 1 - 3t, \qquad x_4 = 2 - 5t, \qquad x_5 = t$$

Solution to (*d*). The last equation in the corresponding system of equations is

$$0x_1 + 0x_2 + 0x_3 = 1$$

Since this equation can never be satisfied, there is no solution to the system.

We have just seen how easy it is to solve a system of linear equations, once its augmented matrix is in reduced row-echelon form. Now we shall give a step-by-step procedure, called ***Gauss-Jordan elimination,**** which can be used to reduce any

**Carl Friedrich Gauss* (1777–1855). Sometimes called the "prince of mathematicians," Gauss made profound contributions to number theory, theory of functions, probability, and statistics. He discovered a way to calculate the orbits of asteroids, made basic discoveries in electromagnetic theory, and invented a telegraph.

matrix to reduced row-echelon form. As we state each step in the procedure, we shall illustrate the idea by reducing the following matrix to reduced row-echelon form.

$$\begin{bmatrix} 0 & 0 & -2 & 0 & 7 & 12 \\ 2 & 4 & -10 & 6 & 12 & 28 \\ 2 & 4 & -5 & 6 & -5 & -1 \end{bmatrix}$$

Step 1. Locate the leftmost column (vertical line) that does not consist entirely of zeros.

$$\begin{bmatrix} 0 & 0 & -2 & 0 & 7 & 12 \\ 2 & 4 & -10 & 6 & 12 & 28 \\ 2 & 4 & -5 & 6 & -5 & -1 \end{bmatrix}$$

↑ **Leftmost nonzero column**

Step 2. Interchange the top row with another row, if necessary, so that the entry at the top of the column found in Step 1 is different from zero.

$$\begin{bmatrix} 2 & 4 & -10 & 6 & 12 & 28 \\ 0 & 0 & -2 & 0 & 7 & 12 \\ 2 & 4 & -5 & 6 & -5 & -1 \end{bmatrix}$$

The first and second rows in the previous matrix were interchanged.

Step 3. If the entry that is now at the top of the column found in Step 1 is a, multiply the first row by $1/a$ in order to introduce a leading 1.

$$\begin{bmatrix} 1 & 2 & -5 & 3 & 6 & 14 \\ 0 & 0 & -2 & 0 & 7 & 12 \\ 2 & 4 & -5 & 6 & -5 & -1 \end{bmatrix}$$

The first row of the previous matrix was multiplied by 1/2.

Step 4. Add suitable multiples of the top row to the rows below so that all entries below leading 1 become zeros.

$$\begin{bmatrix} 1 & 2 & -5 & 3 & 6 & 14 \\ 0 & 0 & -2 & 0 & 7 & 12 \\ 0 & 0 & 5 & 0 & -17 & -29 \end{bmatrix}$$

−2 times the first row of the previous matrix was added to the third row.

Camille Jordan (1838–1922). Jordan was a professor at the École Polytechnique in Paris. He did pioneering work in several branches of mathematics, including matrix theory. He is particularly famous for the Jordan Curve Theorem, which states: A simple closed curve (such as a circle or a square) divides the plane into two nonintersecting connected regions.

Step 5. Now cover the top row in the matrix and begin again with Step 1 applied to the submatrix that remains. Continue in this way until the *entire* matrix is in row-echelon form.

$$\begin{bmatrix} 1 & 2 & -5 & 3 & 6 & 14 \\ 0 & 0 & -2 & 0 & 7 & 12 \\ 0 & 0 & 5 & 0 & -17 & -29 \end{bmatrix}$$

↑ **Leftmost nonzero column in the submatrix** (third column)

$$\begin{bmatrix} 1 & 2 & -5 & 3 & 6 & 14 \\ 0 & 0 & 1 & 0 & -\frac{7}{2} & -6 \\ 0 & 0 & 5 & 0 & -17 & -29 \end{bmatrix}$$

The first row in the submatrix was multiplied by $-1/2$ to introduce a leading 1.

$$\begin{bmatrix} 1 & 2 & -5 & 3 & 6 & 14 \\ 0 & 0 & 1 & 0 & -\frac{7}{2} & -6 \\ 0 & 0 & 0 & 0 & \frac{1}{2} & 1 \end{bmatrix}$$

-5 times the first row of the submatrix was added to the second row of the submatrix to introduce a zero below the leading 1.

$$\begin{bmatrix} 1 & 2 & -5 & 3 & 6 & 14 \\ 0 & 0 & 1 & 0 & -\frac{7}{2} & -6 \\ 0 & 0 & 0 & 0 & \frac{1}{2} & 1 \end{bmatrix}$$

The top row in the submatrix was covered and we returned again to Step 1.

↑ **Leftmost nonzero column in the new submatrix** (fifth column)

$$\begin{bmatrix} 1 & 2 & -5 & 3 & 6 & 14 \\ 0 & 0 & 1 & 0 & -\frac{7}{2} & -6 \\ 0 & 0 & 0 & 0 & 1 & 2 \end{bmatrix}$$

The first (and only) row in the new submatrix was multiplied by 2 to introduce a leading 1.

The *entire* matrix is now in row-echelon form. To find the reduced row-echelon form we need the following additional step.

Step 6. Beginning with the last nonzero row and working upward, add suitable multiples of each row to the rows above to introduce zeros above the leading 1's.

$$\begin{bmatrix} 1 & 2 & -5 & 3 & 6 & 14 \\ 0 & 0 & 1 & 0 & 0 & 1 \\ 0 & 0 & 0 & 0 & 1 & 2 \end{bmatrix}$$

7/2 times the third row of the previous matrix was added to the second row.

$$\begin{bmatrix} 1 & 2 & -5 & 3 & 0 & 2 \\ 0 & 0 & 1 & 0 & 0 & 1 \\ 0 & 0 & 0 & 0 & 1 & 2 \end{bmatrix}$$

−6 times the third row was added to the first row.

$$\begin{bmatrix} 1 & 2 & 0 & 3 & 0 & 7 \\ 0 & 0 & 1 & 0 & 0 & 1 \\ 0 & 0 & 0 & 0 & 1 & 2 \end{bmatrix}$$

5 times the second row was added to the first row.

The last matrix is in reduced row-echelon form.

Example 6

Solve by Gauss-Jordan elimination.

$$\begin{aligned} x_1 + 3x_2 - 2x_3 \qquad\qquad + 2x_5 \qquad\qquad &= 0 \\ 2x_1 + 6x_2 - 5x_3 - 2x_4 + 4x_5 - 3x_6 &= -1 \\ 5x_3 + 10x_4 \qquad\qquad + 15x_6 &= 5 \\ 2x_1 + 6x_2 \qquad\qquad + 8x_4 + 4x_5 + 18x_6 &= 6 \end{aligned}$$

The augmented matrix for the system is

$$\begin{bmatrix} 1 & 3 & -2 & 0 & 2 & 0 & 0 \\ 2 & 6 & -5 & -2 & 4 & -3 & -1 \\ 0 & 0 & 5 & 10 & 0 & 15 & 5 \\ 2 & 6 & 0 & 8 & 4 & 18 & 6 \end{bmatrix}$$

Adding −2 times the first row to the second and fourth rows gives

$$\begin{bmatrix} 1 & 3 & -2 & 0 & 2 & 0 & 0 \\ 0 & 0 & -1 & -2 & 0 & -3 & -1 \\ 0 & 0 & 5 & 10 & 0 & 15 & 5 \\ 0 & 0 & 4 & 8 & 0 & 18 & 6 \end{bmatrix}$$

Multiplying the second row by −1 and then adding −5 times the second row to the third row and −4 times the second row to the fourth row gives

$$\begin{bmatrix} 1 & 3 & -2 & 0 & 2 & 0 & 0 \\ 0 & 0 & 1 & 2 & 0 & 3 & 1 \\ 0 & 0 & 0 & 0 & 0 & 0 & 0 \\ 0 & 0 & 0 & 0 & 0 & 6 & 2 \end{bmatrix}$$

Interchanging the third and fourth rows and then multiplying the third row of the resulting matrix by 1/6 gives the row-echelon form

$$\begin{bmatrix} 1 & 3 & -2 & 0 & 2 & 0 & 0 \\ 0 & 0 & 1 & 2 & 0 & 3 & 1 \\ 0 & 0 & 0 & 0 & 0 & 1 & \frac{1}{3} \\ 0 & 0 & 0 & 0 & 0 & 0 & 0 \end{bmatrix}$$

Adding -3 times the third row to the second row and then adding 2 times the second row of the resulting matrix to the first row yields the reduced row-echelon form

$$\begin{bmatrix} 1 & 3 & 0 & 4 & 2 & 0 & 0 \\ 0 & 0 & 1 & 2 & 0 & 0 & 0 \\ 0 & 0 & 0 & 0 & 0 & 1 & \frac{1}{3} \\ 0 & 0 & 0 & 0 & 0 & 0 & 0 \end{bmatrix}$$

The corresponding system of equations is

$$\begin{aligned} x_1 + 3x_2 \quad + 4x_4 + 2x_5 \quad &= 0 \\ x_3 + 2x_4 \quad &= 0 \\ x_6 &= \tfrac{1}{3} \end{aligned}$$

(We have discarded the last equation, $0x_1 + 0x_2 + 0x_3 + 0x_4 + 0x_5 + 0x_6 = 0$, since it will be satisfied automatically by the solutions of the remaining equations.) Solving for the leading variables, we obtain

$$\begin{aligned} x_1 &= -3x_2 - 4x_4 - 2x_5 \\ x_3 &= -2x_4 \\ x_6 &= \tfrac{1}{3} \end{aligned}$$

If we assign x_2, x_4, and x_5 the arbitrary values r, s, and t respectively, the solution set is given by the formulas

$$x_1 = -3r - 4s - 2t, \qquad x_2 = r, \qquad x_3 = -2s, \qquad x_4 = s, \qquad x_5 = t, \qquad x_6 = \tfrac{1}{3}$$

Example 7

It is often more convenient to solve a system of linear equations by bringing the augmented matrix into row-echelon form without continuing all the way to the reduced row-echelon form. When this is done, the corresponding system of equations can be solved by a technique called ***back-substitution***. We shall illustrate this method using the system of equations in Example 6.

From the computations in Example 6, a row-echelon form of the augmented matrix is

$$\begin{bmatrix} 1 & 3 & -2 & 0 & 2 & 0 & 0 \\ 0 & 0 & 1 & 2 & 0 & 3 & 1 \\ 0 & 0 & 0 & 0 & 0 & 1 & \frac{1}{3} \\ 0 & 0 & 0 & 0 & 0 & 0 & 0 \end{bmatrix}$$

To solve the corresponding system of equations

$$\begin{aligned} x_1 + 3x_2 - 2x_3 \qquad\quad + 2x_5 \qquad\quad &= 0 \\ x_3 + 2x_4 \qquad\quad + 3x_6 &= 1 \\ x_6 &= \tfrac{1}{3} \end{aligned}$$

we proceed as follows:

Step 1. Solve the equations for the leading variables.

$$\begin{aligned} x_1 &= -3x_2 + 2x_3 - 2x_5 \\ x_3 &= 1 - 2x_4 - 3x_6 \\ x_6 &= \tfrac{1}{3} \end{aligned}$$

Step 2. Beginning with the bottom equation and working upward, successively substitute each equation into all the equations above it.

Substituting $x_6 = 1/3$ into the second equation yields

$$\begin{aligned} x_1 &= -3x_2 + 2x_3 - 2x_5 \\ x_3 &= -2x_4 \\ x_6 &= \tfrac{1}{3} \end{aligned}$$

Substituting $x_3 = -2x_4$ into the first equation yields

$$\begin{aligned} x_1 &= -3x_2 - 4x_4 - 2x_5 \\ x_3 &= -2x_4 \\ x_6 &= \tfrac{1}{3} \end{aligned}$$

Step 3. Assign arbitrary values to the nonleading variables.

If we assign x_2, x_4, and x_5 the arbitrary values r, s, and t respectively, the solution set is given by the formulas

$$x_1 = -3r - 4s - 2t, \quad x_2 = r, \quad x_3 = -2s, \quad x_4 = s, \quad x_5 = t, \quad x_6 = \tfrac{1}{3}$$

This agrees with the solution obtained in Example 6.

The method of solving systems of linear equations by reducing the augmented matrix to row-echelon form is called ***Gaussian elimination***.

REMARK. The procedure we have given for reducing a matrix to row-echelon form or reduced row-echelon form is well-suited for computer computation because it is systematic. However, this procedure sometimes introduces fractions, which might otherwise be avoided by varying the steps in the right way. Thus once the basic procedure has been mastered, the reader may wish to vary the steps in specific problems to avoid fractions (see Exercise 13). It can be proved, although we shall not do it, that no matter how the elementary row operations are varied, one will always arrive at the same reduced row-echelon form; that is, the reduced row-echelon form is unique. However, a row-echelon form is *not* unique; by changing the sequence of elementary row operations it is possible to arrive at a different row-echelon form (see Exercise 14).

EXERCISE SET 1.2

1. Which of the following are in reduced row-echelon form?

(a) $\begin{bmatrix} 1 & 0 & 0 \\ 0 & 0 & 0 \\ 0 & 0 & 1 \end{bmatrix}$ (b) $\begin{bmatrix} 0 & 1 & 0 \\ 1 & 0 & 0 \\ 0 & 0 & 0 \end{bmatrix}$ (c) $\begin{bmatrix} 1 & 1 & 0 \\ 0 & 1 & 0 \\ 0 & 0 & 0 \end{bmatrix}$

(d) $\begin{bmatrix} 1 & 2 & 0 & 3 & 0 \\ 0 & 0 & 1 & 1 & 0 \\ 0 & 0 & 0 & 0 & 1 \\ 0 & 0 & 0 & 0 & 0 \end{bmatrix}$ (e) $\begin{bmatrix} 1 & 0 & 0 & 5 \\ 0 & 0 & 1 & 3 \\ 0 & 1 & 0 & 4 \end{bmatrix}$ (f) $\begin{bmatrix} 1 & 0 & 3 & 1 \\ 0 & 1 & 2 & 4 \end{bmatrix}$

2. Which of the following are in row-echelon form?

(a) $\begin{bmatrix} 1 & 2 & 3 \\ 0 & 0 & 0 \\ 0 & 0 & 1 \end{bmatrix}$ (b) $\begin{bmatrix} 1 & -7 & 5 & 5 \\ 0 & 1 & 3 & 2 \end{bmatrix}$ (c) $\begin{bmatrix} 1 & 1 & 0 \\ 0 & 1 & 0 \\ 0 & 0 & 0 \end{bmatrix}$

(d) $\begin{bmatrix} 1 & 3 & 0 & 2 & 0 \\ 1 & 0 & 2 & 2 & 0 \\ 0 & 0 & 0 & 0 & 1 \\ 0 & 0 & 0 & 0 & 0 \end{bmatrix}$ (e) $\begin{bmatrix} 2 & 3 & 4 \\ 0 & 1 & 2 \\ 0 & 0 & 3 \end{bmatrix}$ (f) $\begin{bmatrix} 0 & 0 & 0 \\ 0 & 0 & 0 \\ 0 & 0 & 0 \end{bmatrix}$

3. In each part suppose that the augmented matrix for a system of linear equations has been reduced by row operations to the given reduced row-echelon form. Solve the system.

(a) $\begin{bmatrix} 1 & 0 & 0 & 4 \\ 0 & 1 & 0 & 3 \\ 0 & 0 & 1 & 2 \end{bmatrix}$ (b) $\begin{bmatrix} 1 & 0 & 0 & 3 & 2 \\ 0 & 1 & 0 & -1 & 4 \\ 0 & 0 & 1 & 1 & 2 \end{bmatrix}$

(c) $\begin{bmatrix} 1 & 5 & 0 & 0 & 5 & -1 \\ 0 & 0 & 1 & 0 & 3 & 1 \\ 0 & 0 & 0 & 1 & 4 & 2 \\ 0 & 0 & 0 & 0 & 0 & 0 \end{bmatrix}$ (d) $\begin{bmatrix} 1 & 2 & 0 & 0 \\ 0 & 0 & 1 & 0 \\ 0 & 0 & 0 & 1 \end{bmatrix}$

4. In each part suppose that the augmented matrix for a system of linear equations has been reduced by row operations to the given row-echelon form. Solve the system.

(a) $\begin{bmatrix} 1 & 2 & -4 & 2 \\ 0 & 1 & -2 & -1 \\ 0 & 0 & 1 & 2 \end{bmatrix}$ (b) $\begin{bmatrix} 1 & 0 & 4 & 7 & 10 \\ 0 & 1 & -3 & -4 & -2 \\ 0 & 0 & 1 & 1 & 2 \end{bmatrix}$

(c) $\begin{bmatrix} 1 & 5 & -4 & 0 & -7 & -5 \\ 0 & 0 & 1 & 1 & 7 & 3 \\ 0 & 0 & 0 & 1 & 4 & 2 \\ 0 & 0 & 0 & 0 & 0 & 0 \end{bmatrix}$ (d) $\begin{bmatrix} 1 & 2 & 2 & 2 \\ 0 & 1 & 3 & 3 \\ 0 & 0 & 0 & 1 \end{bmatrix}$

5. Solve each of the following systems by Gauss-Jordan elimination.

(a)
$$\begin{aligned} 2x_1 + x_2 + x_3 &= 8 \\ 3x_1 - 2x_2 - x_3 &= 1 \\ 4x_1 - 7x_2 + 3x_3 &= 10 \end{aligned}$$

(b)
$$\begin{aligned} x_1 + x_2 + x_3 &= 0 \\ -2x_1 + 5x_2 + 2x_3 &= 0 \\ -7x_1 + 7x_2 + x_3 &= 0 \end{aligned}$$

(c)
$$\begin{aligned} x_1 + x_2 - 2x_3 + x_4 + 3x_5 &= 1 \\ 3x_1 + 2x_2 - 4x_3 - 3x_4 - 9x_5 &= 3 \\ 2x_1 - x_2 + 2x_3 + 2x_4 + 6x_5 &= 2 \\ 6x_1 + 2x_2 - 4x_3 \qquad\qquad &= 6 \\ 2x_2 - 4x_3 - 6x_4 - 18x_5 &= 0 \end{aligned}$$

6. Solve each of the systems in Exercise 5 by Gaussian elimination.

7. Solve each of the following systems by Gauss-Jordan elimination.

(a)
$$\begin{aligned} 2x_1 - 3x_2 &= -2 \\ 2x_1 + x_2 &= 1 \\ 3x_1 + 2x_2 &= 1 \end{aligned}$$

(b)
$$\begin{aligned} 3x_1 + 2x_2 - x_3 &= -15 \\ 5x_1 + 3x_2 + 2x_3 &= 0 \\ 3x_1 + x_2 + 3x_3 &= 11 \\ 11x_1 + 7x_2 \qquad &= -30 \end{aligned}$$

(c)
$$\begin{aligned} 4x_1 - 8x_2 &= 12 \\ 3x_1 - 6x_2 &= 9 \\ -2x_1 + 4x_2 &= -6 \end{aligned}$$

8. Solve each of the systems in Exercise 7 by Gaussian elimination.

9. Solve each of the following systems by Gauss-Jordan elimination.

(a)
$$\begin{aligned} 5x_1 + 2x_2 + 6x_3 &= 0 \\ -2x_1 + x_2 + 3x_3 &= 0 \end{aligned}$$

(b)
$$\begin{aligned} x_1 - 2x_2 + x_3 - 4x_4 &= 1 \\ x_1 + 3x_2 + 7x_3 + 2x_4 &= 2 \\ x_1 - 12x_2 - 11x_3 - 16x_4 &= 5 \end{aligned}$$

10. Solve each of the systems in Exercise 9 by Gaussian elimination.

11. Solve the following systems, where a, b, and c are constants.

(a)
$$\begin{aligned} 2x + y &= a \\ 3x + 6y &= b \end{aligned}$$

(b)
$$\begin{aligned} x_1 + x_2 + x_3 &= a \\ 2x_1 \qquad + 2x_3 &= b \\ 3x_2 + 3x_3 &= c \end{aligned}$$

12. For which values of a will the following system have no solutions? Exactly one solution? Infinitely many solutions?

$$\begin{aligned} x + 2y - 3z &= 4 \\ 3x - y + 5z &= 2 \\ 4x + y + (a^2 - 14)z &= a + 2 \end{aligned}$$

13. Reduce

$$\begin{bmatrix} 2 & 1 & 3 \\ 0 & -2 & 7 \\ 3 & 4 & 5 \end{bmatrix}$$

to reduced row-echelon form without introducing any fractions.

14. Find two different row-echelon forms of

$$\begin{bmatrix} 1 & 3 \\ 2 & 7 \end{bmatrix}$$

15. Solve the following system of nonlinear equations for the unknown angles α, β, and γ, where $0 \leq \alpha \leq 2\pi$, $0 \leq \beta \leq 2\pi$ and $0 \leq \gamma < \pi$.

$$\begin{aligned} 2 \sin \alpha - \cos \beta + 3 \tan \gamma &= 3 \\ 4 \sin \alpha + 2 \cos \beta - 2 \tan \gamma &= 2 \\ 6 \sin \alpha - 3 \cos \beta + \tan \gamma &= 9 \end{aligned}$$

16. Describe the possible reduced row-echelon forms of

$$\begin{bmatrix} a & b & c \\ d & e & f \\ g & h & i \end{bmatrix}$$

17. Show that if $ad - bc \neq 0$, then the reduced row-echelon form of

$$\begin{bmatrix} a & b \\ c & d \end{bmatrix} \quad \text{is} \quad \begin{bmatrix} 1 & 0 \\ 0 & 1 \end{bmatrix}$$

18. Use Exercise 17 to show that if $ad - bc \neq 0$, then the system

$$\begin{aligned} ax + by &= k \\ cx + dy &= l \end{aligned}$$

has exactly one solution.

1.3 HOMOGENEOUS SYSTEMS OF LINEAR EQUATIONS

As we have already pointed out, every system of linear equations has either one solution, infinitely many solutions, or no solutions at all. As we progress, there will be situations in which we will not be interested in finding solutions to a given system, but instead will be concerned with deciding how many solutions the system has. In this section we consider several cases in which it is possible to make statements about the number of solutions by inspection.

A system of linear equations is said to be ***homogeneous*** if all the constant terms are zero; that is, the system has the form

$$\begin{aligned} a_{11}x_1 + a_{12}x_2 + \cdots + a_{1n}x_n &= 0 \\ a_{21}x_1 + a_{22}x_2 + \cdots + a_{2n}x_n &= 0 \\ \vdots \qquad \quad \vdots \qquad \qquad \quad \vdots \quad &\quad \vdots \\ a_{m1}x_1 + a_{m2}x_2 + \cdots + a_{mn}x_n &= 0 \end{aligned}$$

Every homogeneous system of linear equations is consistent, since $x_1 = 0$, $x_2 = 0, \ldots, x_n = 0$ is always a solution. This solution is called the ***trivial solution***; if there are other solutions, they are called ***nontrivial solutions***.

Since a homogeneous system of linear equations must be consistent, there is either one solution or infinitely many solutions. Since one of these solutions is the trivial solution, we can make the following statement.

For a homogeneous system of linear equations, exactly one of the following is true.

1. *The system has only the trivial solution.*
2. *The system has infinitely many nontrivial solutions in addition to the trivial solution.*

There is one case in which a homogeneous system is assured of having nontrivial solutions; namely, whenever the system involves more unknowns than equations. To see why, consider the following example of four equations in five unknowns.

Example 8

Solve the following homogeneous system of linear equations by Gauss-Jordan elimination.

$$\begin{aligned} 2x_1 + 2x_2 - x_3 \qquad\qquad + x_5 &= 0 \\ -x_1 - x_2 + 2x_3 - 3x_4 + x_5 &= 0 \\ x_1 + x_2 - 2x_3 \qquad\qquad - x_5 &= 0 \\ x_3 + x_4 + x_5 &= 0 \end{aligned} \tag{1.2}$$

The augmented matrix for the system is

$$\begin{bmatrix} 2 & 2 & -1 & 0 & 1 & 0 \\ -1 & -1 & 2 & -3 & 1 & 0 \\ 1 & 1 & -2 & 0 & -1 & 0 \\ 0 & 0 & 1 & 1 & 1 & 0 \end{bmatrix}$$

Reducing this matrix to reduced row-echelon form, we obtain

$$\begin{bmatrix} 1 & 1 & 0 & 0 & 1 & 0 \\ 0 & 0 & 1 & 0 & 1 & 0 \\ 0 & 0 & 0 & 1 & 0 & 0 \\ 0 & 0 & 0 & 0 & 0 & 0 \end{bmatrix}$$

The corresponding system of equations is

$$\begin{aligned} x_1 + x_2 \qquad + x_5 &= 0 \\ x_3 \qquad + x_5 &= 0 \\ x_4 \qquad &= 0 \end{aligned} \tag{1.3}$$

Solving for the leading variables yields

$$\begin{aligned} x_1 &= -x_2 - x_5 \\ x_3 &= -x_5 \\ x_4 &= 0 \end{aligned}$$

The solution set is therefore given by

$$x_1 = -s - t, \qquad x_2 = s, \qquad x_3 = -t, \qquad x_4 = 0, \qquad x_5 = t$$

Note that the trivial is obtained when $s = t = 0$.

Example 8 illustrates two important points about solving homogeneous systems of linear equations. First, none of the three elementary row operations can alter the final column of zeros in the augmented matrix, so that the system of equations corresponding to the reduced row-echelon form of the augmented matrix must also be a homogeneous system (see system 1.3 in Example 8). Second, depending on whether the reduced row-echelon form of the augmented matrix has any zero rows, the number of equations in the reduced system is the same or less than the number of equations in the original system (compare systems 1.2 and 1.3 in Example 8). Thus if the given homogeneous system has m equations in n unknowns with $m < n$, and if there are r nonzero rows in the reduced row-echelon form of the augmented matrix we will have $r < n$; thus the system of equations corresponding to the reduced row echelon form of the augmented matrix will look like

$$\begin{array}{llll} \cdots x_{k_1} & & & + \Sigma(\ ) = 0 \\ & \cdots x_{k_2} & & + \Sigma(\ ) = 0 \\ & & \cdots \ddots & \quad \vdots \\ & & \qquad x_{k_r} & + \Sigma(\ ) = 0 \end{array} \tag{1.4}$$

where $x_{k_1}, x_{k_2}, \ldots, x_{k_r}$ are the leading variables and $\Sigma(\ )$ denotes sums that involve the $n - r$ remaining variables. Solving for the leading variables gives

$$\begin{aligned} x_{k_1} &= -\Sigma(\ ) \\ x_{k_2} &= -\Sigma(\ ) \\ &\ \ \vdots \\ x_{k_r} &= -\Sigma(\ ) \end{aligned}$$

As in Example 8, we can assign arbitrary values to the variables on the right-hand side and thus obtain infinitely many solutions to the system.

In summary, we have the following important theorem.

Theorem 1. *A homogeneous system of linear equations with more unknowns than equations always has infinitely many solutions.*

EXERCISE SET 1.3

1. Without using pencil and paper, determine which of the following homogeneous systems have nontrivial solutions.

(a) $$\begin{aligned} x_1 + 3x_2 + 5x_3 + x_4 &= 0 \\ 4x_1 - 7x_2 - 3x_3 - x_4 &= 0 \\ 3x_1 + 2x_2 + 7x_3 + 8x_4 &= 0 \end{aligned}$$

(b) $$\begin{aligned} x_1 + 2x_2 + 3x_3 &= 0 \\ x_2 + 4x_3 &= 0 \\ 5x_3 &= 0 \end{aligned}$$

(c) $$\begin{aligned} a_{11}x_1 + a_{12}x_2 + a_{13}x_3 &= 0 \\ a_{21}x_1 + a_{22}x_2 + a_{23}x_3 &= 0 \end{aligned}$$

(d) $$\begin{aligned} x_1 + x_2 &= 0 \\ 2x_1 + 2x_2 &= 0 \end{aligned}$$

In Exercises 2–5 solve the given homogeneous system of linear equations.

2. $$\begin{aligned} 2x_1 + x_2 + 3x_3 &= 0 \\ x_1 + 2x_2 &= 0 \\ x_2 + x_3 &= 0 \end{aligned}$$

3. $$\begin{aligned} 3x_1 + x_2 + x_3 + x_4 &= 0 \\ 5x_1 - x_2 + x_3 - x_4 &= 0 \end{aligned}$$

4. $$\begin{aligned} 2x_1 - 4x_2 + x_3 + x_4 &= 0 \\ x_1 - 5x_2 + 2x_3 &= 0 \\ -2x_2 - 2x_3 - x_4 &= 0 \\ x_1 + 3x_2 + x_4 &= 0 \\ x_1 - 2x_2 - x_3 + x_4 &= 0 \end{aligned}$$

5. $$\begin{aligned} x + 6y - 2z &= 0 \\ 2x - 4y + z &= 0 \end{aligned}$$

6. For which value(s) of λ does the following system of equations have nontrivial solutions?

$$\begin{aligned} (\lambda - 3)x + y &= 0 \\ x + (\lambda - 3)y &= 0 \end{aligned}$$

7. Consider the system of equations

$$\begin{aligned} ax + by &= 0 \\ cx + dy &= 0 \\ ex + fy &= 0 \end{aligned}$$

Discuss the relative positions of the lines $ax + by = 0$, $cx + dy = 0$, and $ex + fy = 0$ when:

(a) the system has only the trivial solution
(b) the system has nontrivial solutions

8. Consider the system of equations

$$\begin{aligned} ax + by &= 0 \\ cx + dy &= 0 \end{aligned}$$

(a) Show that if $x = x_0$, $y = y_0$ is any solution and k is any constant, then $x = kx_0$, $y = ky_0$ is also a solution.
(b) Show that if $x = x_0$, $y = y_0$ and $x = x_1$, $y = y_1$ are any two solutions, then $x = x_0 + x_1$, $y = y_0 + y_1$ is also a solution.

9. Consider the systems of equations

(I) $$\begin{aligned} ax + by &= k \\ cx + dy &= l \end{aligned}$$ (II) $$\begin{aligned} ax + by &= 0 \\ cx + dy &= 0 \end{aligned}$$

(a) Show that if $x = x_1$, $y = y_1$ and $x = x_2$, $y = y_2$ are both solutions of I, then $x = x_1 - x_2$, $y = y_1 - y_2$ is a solution of II.
(b) Show that if $x = x_1$, $y = y_1$ is a solution of I and $x = x_0$, $y = y_0$ is a solution of II, then $x = x_1 + x_0$, $y = y_1 + y_0$ is a solution of I.

10. (a) In the system of equations numbered (1.4), explain why it would be incorrect to denote the leading variables by $x_1, x_2, \ldots, x_r$ rather than $x_{k_1}, x_{k_2}, \ldots, x_{k_r}$ as we have done.
(b) The system of equations numbered (1.3) is a specific case of (1.4). What value does r have in this case? What are $x_{k_1}, x_{k_2}, \ldots, x_{k_r}$ in this case? Write out the sums denoted by $\Sigma(\ )$ in (1.4).

1.4 MATRICES AND MATRIX OPERATIONS

Rectangular arrays of real numbers arise in many contexts other than as augmented matrices for systems of linear equations. In this section we consider such arrays as objects in their own right and develop some of their properties for use in our later work.

Definition. A ***matrix*** is a rectangular array of numbers. The numbers in the array are called the ***entries*** in the matrix.

Example 9

The following are matrices.

$$\begin{bmatrix} 1 & 2 \\ 3 & 0 \\ -1 & 4 \end{bmatrix} \quad [2 \ \ 1 \ \ 0 \ \ -3] \quad \begin{bmatrix} -\sqrt{2} & \pi & e \\ 3 & \frac{1}{2} & 0 \\ 0 & 0 & 0 \end{bmatrix} \quad \begin{bmatrix} 1 \\ 3 \end{bmatrix} \quad [4]$$

As these examples indicate, matrices vary in size. The ***size*** of a matrix is described by specifying the number of rows (horizontal lines) and columns (vertical lines) that occur in the matrix. The first matrix in Example 9 has 3 rows and 2 columns so that its size is 3 by 2 (written 3×2). The first number always indicates the number of rows and the second indicates the number of columns. The remaining matrices in Example 9 thus have sizes 1×4, 3×3, 2×1, and 1×1 respectively.

We shall use capital letters to denote matrices and lowercase letters to denote numerical quantities; thus, we might write

$$A = \begin{bmatrix} 2 & 1 & 7 \\ 3 & 4 & 2 \end{bmatrix} \quad \text{or} \quad C = \begin{bmatrix} a & b & c \\ d & e & f \end{bmatrix}$$

When discussing matrices, it is common to refer to numerical quantities as ***scalars***. In this text *all our scalars will be real numbers.*

If A is a matrix, we will use a_{ij} to denote the entry that occurs in row i and column j of A. Thus a general 3×4 matrix can be written

$$A = \begin{bmatrix} a_{11} & a_{12} & a_{13} & a_{14} \\ a_{21} & a_{22} & a_{23} & a_{24} \\ a_{31} & a_{32} & a_{33} & a_{34} \end{bmatrix}$$

Naturally, if we use B to denote the matrix, then we will use b_{ij} for the entry in row i and column j. Thus a general $m \times n$ matrix might be written

$$B = \begin{bmatrix} b_{11} & b_{12} & \cdots & b_{1n} \\ b_{21} & b_{22} & \cdots & b_{2n} \\ \vdots & \vdots & & \vdots \\ b_{m1} & b_{m2} & \cdots & b_{mn} \end{bmatrix}$$

A matrix A with n rows and n columns is called a ***square matrix of order n***, and the entries $a_{11}, a_{22}, \ldots, a_{nn}$ are said to be on the ***main diagonal*** of A (see Figure 1.2).

$$\begin{bmatrix} a_{11} & a_{12} & \cdots & a_{1n} \\ a_{21} & a_{22} & \cdots & a_{2n} \\ \vdots & \vdots & & \vdots \\ a_{n1} & a_{n2} & \cdots & a_{nn} \end{bmatrix}$$

Figure 1.2

So far we have used matrices to abbreviate the work in solving systems of linear equations. For other applications, however, it is desirable to develop an "arithmetic of matrices" in which matrices can be added and multiplied in a useful way. The remainder of this section will be devoted to developing this arithmetic.

Two matrices are said to be ***equal*** if they have the same size and the corresponding entries in the two matrices are equal.

Example 10

Consider the matrices

$$A = \begin{bmatrix} 2 & 1 \\ 3 & 4 \end{bmatrix} \qquad B = \begin{bmatrix} 2 & 1 \\ 3 & 5 \end{bmatrix} \qquad C = \begin{bmatrix} 2 & 1 & 0 \\ 3 & 4 & 0 \end{bmatrix}$$

Here $A \neq C$ since A and C do not have the same size. For the same reason $B \neq C$. Also, $A \neq B$ since not all the corresponding entries are equal.

Definition. If A and B are any two matrices of the same size, then the ***sum*** $A + B$ is the matrix obtained by adding together the corresponding entries in the two matrices. Matrices of different sizes cannot be added.

Example 11

Consider the matrices

$$A = \begin{bmatrix} 2 & 1 & 0 & 3 \\ -1 & 0 & 2 & 4 \\ 4 & -2 & 7 & 0 \end{bmatrix} \quad B = \begin{bmatrix} -4 & 3 & 5 & 1 \\ 2 & 2 & 0 & -1 \\ 3 & 2 & -4 & 5 \end{bmatrix} \quad C = \begin{bmatrix} 1 & 1 \\ 2 & 2 \end{bmatrix}$$

Then

$$A + B = \begin{bmatrix} -2 & 4 & 5 & 4 \\ 1 & 2 & 2 & 3 \\ 7 & 0 & 3 & 5 \end{bmatrix}$$

while $A + C$ and $B + C$ are undefined.

Definition. If A is any matrix and c is any scalar, then the ***product*** cA is the matrix obtained by multiplying each entry of A by c.

Example 12

If A is the matrix

$$A = \begin{bmatrix} 4 & 2 \\ 1 & 3 \\ -1 & 0 \end{bmatrix}$$

then

$$2A = \begin{bmatrix} 8 & 4 \\ 2 & 6 \\ -2 & 0 \end{bmatrix} \quad \text{and} \quad (-1)A = \begin{bmatrix} -4 & -2 \\ -1 & -3 \\ 1 & 0 \end{bmatrix}$$

If B is any matrix, then $-B$ will denote the product $(-1)B$. If A and B are two matrices of the same size, then $A - B$ is defined to be the sum $A + (-B) = A + (-1)B$.

Example 13

Consider the matrices

$$A = \begin{bmatrix} 2 & 3 & 4 \\ 1 & 2 & 1 \end{bmatrix} \quad \text{and} \quad B = \begin{bmatrix} 0 & 2 & 7 \\ 1 & -3 & 5 \end{bmatrix}$$

From the above definitions

$$-B = \begin{bmatrix} 0 & -2 & -7 \\ -1 & 3 & -5 \end{bmatrix}$$

and

$$A - B = \begin{bmatrix} 2 & 3 & 4 \\ 1 & 2 & 1 \end{bmatrix} + \begin{bmatrix} 0 & -2 & -7 \\ -1 & 3 & -5 \end{bmatrix} = \begin{bmatrix} 2 & 1 & -3 \\ 0 & 5 & -4 \end{bmatrix}$$

Observe that $A - B$ can be obtained directly by subtracting the entries of B from the corresponding entries of A.

Above, we defined the multiplication of a matrix by a scalar. The next question, then, is how do we multiply two matrices together? Perhaps the most natural definition of matrix multiplication would seem to be: "multiply corresponding entries together." Surprisingly, however, this definition would not be very useful for most problems. Experience has led mathematicians to the following less intuitive but more useful definition of matrix multiplication.

Definition. If A is an $m \times r$ matrix and B is an $r \times n$ matrix, then the ***product*** AB is the $m \times n$ matrix whose entries are determined as follows. To find the entry in row i and column j of AB, single out row i from the matrix A and column j from the matrix B. Multiply the corresponding entries from the row and column together and then add up the resulting products.

Example 14

Consider the matrices

$$A = \begin{bmatrix} 1 & 2 & 4 \\ 2 & 6 & 0 \end{bmatrix} \qquad B = \begin{bmatrix} 4 & 1 & 4 & 3 \\ 0 & -1 & 3 & 1 \\ 2 & 7 & 5 & 2 \end{bmatrix}$$

Since A is a 2×3 matrix and B is a 3×4 matrix, the product AB is a 2×4 matrix. To determine, for example, the entry in row 2 and column 3 of AB, we single out row 2 from A and column 3 from B. Then, as illustrated below, we multiply corresponding entries together and add up these products.

$$\begin{bmatrix} 1 & 2 & 4 \\ 2 & 6 & 0 \end{bmatrix} \begin{bmatrix} 4 & 1 & 4 & 3 \\ 0 & -1 & 3 & 1 \\ 2 & 7 & 5 & 2 \end{bmatrix} = \begin{bmatrix} \square & \square & \square & \square \\ \square & \square & \boxed{26} & \square \end{bmatrix}$$

$$(2 \cdot 4) + (6 \cdot 3) + (0 \cdot 5) = 26$$

The entry in row 1 and column 4 of AB is computed as follows.

$$\begin{bmatrix} 1 & 2 & 4 \\ 2 & 6 & 0 \end{bmatrix} \begin{bmatrix} 4 & 1 & 4 & 3 \\ 0 & -1 & 3 & 1 \\ 2 & 7 & 5 & 2 \end{bmatrix} = \begin{bmatrix} \square & \square & \square & \boxed{13} \\ \square & \square & \boxed{26} & \square \end{bmatrix}$$

$$(1 \cdot 3) + (2 \cdot 1) + (4 \cdot 2) = 13$$

The computations for the remaining products are

$$\begin{aligned} (1 \cdot 4) + (2 \cdot 0) + (4 \cdot 2) &= 12 \\ (1 \cdot 1) - (2 \cdot 1) + (4 \cdot 7) &= 27 \\ (1 \cdot 4) + (2 \cdot 3) + (4 \cdot 5) &= 30 \\ (2 \cdot 4) + (6 \cdot 0) + (0 \cdot 2) &= 8 \\ (2 \cdot 1) - (6 \cdot 1) + (0 \cdot 7) &= -4 \\ (2 \cdot 3) + (6 \cdot 1) + (0 \cdot 2) &= 12 \end{aligned} \qquad AB = \begin{bmatrix} 12 & 27 & 30 & 13 \\ 8 & -4 & 26 & 12 \end{bmatrix}$$

The definition of matrix multiplication requires that the number of columns of the first factor A be the same as the number of rows of the second factor B in order to form the product AB. If this condition is not satisfied, the product is undefined. A convenient way to determine whether a product of two matrices is defined is to write down the size of the first factor and, to the right of it, write down the size of the second factor. If, as in Figure 1.3, the inside numbers are the same, then the product is defined. The outside numbers then give the size of the product.

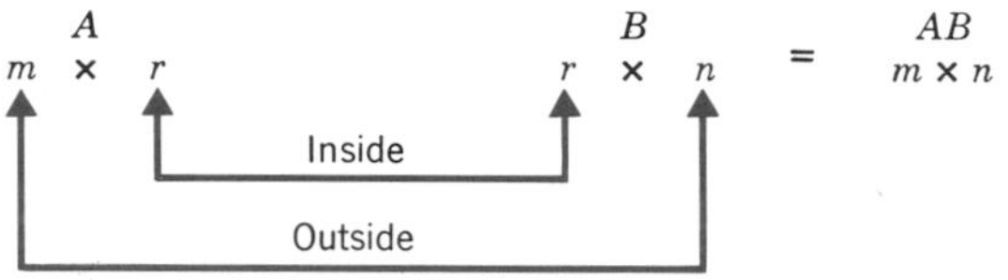

Figure 1.3

Example 15

Suppose that A is a 3×4 matrix, B is a 4×7 matrix, and C is a 7×3 matrix. Then AB is defined and is a 3×7 matrix; CA is defined and is a 7×4 matrix; BC is defined and is a 4×3 matrix. The products AC, CB, and BA are all undefined.

Example 16

If A is a general $m \times r$ matrix and B is a general $r \times n$ matrix, then, as suggested by the shading below, the entry in row i and column j of AB is given by the formula

$$a_{i1}b_{1j} + a_{i2}b_{2j} + a_{i3}b_{3j} + \cdots + a_{ir}b_{rj}$$

$$AB = \begin{bmatrix} a_{11} & a_{12} & \cdots & a_{1r} \\ a_{21} & a_{22} & \cdots & a_{2r} \\ \vdots & \vdots & & \vdots \\ a_{i1} & a_{i2} & \cdots & a_{ir} \\ \vdots & \vdots & & \vdots \\ a_{m1} & a_{m2} & \cdots & a_{mr} \end{bmatrix} \begin{bmatrix} b_{11} & b_{12} & \cdots & b_{1j} & \cdots & b_{1n} \\ b_{21} & b_{22} & \cdots & b_{2j} & \cdots & b_{2n} \\ \vdots & \vdots & & \vdots & & \vdots \\ b_{r1} & b_{r2} & \cdots & b_{rj} & \cdots & b_{rn} \end{bmatrix}$$

Matrix multiplication has an important application to systems of linear equations. Consider any system of m linear equations in n unknowns

$$\begin{aligned} a_{11}x_1 + a_{12}x_2 + \cdots + a_{1n}x_n &= b_1 \\ a_{21}x_1 + a_{22}x_2 + \cdots + a_{2n}x_n &= b_2 \\ \vdots \qquad \vdots \qquad\qquad \vdots \quad &\quad \vdots \\ a_{m1}x_1 + a_{m2}x_2 + \cdots + a_{mn}x_n &= b_m \end{aligned}$$

Since two matrices are equal if and only if their corresponding entries are equal, we can replace the m equations in this system by the single matrix equation

$$\begin{bmatrix} a_{11}x_1 + a_{12}x_2 + \cdots + a_{1n}x_n \\ a_{21}x_1 + a_{22}x_2 + \cdots + a_{2n}x_n \\ \vdots \\ a_{m1}x_1 + a_{m2}x_2 + \cdots + a_{mn}x_n \end{bmatrix} = \begin{bmatrix} b_1 \\ b_2 \\ \vdots \\ b_m \end{bmatrix}$$

The $m \times 1$ matrix on the left side of this equation can be written as a product to give

$$\begin{bmatrix} a_{11} & a_{12} & \cdots & a_{1n} \\ a_{21} & a_{22} & \cdots & a_{2n} \\ \vdots & \vdots & & \vdots \\ a_{m1} & a_{m2} & \cdots & a_{mn} \end{bmatrix} \begin{bmatrix} x_1 \\ x_2 \\ \vdots \\ x_n \end{bmatrix} = \begin{bmatrix} b_1 \\ b_2 \\ \vdots \\ b_m \end{bmatrix}$$

If we designate these matrices by A, X, and B, respectively, the original system of m equations in n unknowns has been replaced by the single matrix equation

$$AX = B \tag{1.5}$$

Some of our later work will be devoted to solving matrix equations like this for the unknown matrix X. As a consequence of this matrix approach, we will obtain effective new methods for solving systems of linear equations. The matrix A in (1.5) is called the ***coefficient matrix*** for the system.

Example 17

At times it is helpful to be able to find a particular row or column in a product AB without computing the entire product. We leave it as an exercise to show that the entries in the jth column of AB are the entries in the product AB_j, where B_j is the matrix formed using only the jth column of B. Thus, if A and B are the matrices in Example 14, the second column of AB can be obtained by the computation

$$\begin{bmatrix} 1 & 2 & 4 \\ 2 & 6 & 0 \end{bmatrix} \begin{bmatrix} 1 \\ -1 \\ 7 \end{bmatrix} = \begin{bmatrix} 27 \\ -4 \end{bmatrix}$$

(↑ under the column: **second column of *B***; ↑ under the result: **second column of *AB***)

Similarly, the entries in the ith row of AB are the entries in the product A_iB, where A_i is the matrix formed by using only the ith row of A. Thus the first row in the product AB of Example 14 can be obtained by the computation

$$[1 \quad 2 \quad 4] \begin{bmatrix} 4 & 1 & 4 & 3 \\ 0 & -1 & 3 & 1 \\ 2 & 7 & 5 & 2 \end{bmatrix} = [12 \quad 27 \quad 30 \quad 13]$$

(arrow to $[1\ 2\ 4]$: **first row of *A***; arrow to $[12\ 27\ 30\ 13]$: **first row of *AB***)

EXERCISE SET 1.4

1. Let A and B be 4×5 matrices and let C, D, and E be 5×2, 4×2, and 5×4 matrices, respectively. Determine which of the following matrix expressions are defined. For those that are defined, give the size of the resulting matrix.

(a) BA (b) $AC + D$ (c) $AE + B$
(d) $AB + B$ (e) $E(A + B)$ (f) $E(AC)$

2. (a) Show that if AB and BA are both defined, then AB and BA are square matrices.
(b) Show that if A is an $m \times n$ matrix and $A(BA)$ is defined, then B is an $n \times m$ matrix.

3. Solve the following matrix equation for a, b, c, and d.

$$\begin{bmatrix} a - b & b + c \\ 3d + c & 2a - 4d \end{bmatrix} = \begin{bmatrix} 8 & 1 \\ 7 & 6 \end{bmatrix}$$

4. Consider the matrices

$$A = \begin{bmatrix} 3 & 0 \\ -1 & 2 \\ 1 & 1 \end{bmatrix} \quad B = \begin{bmatrix} 4 & -1 \\ 0 & 2 \end{bmatrix} \quad C = \begin{bmatrix} 1 & 4 & 2 \\ 3 & 1 & 5 \end{bmatrix}$$

$$D = \begin{bmatrix} 1 & 5 & 2 \\ -1 & 0 & 1 \\ 3 & 2 & 4 \end{bmatrix} \quad E = \begin{bmatrix} 6 & 1 & 3 \\ -1 & 1 & 2 \\ 4 & 1 & 3 \end{bmatrix}$$

Compute
(a) AB (b) $D + E$ (c) $D - E$
(d) DE (e) ED (f) $-7B$

5. Using the matrices from Exercise 4, compute (where possible)
(a) $3C - D$ (b) $(3E)D$
(c) $(AB)C$ (d) $A(BC)$
(e) $(4B)C + 2B$ (f) $D + E^2$ (where $E^2 = EE$)

6. Let

$$A = \begin{bmatrix} 3 & -2 & 7 \\ 6 & 5 & 4 \\ 0 & 4 & 9 \end{bmatrix} \quad \text{and} \quad B = \begin{bmatrix} 6 & -2 & 4 \\ 0 & 1 & 3 \\ 7 & 7 & 5 \end{bmatrix}$$

Use the method of Example 17 to find
(a) the first row of AB (b) the third row of AB
(c) the second column of AB (d) the first column of BA
(e) the third row of AA (f) the third column of AA

7. Let C, D, and E be the matrices in Exercise 4. Using as few computations as possible, determine the entry in row 2 and column 3 of $C(DE)$.

8. (a) Show that if A has a row of zeros and B is any matrix for which AB is defined, then AB also has a row of zeros.
(b) Find a similar result involving a column of zeros.

9. Let A be any $m \times n$ matrix and let 0 be the $m \times n$ matrix each of whose entries is zero. Show that if $kA = 0$, then $k = 0$ or $A = 0$.

10. Let I be the $n \times n$ matrix whose entry in row i and column j is

$$\begin{cases} 1 & \text{if} \quad i = j \\ 0 & \text{if} \quad i \neq j \end{cases}$$

Show that $AI = IA = A$ for every $n \times n$ matrix A.

11. A square matrix is called a ***diagonal matrix*** if all entries off the main diagonal are zeros. Show that the product of diagonal matrices is again a diagonal matrix. State a rule for multiplying diagonal matrices.

12. (a) Show that the entries in the jth column of AB are the entries in the product AB_j, where B_j is the matrix formed from the jth column of B.
(b) Show that the entries in the ith row of AB are the entries in the product A_iB, where A_i is the matrix formed from the ith row of A.

1.5 RULES OF MATRIX ARITHMETIC

Although many of the rules of arithmetic for real numbers also hold for matrices, there are some exceptions. One of the most important exceptions occurs in the multiplication of matrices. For real numbers a and b, we always have $ab = ba$. This property is often called the *commutative law for multiplication.* For matrices, however, AB and BA need not be equal. Equality can fail to hold for three reasons. It can happen, for example, that AB is defined but BA is undefined. This is the case if A is a 2×3 matrix and B is a 3×4 matrix. Also it can happen that AB and BA are both defined but have different sizes. This is the situation if A is a 2×3 matrix and B is a 3×2 matrix. Finally, as our next example shows, it is possible to have $AB \neq BA$ even if both AB and BA are defined and have the same size.

Example 18

Consider the matrices

$$A = \begin{bmatrix} -1 & 0 \\ 2 & 3 \end{bmatrix} \qquad B = \begin{bmatrix} 1 & 2 \\ 3 & 0 \end{bmatrix}$$

Multiplying gives

$$AB = \begin{bmatrix} -1 & -2 \\ 11 & 4 \end{bmatrix} \qquad BA = \begin{bmatrix} 3 & 6 \\ -3 & 0 \end{bmatrix}$$

Thus $AB \neq BA$.

Although the commutative law for multiplication is not valid in matrix arithmetic, many familiar laws of arithmetic are valid for matrices. Some of the most important ones and their names are summarized in the following theorem.

Theorem 2. *Assuming that the sizes of the matrices are such that the indicated operations can be performed, the following rules of matrix arithmetic are valid.*

(*a*) $A + B = B + A$ (*Commutative law for addition*)
(*b*) $A + (B + C) = (A + B) + C$ (*Associative law for addition*)
(*c*) $A(BC) = (AB)C$ (*Associative law for multiplication*)
(*d*) $A(B + C) = AB + AC$ (*Distributive law*)
(*e*) $(B + C)A = BA + CA$ (*Distributive law*)
(*f*) $A(B - C) = AB - AC$
(*g*) $(B - C)A = BA - CA$
(*h*) $a(B + C) = aB + aC$
(*i*) $a(B - C) = aB - aC$
(*j*) $(a + b)C = aC + bC$
(*k*) $(a - b)C = aC - bC$
(*l*) $(ab)C = a(bC)$
(*m*) $a(BC) = (aB)C = B(aC)$

Each of the equations in this theorem asserts an equality between matrices. To prove one of these equalities, it is necessary to show that the matrix on the left side has the same size as the matrix on the right side, and that corresponding entries on the two sides are equal. To illustrate, we shall prove (*h*). Some of the remaining proofs are given as exercises.

Proof of (*h*). Since the left side involves the operation $B + C$, B and C must have the same size, say $m \times n$. It follows that $a(B + C)$ and $aB + aC$ are also $m \times n$ matrices, and thus have the same size.

Let l_{ij} be any entry on the left side, and let r_{ij} be the corresponding entry on the right side. To complete the proof, we must show $l_{ij} = r_{ij}$. If we let a_{ij}, b_{ij}, and c_{ij} be the entries in the ith row and jth column of A, B, and C, respectively, then from the definitions of the matrix operations

$$l_{ij} = a(b_{ij} + c_{ij}) \qquad \text{and} \qquad r_{ij} = ab_{ij} + ac_{ij}$$

Since $a(b_{ij} + c_{ij}) = ab_{ij} + ac_{ij}$, we have $l_{ij} = r_{ij}$, which completes the proof. ▮

Although the operations of matrix addition and matrix multiplication were defined for pairs of matrices, associative laws (*b*) and (*c*) enable us to denote sums and products of three matrices as $A + B + C$ and ABC without inserting any parentheses. This is justified by the fact that no matter how parentheses are inserted, the associative laws guarantee that the same end result will be obtained. Without going into details, we observe that similar results are valid for sums or products involving four or more matrices. In general, *given any sum or any product of matrices, pairs of parentheses can be inserted or deleted anywhere within the expression without affecting the end result.*

Example 19

As an illustration of the associative law for matrix multiplication, consider

$$A = \begin{bmatrix} 1 & 2 \\ 3 & 4 \\ 0 & 1 \end{bmatrix} \qquad B = \begin{bmatrix} 4 & 3 \\ 2 & 1 \end{bmatrix} \qquad C = \begin{bmatrix} 1 & 0 \\ 2 & 3 \end{bmatrix}$$

Then

$$AB = \begin{bmatrix} 1 & 2 \\ 3 & 4 \\ 0 & 1 \end{bmatrix} \begin{bmatrix} 4 & 3 \\ 2 & 1 \end{bmatrix} = \begin{bmatrix} 8 & 5 \\ 20 & 13 \\ 2 & 1 \end{bmatrix}$$

so that

$$(AB)C = \begin{bmatrix} 8 & 5 \\ 20 & 13 \\ 2 & 1 \end{bmatrix} \begin{bmatrix} 1 & 0 \\ 2 & 3 \end{bmatrix} = \begin{bmatrix} 18 & 15 \\ 46 & 39 \\ 4 & 3 \end{bmatrix}$$

On the other hand

$$BC = \begin{bmatrix} 4 & 3 \\ 2 & 1 \end{bmatrix} \begin{bmatrix} 1 & 0 \\ 2 & 3 \end{bmatrix} = \begin{bmatrix} 10 & 9 \\ 4 & 3 \end{bmatrix}$$

so that

$$A(BC) = \begin{bmatrix} 1 & 2 \\ 3 & 4 \\ 0 & 1 \end{bmatrix} \begin{bmatrix} 10 & 9 \\ 4 & 3 \end{bmatrix} = \begin{bmatrix} 18 & 15 \\ 46 & 39 \\ 4 & 3 \end{bmatrix}$$

Thus $(AB)C = A(BC)$, as guaranteed by Theorem 2(*c*).

A matrix, all of whose entries are zero, such as

$$\begin{bmatrix} 0 & 0 \\ 0 & 0 \end{bmatrix}, \quad \begin{bmatrix} 0 & 0 & 0 \\ 0 & 0 & 0 \\ 0 & 0 & 0 \end{bmatrix}, \quad \begin{bmatrix} 0 & 0 & 0 & 0 \\ 0 & 0 & 0 & 0 \end{bmatrix}, \quad \begin{bmatrix} 0 \\ 0 \\ 0 \\ 0 \end{bmatrix}, \quad [0]$$

is called a ***zero matrix***. Zero matrices will be denoted by 0; if it is important to emphasize the size, we shall write $0_{m \times n}$ for the $m \times n$ zero matrix.

If A is any matrix and 0 is the zero matrix with the same size, it is obvious that $A + 0 = A$. The matrix 0 plays much the same role in this matrix equation as the number 0 plays in the numerical equation $a + 0 = a$.

Since we already know that some of the rules of arithmetic for real numbers do not carry over to matrix arithmetic, it would be foolhardy to assume that all the properties of the real number zero carry over to zero matrices. For example, consider the following two standard results in the arithmetic of real numbers.

(i) If $ab = ac$ and $a \neq 0$, then $b = c$. (This is called the *cancellation law*.)
(ii) If $ad = 0$, then at least one of the factors on the left is 0.

As the next example shows, the corresponding results are false in matrix arithmetic.

Example 20

Consider the matrices

$$A = \begin{bmatrix} 0 & 1 \\ 0 & 2 \end{bmatrix} \quad B = \begin{bmatrix} 1 & 1 \\ 3 & 4 \end{bmatrix} \quad C = \begin{bmatrix} 2 & 5 \\ 3 & 4 \end{bmatrix} \quad D = \begin{bmatrix} 3 & 7 \\ 0 & 0 \end{bmatrix}$$

Here

$$AB = AC = \begin{bmatrix} 3 & 4 \\ 6 & 8 \end{bmatrix}$$

Although $A \neq 0$, it is *incorrect* to cancel the A from both sides of the equation $AB = AC$ and write $B = C$. Thus the cancellation law fails to hold for matrices.

Also, $AD = 0$; yet $A \neq 0$ and $D \neq 0$ so that result (ii) listed above does not carry over to matrix arithmetic.

In spite of these negative examples, a number of familiar properties of the real number 0 carry over to zero matrices. Some of the more important ones are summarized in the next theorem. The proofs are left as exercises.

Theorem 3. *Assuming that the sizes of the matrices are such that the indicated operations can be performed, the following rules of matrix arithmetic are valid.*

(*a*) $A + 0 = 0 + A = A$
(*b*) $A - A = 0$
(*c*) $0 - A = -A$
(*d*) $A0 = 0; \quad 0A = 0$

As an application of our results on matrix arithmetic, we prove the following theorem, which was anticipated earlier in the text.

Theorem 4. *Every system of linear equations has either no solutions, exactly one solution, or infinitely many solutions.*

Proof. If $AX = B$ is a system of linear equations, exactly one of the following is true: (a) the system has no solutions, (b) the system has exactly one solution, or (c) the system has more than one solution. The proof will be complete if we can show that the system has infinitely many solutions in case (c).

Assume that $AX = B$ has more than one solution; let X_1 and X_2 be two different solutions. Thus, $AX_1 = B$ and $AX_2 = B$. Subtracting these equations gives $AX_1 - AX_2 = 0$ or $A(X_1 - X_2) = 0$. If we let $X_0 = X_1 - X_2$ and let k be any scalar, then

$$\begin{aligned} A(X_1 + kX_0) &= AX_1 + A(kX_0) \\ &= AX_1 + k(AX_0) \\ &= B + k0 \\ &= B + 0 \\ &= B \end{aligned}$$

But this says that $X_1 + kX_0$ is a solution of $AX = B$. Since there are infinitely many choices for k, $AX = B$ has infinitely many solutions. ∎

Of special interest are square matrices with 1's on the main diagonal and 0's off the main diagonal, such as

$$\begin{bmatrix} 1 & 0 \\ 0 & 1 \end{bmatrix}, \quad \begin{bmatrix} 1 & 0 & 0 \\ 0 & 1 & 0 \\ 0 & 0 & 1 \end{bmatrix}, \quad \begin{bmatrix} 1 & 0 & 0 & 0 \\ 0 & 1 & 0 & 0 \\ 0 & 0 & 1 & 0 \\ 0 & 0 & 0 & 1 \end{bmatrix}, \quad \text{etc.}$$

A matrix of this kind is called an ***identity matrix*** and is denoted by I. If it is important to emphasize the size, we shall write I_n for the $n \times n$ identity matrix.

If A is an $m \times n$ matrix, then, as illustrated in the next example, $AI_n = A$ and $I_m A = A$. Thus, an identity matrix plays much the same role in matrix arithmetic as the number 1 plays in the numerical relationships $a \cdot 1 = a$ and $1 \cdot a = a$.

Example 21

Consider the matrix

$$A = \begin{bmatrix} a_{11} & a_{12} & a_{13} \\ a_{21} & a_{22} & a_{23} \end{bmatrix}$$

Then

$$I_2 A = \begin{bmatrix} 1 & 0 \\ 0 & 1 \end{bmatrix} \begin{bmatrix} a_{11} & a_{12} & a_{13} \\ a_{21} & a_{22} & a_{23} \end{bmatrix} = \begin{bmatrix} a_{11} & a_{12} & a_{13} \\ a_{21} & a_{22} & a_{23} \end{bmatrix} = A$$

and

$$AI_3 = \begin{bmatrix} a_{11} & a_{12} & a_{13} \\ a_{21} & a_{22} & a_{23} \end{bmatrix} \begin{bmatrix} 1 & 0 & 0 \\ 0 & 1 & 0 \\ 0 & 0 & 1 \end{bmatrix} = \begin{bmatrix} a_{11} & a_{12} & a_{13} \\ a_{21} & a_{22} & a_{23} \end{bmatrix} = A$$

If A is any square matrix, and if a matrix B can be found such that $AB = BA = I$, then A is said to be ***invertible*** and B is called an ***inverse*** of A.

Example 22

The matrix

$$B = \begin{bmatrix} 3 & 5 \\ 1 & 2 \end{bmatrix} \quad \text{is an inverse of} \quad A = \begin{bmatrix} 2 & -5 \\ -1 & 3 \end{bmatrix}$$

since

$$AB = \begin{bmatrix} 2 & -5 \\ -1 & 3 \end{bmatrix} \begin{bmatrix} 3 & 5 \\ 1 & 2 \end{bmatrix} = \begin{bmatrix} 1 & 0 \\ 0 & 1 \end{bmatrix} = I$$

and

$$BA = \begin{bmatrix} 3 & 5 \\ 1 & 2 \end{bmatrix} \begin{bmatrix} 2 & -5 \\ -1 & 3 \end{bmatrix} = \begin{bmatrix} 1 & 0 \\ 0 & 1 \end{bmatrix} = I$$

Example 23

The matrix

$$A = \begin{bmatrix} 1 & 4 & 0 \\ 2 & 5 & 0 \\ 3 & 6 & 0 \end{bmatrix}$$

is not invertible. To see why, let

$$B = \begin{bmatrix} b_{11} & b_{12} & b_{13} \\ b_{21} & b_{22} & b_{23} \\ b_{31} & b_{32} & b_{33} \end{bmatrix}$$

be any 3×3 matrix. From Example 17 the third column of BA is

$$\begin{bmatrix} b_{11} & b_{12} & b_{13} \\ b_{21} & b_{22} & b_{23} \\ b_{31} & b_{32} & b_{33} \end{bmatrix} \begin{bmatrix} 0 \\ 0 \\ 0 \end{bmatrix} = \begin{bmatrix} 0 \\ 0 \\ 0 \end{bmatrix}$$

Thus

$$BA \neq I = \begin{bmatrix} 1 & 0 & 0 \\ 0 & 1 & 0 \\ 0 & 0 & 1 \end{bmatrix}$$

It is reasonable to ask whether an invertible matrix can have more than one inverse. The next theorem shows the answer is no—an invertible matrix has exactly one inverse.

Theorem 5. *If B and C are both inverses of the matrix A, then $B = C$.*

Proof. Since B is an inverse of A, $BA = I$. Multiplying both sides on the right by C gives $(BA)C = IC = C$. But $(BA)C = B(AC) = BI = B$, so that $B = C$. ▮

As a consequence of this important result, we can now speak of "the" inverse of an invertible matrix. If A is invertible, then its inverse will be denoted by the symbol A^{-1}. Thus

$$AA^{-1} = I \quad \text{and} \quad A^{-1}A = I$$

The inverse of A plays much the same role in matrix arithmetic that reciprocal a^{-1} plays in the numerical relationships $aa^{-1} = 1$ and $a^{-1}a = 1$.

Example 24

Consider the 2×2 matrix

$$A = \begin{bmatrix} a & b \\ c & d \end{bmatrix}$$

If $ad - bc \neq 0$, then

$$A^{-1} = \frac{1}{ad - bc}\begin{bmatrix} d & -b \\ -c & a \end{bmatrix} = \begin{bmatrix} \dfrac{d}{ad - bc} & -\dfrac{b}{ad - bc} \\ -\dfrac{c}{ad - bc} & \dfrac{a}{ad - bc} \end{bmatrix}$$

since $AA^{-1} = I_2$ and $A^{-1}A = I_2$ (verify). In the next section we shall show how to find inverses of invertible matrices whose sizes are greater than 2×2.

Theorem 6. *If A and B are invertible matrices of the same size, then*

(a) AB is invertible
(b) $(AB)^{-1} = B^{-1}A^{-1}$

Proof. If we can show that $(AB)(B^{-1}A^{-1}) = (B^{-1}A^{-1})(AB) = I$, then we will have simultaneously established that AB is invertible and that $(AB)^{-1} = B^{-1}A^{-1}$. But $(AB)(B^{-1}A^{-1}) = A(BB^{-1})A^{-1} = AIA^{-1} = AA^{-1} = I$. Similarly $(B^{-1}A^{-1})(AB) = I$. ∎

Although we will not prove it, this result can be extended to include three or more factors. Thus we can state the following general rule.

A product of invertible matrices is always invertible, and the inverse of the product is the product of the inverses in the reverse order.

Example 25

Consider the matrices

$$A = \begin{bmatrix} 1 & 2 \\ 1 & 3 \end{bmatrix} \quad B = \begin{bmatrix} 3 & 2 \\ 2 & 2 \end{bmatrix} \quad AB = \begin{bmatrix} 7 & 6 \\ 9 & 8 \end{bmatrix}$$

Applying the formula given in Example 24, we obtain

$$A^{-1} = \begin{bmatrix} 3 & -2 \\ -1 & 1 \end{bmatrix} \quad B^{-1} = \begin{bmatrix} 1 & -1 \\ -1 & \frac{3}{2} \end{bmatrix} \quad (AB)^{-1} = \begin{bmatrix} 4 & -3 \\ -\frac{9}{2} & \frac{7}{2} \end{bmatrix}$$

Also

$$B^{-1}A^{-1} = \begin{bmatrix} 1 & -1 \\ -1 & \frac{3}{2} \end{bmatrix}\begin{bmatrix} 3 & -2 \\ -1 & 1 \end{bmatrix} = \begin{bmatrix} 4 & -3 \\ -\frac{9}{2} & \frac{7}{2} \end{bmatrix}$$

Therefore $(AB)^{-1} = B^{-1}A^{-1}$ as guaranteed by Theorem 6.

If A is a square matrix and n is a positive integer, we define

$$A^n = \underbrace{AA \cdots A}_{n\text{-factors}}$$

$$A^0 = I$$

If, in addition, A is invertible, we define

$$A^{-n} = (A^{-1})^n = \underbrace{A^{-1}A^{-1}\cdots A^{-1}}_{n\text{-factors}}$$

Theorem 7. *If A is an invertible matrix, then:*

(a) A^{-1} is invertible and $(A^{-1})^{-1} = A$.
(b) A^n is invertible and $(A^n)^{-1} = (A^{-1})^n$ for $n = 0, 1, 2, \ldots$.
(c) For any nonzero scalar k, kA is invertible and $(kA)^{-1} = \frac{1}{k}A^{-1}$.

Proof.
(*a*) Since $AA^{-1} = A^{-1}A = I$, A^{-1} is invertible and $(A^{-1})^{-1} = A$.
(*b*) This part is left as an exercise.
(*c*) If k is any nonzero scalar, results (*l*) and (*m*) of Theorem 2 enable us to write

$$(kA)\left(\frac{1}{k}A^{-1}\right) = \frac{1}{k}(kA)A^{-1} = \left(\frac{1}{k}k\right)AA^{-1} = (1)I = I$$

Similarly $\left(\frac{1}{k}A^{-1}\right)(kA) = I$ so that kA is invertible and $(kA)^{-1} = \frac{1}{k}A^{-1}$. ∎

We conclude this section by observing that if A is a square matrix and r and s are integers, then the following familiar laws of exponents are valid:

$$A^rA^s = A^{r+s} \qquad (A^r)^s = A^{rs}$$

The proofs are left as exercises.

EXERCISE SET 1.5

1. Let

$$A = \begin{bmatrix} 3 & 2 \\ -1 & 3 \end{bmatrix} \quad B = \begin{bmatrix} 4 & 0 \\ 1 & 5 \end{bmatrix} \quad C = \begin{bmatrix} 0 & -1 \\ 4 & 6 \end{bmatrix} \quad a = -3 \quad b = 2$$

Show
(a) $A + (B + C) = (A + B) + C$ (b) $(AB)C = A(BC)$
(c) $(a + b)C = aC + bC$ (d) $a(B - C) = aB - aC$

2. Using the matrices and scalars in Exercise 1, show
(a) $a(BC) = (aB)C = B(aC)$ (b) $A(B - C) = AB - AC$.

3. Use the formula given in Example 24 to compute the inverses of the following matrices.

$$A = \begin{bmatrix} 3 & 1 \\ 5 & 2 \end{bmatrix} \quad B = \begin{bmatrix} 2 & -3 \\ 4 & 4 \end{bmatrix} \quad C = \begin{bmatrix} 2 & 0 \\ 0 & 3 \end{bmatrix}$$

4. Verify that the matrices A and B in Exercise 3 satisfy the relationship $(AB)^{-1} = B^{-1}A^{-1}$.

5. Let A and B be square matrices of the same size. Is $(AB)^2 = A^2B^2$ a valid matrix identity? Justify your answer.

6. Let A be an invertible matrix whose inverse is

$$\begin{bmatrix} 3 & 4 \\ 5 & 6 \end{bmatrix}$$

Find the matrix A.

7. Let A be an invertible matrix, and suppose that the inverse of $7A$ is

$$\begin{bmatrix} -1 & 2 \\ 4 & -7 \end{bmatrix}$$

Find the matrix A.

8. Let A be the matrix

$$\begin{bmatrix} 1 & 0 \\ 2 & 3 \end{bmatrix}$$

Compute A^3, A^{-3}, and $A^2 - 2A + I$.

9. Let A be the matrix

$$\begin{bmatrix} 1 & 1 & 0 \\ 0 & 1 & 1 \\ 1 & 0 & 1 \end{bmatrix}$$

Determine if A is invertible, and if so, find its inverse. (***Hint.*** Solve $AX = I$ by equating corresponding entries on the two sides.)

10. Find the inverse of

$$\begin{bmatrix} \cos\theta & \sin\theta \\ -\sin\theta & \cos\theta \end{bmatrix}$$

11. (a) Find 2×2 matrices A and B such that

$$(A + B)^2 \neq A^2 + 2AB + B^2$$

(b) Show that, if A and B are square matrices such that $AB = BA$, then

$$(A + B)^2 = A^2 + 2AB + B^2$$

(c) Find an expansion of $(A + B)^2$ that is valid for all square matrices A and B having the same size.

12. Consider the matrix

$$A = \begin{bmatrix} a_{11} & 0 & 0 & \cdots & 0 \\ 0 & a_{22} & 0 & \cdots & 0 \\ \vdots & \vdots & \vdots & & \vdots \\ 0 & 0 & 0 & \cdots & a_{nn} \end{bmatrix}$$

where $a_{11}a_{22} \cdots a_{nn} \neq 0$. Show that A is invertible and find its inverse.

13. Assume that A is a square matrix which satisfies $A^2 - 3A + I = 0$. Show that $A^{-1} = 3I - A$.

14. (a) Show that a matrix with a row of zeros cannot have an inverse.
(b) Show that a matrix with a column of zeros cannot have an inverse.

15. Is the sum of two invertible matrices necessarily invertible?

16. Let A and B be square matrices such that $AB = 0$. Show that A cannot be invertible unless $B = 0$.

17. In Theorem 3 why didn't we write part (d) as $A0 = 0 = 0A$?

18. The real equation $a^2 = 1$ has exactly two solutions. Find at least eight different 3×3 matrices that satisfy the matrix equation $A^2 = I_3$. (***Hint.*** Look for solutions in which all the entries off the main diagonal are zero.)

19. Let $AX = B$ be any consistent system of linear equations, and let X_1 be a fixed solution. Show that every solution to the system can be written in the form $X = X_1 + X_0$ where X_0 is a solution to $AX = 0$. Show also that every matrix of this form is a solution.

20. Apply parts (d) and (m) of Theorem 2 to the matrices A, B, and $(-1)C$ to derive the result in part (f).

21. Prove part (b) of Theorem 2.

22. Prove Theorem 3.

23. Prove part (c) of Theorem 2.

24. Prove part (c) of Theorem 7.

25. Consider the laws of exponents $A^rA^s = A^{r+s}$ and $(A^r)^s = A^{rs}$.
(a) Show that if A is any square matrix, these laws are valid for all nonnegative integer values of r and s.
(b) Show that if A is invertible, then these laws are valid for all integer values of r and s.

26. Show that if A is invertible and k is any nonzero scalar, then $(kA)^n = k^nA^n$ for all integer values of n.

27. (a) Show that if A is invertible and $AB = AC$ then $B = C$.
(b) Explain why part (a) and Example 20 do not contradict one another.

1.6 ELEMENTARY MATRICES AND A METHOD FOR FINDING A^{-1}

In this section we shall develop a simple scheme or algorithm for finding the inverse of an invertible matrix.

Definition. An $n \times n$ matrix is called an ***elementary matrix*** if it can be obtained from the $n \times n$ identity matrix I_n by performing a single elementary row operation.

Example 26

Listed below are four elementary matrices and the operations that produce them.

(i) $\begin{bmatrix} 1 & 0 \\ 0 & -3 \end{bmatrix}$ (ii) $\begin{bmatrix} 1 & 0 & 0 & 0 \\ 0 & 0 & 0 & 1 \\ 0 & 0 & 1 & 0 \\ 0 & 1 & 0 & 0 \end{bmatrix}$ (iii) $\begin{bmatrix} 1 & 0 & 3 \\ 0 & 1 & 0 \\ 0 & 0 & 1 \end{bmatrix}$ (iv) $\begin{bmatrix} 1 & 0 & 0 \\ 0 & 1 & 0 \\ 0 & 0 & 1 \end{bmatrix}$

(i)	(ii)	(iii)	(iv)
Multiply the second row of I_2 by -3	Interchange the second and fourth rows of I_4	Add 3 times the third row of I_3 to the first row	Multiply the first row of I_3 by 1

When a matrix A is multiplied on the *left* by an elementary matrix E the effect is to perform an elementary row operation on A. This is the content of the following theorem, which we state without proof.

Theorem 8. *If the elementary matrix E results from performing a certain row operation on I_m and if A is an $m \times n$ matrix, then the product EA is the matrix that results when this same row operation is performed on A.*

The following example illustrates this idea.

Example 27

Consider the matrix

$$A = \begin{bmatrix} 1 & 0 & 2 & 3 \\ 2 & -1 & 3 & 6 \\ 1 & 4 & 4 & 0 \end{bmatrix}$$

and consider the elementary matrix

$$E = \begin{bmatrix} 1 & 0 & 0 \\ 0 & 1 & 0 \\ 3 & 0 & 1 \end{bmatrix}$$

which results from adding 3 times the first row of I_3 to the third row. The product EA is

$$EA = \begin{bmatrix} 1 & 0 & 2 & 3 \\ 2 & -1 & 3 & 6 \\ 4 & 4 & 10 & 9 \end{bmatrix}$$

which is precisely the same matrix that results when we add 3 times the first row of A to the third row.

REMARK. Theorem 8 is primarily of theoretical interest and will be used for developing some results about matrices and systems of linear equations. From a computational viewpoint, it is preferable to perform row operations directly rather than multiply on the left by an elementary matrix.

If an elementary row operation is applied to an identity matrix I to produce an elementary matrix E, then there is a second row operation which, when applied to E, produces I back again. For example, if E is obtained by multiplying the ith row of I by a nonzero constant c, then I can be recovered if the ith row of E is multiplied by $1/c$. The various possibilities are listed in Figure 1.4.

Row operation on I that produces E	*Row operation on E that reproduces I*
Multiply row i by $c \neq 0$.	Multiply row i by $1/c$
Interchange rows i and j.	Interchange rows i and j.
Add c times row i to row j.	Add $-c$ times row i to row j

Figure 1.4

The operations on the right side of this table are called the ***inverse operations*** of the corresponding operations on the left.

Example 28

Using the results in Figure 1.4, the first three elementary matrices given in Example 26 can be restored to identity matrices by applying the following row operations: multiply the second row in (i) by $-1/3$; interchange the second and fourth rows in (ii); add -3 times the third row of (iii) to the first row.

The next theorem gives an important property of elementary matrices.

Theorem 9. *Every elementary matrix is invertible, and the inverse is also an elementary matrix.*

Proof. If E is any elementary matrix, then by the observations above, I can be obtained from E by performing a single elementary row operation. Let E_0 be the elementary matrix that results when this row operation is performed on I. Applying Theorem 8, we find that

$$E_0E = I \tag{1.6}$$

To complete the proof, we shall show that

$$EE_0 = I$$

Since E_0 is an elementary matrix, there is an elementary row operation that, when applied to E_0, produces I. Let E_1 be the elementary matrix that results when this row operation is performed on I. Applying Theorem 8 again we obtain

$$E_1 E_0 = I \tag{1.7}$$

Multiplying both sides of equation 1.6 on the left by E_1 gives $E_1 E_0 E = E_1$ or $IE = E_1$ or $E = E_1$. Substituting E for E_1 in equation 1.7 gives $EE_0 = I$, which completes the proof. ▮

If a matrix B can be obtained from a matrix A by performing a finite sequence of elementary row operations then obviously we can get from B back to A by performing the inverses of these elementary row operations in reverse order. Matrices that can be obtained from one another by a finite sequence of elementary row operations are said to be ***row equivalent***.

The next theorem establishes some fundamental relationships between $n \times n$ matrices and systems of n linear equations in n unknowns. These results are extremely important and will be used many times in later sections.

Theorem 10. *If A is an $n \times n$ matrix, then the following statements are equivalent, that is, are all true or all false.*

(*a*) *A is invertible.*
(*b*) *$AX = 0$ has only the trivial solution.*
(*c*) *A is row equivalent to I_n.*

Proof. We shall prove the equivalence by establishing the following chain of implications: $(a) \Rightarrow (b) \Rightarrow (c) \Rightarrow (a)$.

$(a) \Rightarrow (b)$: Assume A is invertible and let X_0 be any solution to $AX = 0$. Thus $AX_0 = 0$. Multiplying both sides of this equation by A^{-1} gives $A^{-1}(AX_0) = A^{-1}0$, or $(A^{-1}A)X_0 = 0$, or $IX_0 = 0$, or $X_0 = 0$. Thus $AX = 0$ has only the trivial solution.

$(b) \Rightarrow (c)$: Let $AX = 0$ be the matrix form of the system

$$\begin{aligned} a_{11}x_1 + a_{12}x_2 + \cdots + a_{1n}x_n &= 0 \\ a_{21}x_1 + a_{22}x_2 + \cdots + a_{2n}x_n &= 0 \\ \vdots \qquad\quad \vdots \qquad\qquad\quad \vdots \quad &\quad \vdots \\ a_{n1}x_1 + a_{n2}x_2 + \cdots + a_{nn}x_n &= 0 \end{aligned} \tag{1.8}$$

and assume the system has only the trivial solution. If we solve by Gauss-Jordan elimination, then the system of equations corresponding to the reduced row-echelon form of the augmented matrix will be

$$\begin{array}{cccc} x_1 & & & = 0 \\ & x_2 & & = 0 \\ & & \ddots & \\ & & x_n & = 0 \end{array} \tag{1.9}$$

Thus the augmented matrix

$$\begin{bmatrix} a_{11} & a_{12} & \cdots & a_{1n} & 0 \\ a_{21} & a_{22} & \cdots & a_{2n} & 0 \\ \vdots & \vdots & & \vdots & \vdots \\ a_{n1} & a_{n2} & \cdots & a_{nn} & 0 \end{bmatrix}$$

for (1.8) can be reduced to the augmented matrix

$$\begin{bmatrix} 1 & 0 & 0 & \cdots & 0 & 0 \\ 0 & 1 & 0 & \cdots & 0 & 0 \\ 0 & 0 & 1 & \cdots & 0 & 0 \\ \vdots & \vdots & \vdots & & \vdots & \vdots \\ 0 & 0 & 0 & \cdots & 1 & 0 \end{bmatrix}$$

for (1.9) by a sequence of elementary row operations. If we disregard the last column (of zeros) in each of these matrices, we can conclude that A can be reduced to I_n by a sequence of elementary row operations; that is, A is row equivalent to I_n.

(c) $\Rightarrow$ *(a)*: Assume A is row equivalent to I_n, so that A can be reduced to I_n by a finite sequence of elementary row operations. By Theorem 8 each of these operations can be accomplished by multiplying on the left by an appropriate elementary matrix. Thus, we can find elementary matrices $E_1, E_2, \ldots, E_k$ such that

$$E_k \cdots E_2 E_1 A = I_n \tag{1.10}$$

By Theorem 9, $E_1, E_2, \ldots, E_k$ are invertible. Multiplying both sides of equation (1.10) on the left successively by $E_k^{-1}, \ldots, E_2^{-1}, E_1^{-1}$ we obtain

$$A = E_1^{-1} E_2^{-1} \cdots E_k^{-1} I_n = E_1^{-1} E_2^{-1} \cdots E_k^{-1} \tag{1.11}$$

Since (1.11) expresses A as a product of invertible matrices, we can conclude that A is invertible. ▮

REMARK. Because I_n is in reduced row-echelon form and because the reduced row-echelon form of a matrix A is unique, part (c) of Theorem 10 is equivalent to stating that I_n is the reduced row-echelon form of A.

As our first application of this theorem, we shall establish a method for determining the inverse of an invertible matrix.

Inverting the left and right sides of (1.11) yields $A^{-1} = E_k \cdots E_2 E_1$, or equivalently

$$A^{-1} = E_k \cdots E_2 E_1 I_n \tag{1.12}$$

which tells us that A^{-1} can be obtained by multiplying I_n successively on the left by the elementary matrices $E_1, E_2, \ldots, E_k$. Since each multiplication on the left by one of these elementary matrices performs a row operation, it follows, by comparing equations (1.10) and (1.12), that *the sequence of row operations that reduces A* *to I_n will reduce I_n to A^{-1}.* Thus, to find the inverse of an invertible matrix A, we must find a sequence of elementary row operations that reduces A to the identity and then perform this same sequence of operations on I_n to obtain A^{-1}. A simple method for carrying out this procedure is given in the following example.

Example 29

Find the inverse of

$$A = \begin{bmatrix} 1 & 2 & 3 \\ 2 & 5 & 3 \\ 1 & 0 & 8 \end{bmatrix}$$

We wish to reduce A to the identity matrix by row operations and simultaneously apply these operations to I to produce A^{-1}. This can be accomplished by adjoining the identity matrix to the right of A and applying row operations to both sides until the left side is reduced to I. The final matrix will then have the form $[I \mid A^{-1}]$. The computations can be carried out as follows.

$$\left[\begin{array}{rrr|rrr} 1 & 2 & 3 & 1 & 0 & 0 \\ 2 & 5 & 3 & 0 & 1 & 0 \\ 1 & 0 & 8 & 0 & 0 & 1 \end{array}\right]$$

$$\left[\begin{array}{rrr|rrr} 1 & 2 & 3 & 1 & 0 & 0 \\ 0 & 1 & -3 & -2 & 1 & 0 \\ 0 & -2 & 5 & -1 & 0 & 1 \end{array}\right]$$

We added −2 times the first row to the second and −1 times the first row to the third.

$$\left[\begin{array}{rrr|rrr} 1 & 2 & 3 & 1 & 0 & 0 \\ 0 & 1 & -3 & -2 & 1 & 0 \\ 0 & 0 & -1 & -5 & 2 & 1 \end{array}\right]$$

We added 2 times the second row to the third.

$$\left[\begin{array}{rrr|rrr} 1 & 2 & 3 & 1 & 0 & 0 \\ 0 & 1 & -3 & -2 & 1 & 0 \\ 0 & 0 & 1 & 5 & -2 & -1 \end{array}\right]$$

We multiplied the third row by −1.

$$\left[\begin{array}{rrr|rrr} 1 & 2 & 0 & -14 & 6 & 3 \\ 0 & 1 & 0 & 13 & -5 & -3 \\ 0 & 0 & 1 & 5 & -2 & -1 \end{array}\right]$$

We added 3 times the third row to the second and −3 times the third row to the first.

$$\left[\begin{array}{rrr|rrr} 1 & 0 & 0 & -40 & 16 & 9 \\ 0 & 1 & 0 & 13 & -5 & -3 \\ 0 & 0 & 1 & 5 & -2 & -1 \end{array}\right]$$

We added −2 times the second row to the first.

Thus

$$A^{-1} = \begin{bmatrix} -40 & 16 & 9 \\ 13 & -5 & -3 \\ 5 & -2 & -1 \end{bmatrix}$$

Often it will not be known in advance whether a given matrix is invertible. If the procedure used in this example is attempted on a matrix that is not invertible, then, by part (c) of Theorem 10, it will be impossible to reduce the left side to I by row operations. At some point in the computation a row of zeros will occur on the left side. It can then be concluded that the given matrix is not invertible and the computations can be stopped.

Example 30

Consider the matrix

$$A = \begin{bmatrix} 1 & 6 & 4 \\ 2 & 4 & -1 \\ -1 & 2 & 5 \end{bmatrix}$$

Applying the procedure of Example 29 yields

$$\left[\begin{array}{rrr|rrr} 1 & 6 & 4 & 1 & 0 & 0 \\ 2 & 4 & -1 & 0 & 1 & 0 \\ -1 & 2 & 5 & 0 & 0 & 1 \end{array}\right]$$

$$\left[\begin{array}{rrr|rrr} 1 & 6 & 4 & 1 & 0 & 0 \\ 0 & -8 & -9 & -2 & 1 & 0 \\ 0 & 8 & 9 & 1 & 0 & 1 \end{array}\right]$$

We added -2 times the first row to the second and added the first row to the third.

$$\left[\begin{array}{rrr|rrr} 1 & 6 & 4 & 1 & 0 & 0 \\ 0 & -8 & -9 & -2 & 1 & 0 \\ 0 & 0 & 0 & -1 & 1 & 1 \end{array}\right]$$

We added the second row to the third.

Since we have obtained a row of zeros on the left side, A is not invertible.

Example 31

In Example 29 we showed that

$$A = \begin{bmatrix} 1 & 2 & 3 \\ 2 & 5 & 3 \\ 1 & 0 & 8 \end{bmatrix}$$

is an invertible matrix. From Theorem 10 we can now conclude that the system of equations

$$\begin{aligned} x_1 + 2x_2 + 3x_3 &= 0 \\ 2x_1 + 5x_2 + 3x_3 &= 0 \\ x_1 \qquad\quad + 8x_3 &= 0 \end{aligned}$$

has only the trivial solution.

EXERCISE SET 1.6

1. Which of the following are elementary matrices?

(a) $\begin{bmatrix} 2 & 0 \\ 0 & 1 \end{bmatrix}$ (b) $\begin{bmatrix} 1 & 0 \\ 3 & 1 \end{bmatrix}$ (c) $\begin{bmatrix} 2 & 0 \\ 0 & 2 \end{bmatrix}$

(d) $\begin{bmatrix} 0 & 1 & 0 \\ 1 & 0 & 0 \\ 0 & 0 & 1 \end{bmatrix}$ (e) $\begin{bmatrix} 0 & 1 & 0 \\ 0 & 0 & 1 \\ 0 & 0 & 1 \end{bmatrix}$

(f) $\begin{bmatrix} 1 & 0 & 0 \\ 0 & 1 & -3 \\ 0 & 0 & 1 \end{bmatrix}$ (g) $\begin{bmatrix} 1 & 0 & 0 & 0 \\ 0 & 1 & 0 & 0 \\ 0 & 1 & 1 & 0 \\ 0 & 0 & 0 & 1 \end{bmatrix}$

2. Determine the row operation that will restore the given elementary matrix to an identity matrix.

(a) $\begin{bmatrix} 1 & 0 \\ 5 & 1 \end{bmatrix}$ (b) $\begin{bmatrix} 0 & 0 & 1 \\ 0 & 1 & 0 \\ 1 & 0 & 0 \end{bmatrix}$ (c) $\begin{bmatrix} 1 & 0 & 0 & 0 \\ 0 & 8 & 0 & 0 \\ 0 & 0 & 1 & 0 \\ 0 & 0 & 0 & 1 \end{bmatrix}$

3. Consider the matrices

$$A = \begin{bmatrix} 1 & 2 & 3 \\ 4 & 5 & 6 \\ 7 & 8 & 9 \end{bmatrix} \quad B = \begin{bmatrix} 7 & 8 & 9 \\ 4 & 5 & 6 \\ 1 & 2 & 3 \end{bmatrix} \quad C = \begin{bmatrix} 1 & 2 & 3 \\ 4 & 5 & 6 \\ 9 & 12 & 15 \end{bmatrix}$$

Find elementary matrices E_1, E_2, E_3, and E_4 such that
(a) $E_1A = B$ (b) $E_2B = A$ (c) $E_3A = C$ (d) $E_4C = A$

4. In Exercise 3 is it possible to find an elementary matrix E such that $EB = C$? Justify your answer.

In Exercises 5–7 use the method shown in Examples 29 and 30 to find the inverse of the given matrix if the matrix is invertible.

5. (a) $\begin{bmatrix} 1 & 2 \\ 3 & 5 \end{bmatrix}$ (b) $\begin{bmatrix} -2 & 3 \\ 3 & -5 \end{bmatrix}$ (c) $\begin{bmatrix} 8 & -6 \\ -4 & 3 \end{bmatrix}$

6. (a) $\begin{bmatrix} 3 & 4 & -1 \\ 1 & 0 & 3 \\ 2 & 5 & -4 \end{bmatrix}$ (b) $\begin{bmatrix} 3 & 1 & 5 \\ 2 & 4 & 1 \\ -4 & 2 & -9 \end{bmatrix}$ (c) $\begin{bmatrix} 1 & 0 & 1 \\ 0 & 1 & 1 \\ 1 & 1 & 0 \end{bmatrix}$

(d) $\begin{bmatrix} 2 & 6 & 6 \\ 2 & 7 & 6 \\ 2 & 7 & 7 \end{bmatrix}$ (e) $\begin{bmatrix} 1 & 0 & 1 \\ -1 & 1 & 1 \\ 0 & 1 & 0 \end{bmatrix}$ (f) $\begin{bmatrix} \frac{1}{5} & \frac{1}{5} & \frac{1}{5} \\ \frac{1}{5} & \frac{1}{5} & -\frac{4}{5} \\ -\frac{2}{5} & \frac{1}{10} & \frac{1}{10} \end{bmatrix}$

7. (a) $\begin{bmatrix} \frac{1}{\sqrt{2}} & \frac{1}{\sqrt{2}} & 0 \\ -\frac{1}{\sqrt{2}} & \frac{1}{\sqrt{2}} & 0 \\ 0 & 0 & 1 \end{bmatrix}$ (b) $\begin{bmatrix} 1 & 0 & 0 & 0 \\ 1 & 2 & 0 & 0 \\ 1 & 2 & 4 & 0 \\ 1 & 2 & 4 & 8 \end{bmatrix}$ (c) $\begin{bmatrix} 5 & 11 & 7 & 3 \\ 2 & 1 & 4 & -5 \\ 3 & -2 & 8 & 7 \\ 0 & 0 & 0 & 0 \end{bmatrix}$

8. Show that the matrix

$$A = \begin{bmatrix} \cos\theta & \sin\theta & 0 \\ -\sin\theta & \cos\theta & 0 \\ 0 & 0 & 1 \end{bmatrix}$$

is invertible for all values of θ and find A^{-1}.

9. Consider the matrix

$$A = \begin{bmatrix} 1 & 0 \\ 3 & 4 \end{bmatrix}$$

(a) Find elementary matrices E_1 and E_2 such that $E_2E_1A = I$.
(b) Write A^{-1} as a product of two elementary matrices.
(c) Write A as a product of two elementary matrices.

10. Perform the following row operations on

$$A = \begin{bmatrix} 3 & 1 & 0 \\ -2 & 1 & 4 \\ 3 & 5 & 5 \end{bmatrix}$$

by multiplying A on the left by a suitable elementary matrix. Check your answer in each case by performing the row operation directly on A.
(a) Interchange the first and third rows.
(b) Multiply the second row by 1/3.
(c) Add twice the second row to the first row.

11. Express the matrix

$$A = \begin{bmatrix} 1 & 3 & 3 & 8 \\ -2 & -5 & 1 & -8 \\ 0 & 1 & 7 & 8 \end{bmatrix}$$

in the form $A = EFR$, where E and F are elementary matrices, and R is in row-echelon form.

12. Show that if

$$A = \begin{bmatrix} 1 & 0 & 0 \\ 0 & 1 & 0 \\ a & b & c \end{bmatrix}$$

is an elementary matrix, then at least one entry in the third row must be a zero.

13. Find the inverse of each of the following 4×4 matrices, where k_1, k_2, k_3, k_4 and k are all nonzero.

(a) $\begin{bmatrix} k_1 & 0 & 0 & 0 \\ 0 & k_2 & 0 & 0 \\ 0 & 0 & k_3 & 0 \\ 0 & 0 & 0 & k_4 \end{bmatrix}$ (b) $\begin{bmatrix} 0 & 0 & 0 & k_1 \\ 0 & 0 & k_2 & 0 \\ 0 & k_3 & 0 & 0 \\ k_4 & 0 & 0 & 0 \end{bmatrix}$ (c) $\begin{bmatrix} k & 0 & 0 & 0 \\ 1 & k & 0 & 0 \\ 0 & 1 & k & 0 \\ 0 & 0 & 1 & k \end{bmatrix}$

14. Prove that if A is an $m \times n$ matrix, there is an invertible matrix C such that CA is in reduced row-echelon form.

15. Prove that if A is an invertible matrix and B is row equivalent to A, then B is also invertible.

1.7 FURTHER RESULTS ON SYSTEMS OF EQUATIONS AND INVERTIBILITY

In this section we shall establish more results about systems of linear equations and invertibility of matrices. Our work will lead to a method for solving n equations in n unknowns that is more efficient than Gaussian elimination for certain kinds of problems.

Theorem 11. *If A is an invertible $n \times n$ matrix, then for each $n \times 1$ matrix B, the system of equations $AX = B$ has exactly one solution, namely, $X = A^{-1}B$.*

Proof. Since $A(A^{-1}B) = B$, $X = A^{-1}B$ is a solution of $AX = B$. To show that this is the only solution, we will assume that X_0 is an arbitrary solution, and then show that X_0 must be the solution $A^{-1}B$.

If X_0 is any solution, then $AX_0 = B$. Multiplying both sides by A^{-1}, we obtain $X_0 = A^{-1}B$. ▮

Example 32

Consider the system of linear equations

$$\begin{aligned} x_1 + 2x_2 + 3x_3 &= 5 \\ 2x_1 + 5x_2 + 3x_3 &= 3 \\ x_1 \qquad\quad + 8x_3 &= 17 \end{aligned}$$

In matrix form this system can be written as $AX = B$, where

$$A = \begin{bmatrix} 1 & 2 & 3 \\ 2 & 5 & 3 \\ 1 & 0 & 8 \end{bmatrix} \qquad X = \begin{bmatrix} x_1 \\ x_2 \\ x_3 \end{bmatrix} \qquad B = \begin{bmatrix} 5 \\ 3 \\ 17 \end{bmatrix}$$

In Example 29 we showed that A is invertible and

$$A^{-1} = \begin{bmatrix} -40 & 16 & 9 \\ 13 & -5 & -3 \\ 5 & -2 & -1 \end{bmatrix}$$

By Theorem 11 the solution of the system is

$$X = A^{-1}B = \begin{bmatrix} -40 & 16 & 9 \\ 13 & -5 & -3 \\ 5 & -2 & -1 \end{bmatrix} \begin{bmatrix} 5 \\ 3 \\ 17 \end{bmatrix} = \begin{bmatrix} 1 \\ -1 \\ 2 \end{bmatrix}$$

or $x_1 = 1, x_2 = -1, x_3 = 2$.

The technique illustrated in this example only applies when the coefficient matrix A is square, that is, when the system has as many equations as unknowns. However, many problems in science and engineering involve systems of this type. The method is particularly useful when it is necessary to solve a whole series of systems

$$AX = B_1, AX = B_2, \ldots, AX = B_k$$

each of which has the same square coefficient matrix A. In this case the solutions

$$X = A^{-1}B_1, X = A^{-1}B_2, \ldots, X = A^{-1}B_k$$

can be obtained using one matrix inversion and k matrix multiplications. This procedure is more efficient than separately applying Gaussian elimination to each of the k systems.

We digress for a moment to illustrate how this situation can arise in applications. In certain applied problems, physical systems are considered that can be described as *black boxes*. This term indicates that the system has been stripped to its bare essentials. As illustrated in Figure 1.5, one imagines simply that if a certain input is applied to the system, then a certain output for the system will result. The internal workings of the system are either unknown or unimportant for the problem—hence, the term *black box*. For many important black-box systems, both the input and the output can be described mathematically as matrices having a single column. For example, if the black box consists of certain electronic circuitry, then the input might be an $n \times 1$ matrix whose entries are n voltages read across certain input terminals, and the output might be an $n \times 1$ matrix whose entries are the resulting currents in n output wires. Mathematically speaking, such a system does nothing more than transform an $n \times 1$ input matrix into an

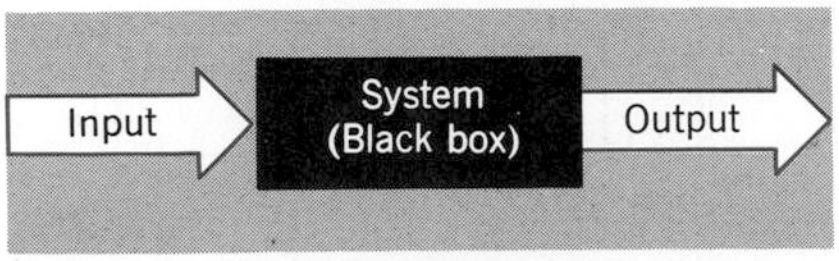

Figure 1.5

$n \times 1$ output matrix. For a large class of black-box systems an input matrix C is related to the output matrix B by a matrix equation

$$AC = B$$

where A is an $n \times n$ matrix whose entries are physical parameters determined by the system. A system of this kind is an example of what is called a *linear physical system.* In applications it is often important to determine what input must be applied to the system to achieve a certain desired output. For a linear physical system of the type we have just discussed, this amounts to solving the equation $AX = B$ for the unknown input X, given the desired output B. Thus if we have a succession of different output matrices $B_1, \ldots, B_k$, and we want to determine the input matrices that produce these given outputs, we must successively solve the k systems of linear equations

$$AX = B_j \qquad j = 1, 2, \ldots, k$$

each of which has the same square coefficient matrix A.

The next theorem simplifies the problem of showing a matrix is invertible. Up to now, to show an $n \times n$ matrix A is invertible, it was necessary to find an $n \times n$ matrix B such that

$$AB = I \qquad \text{and} \qquad BA = I$$

The next theorem shows that if we produce an $n \times n$ matrix B satisfying *either* condition, then the other condition holds automatically.

Theorem 12. *Let A be a square matrix.*

(a) *If B is a square matrix satisfying $BA = I$, then $B = A^{-1}$.*
(b) *If B is a square matrix satisfying $AB = I$, then $B = A^{-1}$.*

Proof. We shall prove (a) and leave (b) as an exercise.

(a) Assume $BA = I$. If we can show that A is invertible, the proof can be completed by multiplying $BA = I$ on both sides by A^{-1} to obtain

$$BAA^{-1} = IA^{-1} \qquad \text{or} \qquad BI = IA^{-1} \qquad \text{or} \qquad B = A^{-1}$$

To show that A is invertible, it suffices to show that the system $AX = 0$ has only the trivial solution (see Theorem 10). However, if we multiply both sides of $AX = 0$ on the left by B, we obtain $BAX = B0$ or $IX = 0$ or $X = 0$. The system of equations $AX = 0$ therefore has only the trivial solution. ∎

We are now in a position to add a fourth statement equivalent to the three given in Theorem 10.

Theorem 13. *If A is an $n \times n$ matrix, then the following statements are equivalent.*

(a) *A is invertible.*
(b) *$AX = 0$ has only the trivial solution.*

(*c*) *A is row equivalent to I_n.*
(*d*) *$AX = B$ is consistent for every $n \times 1$ matrix B.*

Proof. Since we proved in Theorem 10 that (*a*), (*b*), and (*c*) are equivalent, it will be sufficient to prove (*a*) ⇒ (*d*) and (*d*) ⇒ (*a*).

(*a*) ⇒ (*d*): If A is invertible and B is any $n \times 1$ matrix then $X = A^{-1}B$ is a solution of $AX = B$ by Theorem 11. Thus $AX = B$ is consistent.

(*d*) ⇒ (*a*): If the system $AX = B$ is consistent for every $n \times 1$ matrix B then, in particular, the systems

$$AX = \begin{bmatrix} 1 \\ 0 \\ 0 \\ \vdots \\ 0 \end{bmatrix}, \qquad AX = \begin{bmatrix} 0 \\ 1 \\ 0 \\ \vdots \\ 0 \end{bmatrix}, \qquad \ldots, \qquad AX = \begin{bmatrix} 0 \\ 0 \\ 0 \\ \vdots \\ 1 \end{bmatrix}$$

will be consistent. Let X_1 be a solution of the first system, X_2 a solution of the second system, . . . , and X_n a solution of the last system, and let us form an $n \times n$ matrix C having these solutions as columns. Thus C has the form

$$C = [X_1 \mid X_2 \mid \cdots \mid X_n]$$

As discussed in Example 17, the successive columns of the product AC will be

$$AX_1, AX_2, \ldots, AX_n$$

Thus

$$AC = [AX_1 \mid AX_2 \mid \cdots \mid AX_n] = \begin{bmatrix} 1 & 0 & \cdots & 0 \\ 0 & 1 & \cdots & 0 \\ 0 & 0 & \cdots & 0 \\ \vdots & \vdots & & \vdots \\ 0 & 0 & \cdots & 1 \end{bmatrix} = I$$

By part (*b*) of Theorem 12 it follows that $C = A^{-1}$. Thus A is invertible. ∎

In our later work the following fundamental problem will occur over and over again in various contexts.

A Fundamental Problem. Let A be a fixed $m \times n$ matrix. Find all $m \times 1$ matrices B such that the system of equations $AX = B$ is consistent.

If A is an invertible matrix, Theorem 11 completely solves this problem by asserting that for *every* $m \times 1$ matrix B, $AX = B$ has the unique solution $X = A^{-1}B$. If A is not square, or if A is square but not invertible, then Theorem 11 does not apply. In these cases we would like to determine what conditions, if any, the matrix B must satisfy in order for $AX = B$ to be consistent. The following example illustrates how Gaussian elimination can be used to determine such conditions.

Example 33

What conditions must b_1, b_2, and b_3 satisfy in order for the system of equations

$$\begin{aligned} x_1 + x_2 + 2x_3 &= b_1 \\ x_1 + x_3 &= b_2 \\ 2x_1 + x_2 + 3x_3 &= b_3 \end{aligned}$$

to be consistent?

Solution. The augmented matrix is

$$\begin{bmatrix} 1 & 1 & 2 & b_1 \\ 1 & 0 & 1 & b_2 \\ 2 & 1 & 3 & b_3 \end{bmatrix}$$

which can be reduced to row-echelon form as follows.

$$\begin{bmatrix} 1 & 1 & 2 & b_1 \\ 0 & -1 & -1 & b_2 - b_1 \\ 0 & -1 & -1 & b_3 - 2b_1 \end{bmatrix}$$

1 times the first row was added to the second and -2 times the first row was added to the third.

$$\begin{bmatrix} 1 & 1 & 2 & b_1 \\ 0 & 1 & 1 & b_1 - b_2 \\ 0 & -1 & -1 & b_3 - 2b_1 \end{bmatrix}$$

The second row was multiplied by -1.

$$\begin{bmatrix} 1 & 1 & 1 & b_2 \\ 0 & 1 & 1 & b_1 - b_2 \\ 0 & 0 & 0 & b_3 - b_2 - b_1 \end{bmatrix}$$

The second row was added to the third.

It is now evident from the third row in the matrix that the system has a solution if and only if b_1, b_2, and b_3 satisfy the condition

$$b_3 - b_2 - b_1 = 0 \qquad \text{or} \qquad b_3 = b_1 + b_2$$

To express this condition another way, $AX = B$ is consistent if and only if B is a matrix of the form

$$B = \begin{bmatrix} b_1 \\ b_2 \\ b_1 + b_2 \end{bmatrix}$$

where b_1 and b_2 are arbitrary.

EXERCISE SET 1.7

In Exercises 1–6 solve the system using the method of Example 32.

1. $\begin{aligned} x_1 + 2x_2 &= 7 \\ 2x_1 + 5x_2 &= -3 \end{aligned}$

2. $\begin{aligned} 3x_1 - 6x_2 &= 8 \\ 2x_1 + 5x_2 &= 1 \end{aligned}$

3. $\begin{aligned} x_1 + 2x_2 + 2x_3 &= -1 \\ x_1 + 3x_2 + x_3 &= 4 \\ x_1 + 3x_2 + 2x_3 &= 3 \end{aligned}$

4. $\begin{aligned} 2x_1 + x_2 + x_3 &= 7 \\ 3x_1 + 2x_2 + x_3 &= -3 \\ x_2 + x_3 &= 5 \end{aligned}$

5. $\begin{aligned} \tfrac{1}{5}x + \tfrac{1}{5}y + \tfrac{1}{5}z &= 1 \\ \tfrac{1}{5}x + \tfrac{1}{5}y - \tfrac{4}{5}z &= 2 \\ -\tfrac{2}{5}x + \tfrac{1}{10}y + \tfrac{1}{10}z &= 0 \end{aligned}$

6. $\begin{aligned} 3w + x + 7y + 9z &= 4 \\ w + x + 4y + 4z &= 7 \\ -w \qquad - 2y - 3z &= 0 \\ -2w - x - 4y - 6z &= 6 \end{aligned}$

7. Solve the system

$$\begin{aligned} x_1 + 2x_2 + x_3 &= b_1 \\ x_1 - x_2 + x_3 &= b_2 \\ x_1 + x_2 \qquad &= b_3 \end{aligned}$$

when

(a) $b_1 = -1, b_2 = 3, b_3 = 4$ (b) $b_1 = 5, b_2 = 0, b_3 = 0$

(c) $b_1 = -1, b_2 = -1, b_3 = 3$ (d) $b_1 = \frac{1}{2}, b_2 = 3, b_3 = \frac{1}{7}$

8. What conditions must the b's satisfy in order for the given system to be consistent?

(a) $\begin{aligned} x_1 - x_2 + 3x_3 &= b_1 \\ 3x_1 - 3x_2 + 9x_3 &= b_2 \\ -2x_1 + 2x_2 - 6x_3 &= b_3 \end{aligned}$

(b) $\begin{aligned} 2x_1 + 3x_2 - x_3 + x_4 &= b_1 \\ x_1 + 5x_2 + x_3 - 2x_4 &= b_2 \\ -x_1 + 2x_2 + 2x_3 - 3x_4 &= b_3 \\ 3x_1 + x_2 - 3x_3 + 4x_4 &= b_4 \end{aligned}$

9. Consider the matrices

$$A = \begin{bmatrix} 2 & 2 & 3 \\ 1 & 2 & 1 \\ 2 & -2 & 1 \end{bmatrix} \quad \text{and} \quad X = \begin{bmatrix} x_1 \\ x_2 \\ x_3 \end{bmatrix}$$

(a) Show that the equation $AX = X$ can be rewritten as $(A - I)X = 0$ and use this result to solve $AX = X$ for X.

(b) Solve $AX = 4X$.

10. Without using pencil and paper, determine if the following matrices are invertible.

(a) $\begin{bmatrix} 2 & 1 & -3 & 1 \\ 0 & 5 & 4 & 3 \\ 0 & 0 & 1 & 2 \\ 0 & 0 & 0 & 3 \end{bmatrix}$ (b) $\begin{bmatrix} 5 & 1 & 4 & 1 \\ 0 & 0 & 2 & -1 \\ 0 & 0 & 1 & 1 \\ 0 & 0 & 0 & 7 \end{bmatrix}$

Hint. Consider the associated homogeneous systems

$$\begin{aligned} 2x_1 + x_2 - 3x_3 + x_4 &= 0 \\ 5x_2 + 4x_3 + 3x_4 &= 0 \\ x_3 + 2x_4 &= 0 \\ 3x_4 &= 0 \end{aligned} \quad \text{and} \quad \begin{aligned} 5x_1 + x_2 + 4x_3 + x_4 &= 0 \\ 2x_3 - x_4 &= 0 \\ x_3 + x_4 &= 0 \\ 7x_4 &= 0 \end{aligned}$$

11. Let $AX = 0$ be a homogeneous system of n linear equations in n unknowns that has only the trivial solution. Show that if k is any positive integer, then the system $A^kX = 0$ also has only the trivial solution.

12. Let $AX = 0$ be a homogeneous system of n linear equations in n unknowns, and let Q be an invertible matrix. Show that $AX = 0$ has just the trivial solution if and only if $(QA)X = 0$ has just the trivial solution.

13. Show that an $n \times n$ matrix A is invertible if and only if it can be written as a product of elementary matrices.

14. Use part (a) of Theorem 12 to prove part (b).

2 Determinants

2.1 THE DETERMINANT FUNCTION

We are all familiar with functions like $f(x) = \sin x$ and $f(x) = x^2$, which associate a real number $f(x)$ with a real value of the variable x. Since both x and $f(x)$ assume only real values, such functions can be described as real-valued functions of a real variable. In this section we initiate the study of real-valued functions of a matrix variable, that is, functions that associate a real number $f(X)$ with a matrix X. Our primary effort will be devoted to the study of one such function called the determinant function. Our work on the determinant function will have important applications to the theory of systems of linear equations and will also lead us to an explicit formula for the inverse of an invertible matrix.

Before we shall be able to define the determinant function, it will be necessary to establish some results concerning permutations.

Definition. A ***permutation*** of the set of integers $\{1, 2, \ldots, n\}$ is an arrangement of these integers in some order without omissions or repetitions.

Example 1

There are six different permutations of the set of integers $\{1, 2, 3\}$. These are

$$
\begin{array}{ccc}
(1, 2, 3) & (2, 1, 3) & (3, 1, 2) \\
(1, 3, 2) & (2, 3, 1) & (3, 2, 1)
\end{array}
$$

One convenient method of systematically listing permutations is to use a *permutation tree*. This method will be illustrated in our next example.

Example 2

List all permutations of the set of integers $\{1, 2, 3, 4\}$.

Solution.

Consider the Figure 2.1. The four dots labeled 1, 2, 3, 4 at the top of the figure represent the possible choices for the first number in the permutation. The three branches emanating from these dots represent the possible choices for the second position in the permutation. Thus, if the permutation begins (2, –, –, –), the three possibilities for the second position are 1, 3, and 4. The two branches emanating from each dot in the second position represent the possible choices for the third position. Thus, if the permutation begins (2, 3, –, –), the two possible choices for the third position are 1 and 4. Finally, the single branch emanating from each dot in the third position represents the only possible choice for the fourth position. Thus, if the permutation begins (2, 3, 4, –), the only choice for the fourth position is 1. The different permutations can now be listed by tracing out all the possible paths through the "tree" from the first position to the last position. We obtain the following list by this process.

(1, 2, 3, 4)	(2, 1, 3, 4)	(3, 1, 2, 4)	(4, 1, 2, 3)
(1, 2, 4, 3)	(2, 1, 4, 3)	(3, 1, 4, 2)	(4, 1, 3, 2)
(1, 3, 2, 4)	(2, 3, 1, 4)	(3, 2, 1, 4)	(4, 2, 1, 3)
(1, 3, 4, 2)	(2, 3, 4, 1)	(3, 2, 4, 1)	(4, 2, 3, 1)
(1, 4, 2, 3)	(2, 4, 1, 3)	(3, 4, 1, 2)	(4, 3, 1, 2)
(1, 4, 3, 2)	(2, 4, 3, 1)	(3, 4, 2, 1)	(4, 3, 2, 1)

From this example we see that there are 24 permutations of $\{1, 2, 3, 4\}$. This result could have been anticipated without actually listing the permutations by arguing as follows. Since the first position can be filled in four ways and then the second position in three ways, there are $4 \cdot 3$ ways of filling the first two positions. Since the third position can then be filled in two ways, there are $4 \cdot 3 \cdot 2$ ways of filling the first three positions. Finally, since the last position can then be filled in only one way, there are $4 \cdot 3 \cdot 2 \cdot 1 = 24$ ways of filling all four positions. In general, the set $\{1, 2, \ldots, n\}$ will have $n(n-1)(n-2) \cdots 2 \cdot 1 = n!$ different permutations.

To denote a general permutation of the set $\{1, 2, \ldots, n\}$, we shall write $(j_1, j_2, \ldots, j_n)$. Here, j_1 is the first integer in the permutation, j_2 is the second, etc. An ***inversion*** is said to occur in a permutation $(j_1, j_2, \ldots, j_n)$ whenever a larger

Figure 2.1

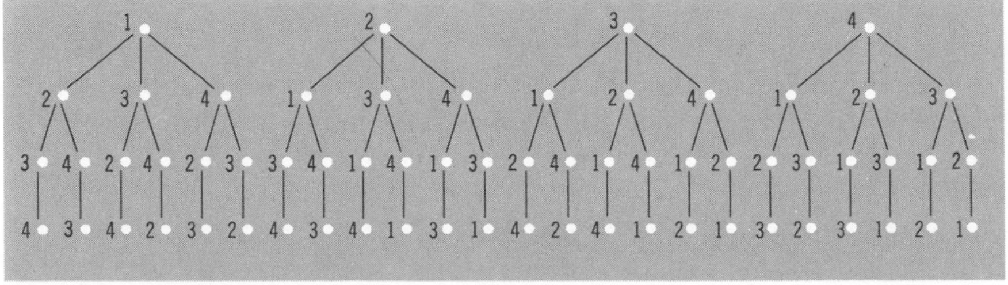

integer precedes a smaller one. The total number of inversions occurring in a permutation can be obtained as follows: (1) Find the number of integers that are less than j_1 and that follow j_1 in the permutation; (2) find the number of integers that are less than j_2 and that follow j_2 in the permutation. Continue this counting process for $j_3, \ldots, j_{n-1}$. The sum of these numbers will be the total number of inversions in the permutation.

Example 3

Determine the number of inversions in the following permutations:

(i) (6, 1, 3, 4, 5, 2) (ii) (2, 4, 1, 3) (iii) (1, 2, 3, 4)

(i) The number of inversions is $5 + 0 + 1 + 1 + 1 = 8$.
(ii) The number of inversions is $1 + 2 + 0 = 3$.
(iii) There are no inversions in this permutation.

Definition. A permutation is called ***even*** if the total number of inversions is an even integer and is called ***odd*** if the total number of inversions is an odd integer.

Example 4

The following table classifies the various permutations of {1, 2, 3} as even or odd.

Permutation	*Number of inversions*	*Classification*
(1, 2, 3)	0	even
(1, 3, 2)	1	odd
(2, 1, 3)	1	odd
(2, 3, 1)	2	even
(3, 1, 2)	2	even
(3, 2, 1)	3	odd

Consider the $n \times n$ matrix

$$A = \begin{bmatrix} a_{11} & a_{12} & \cdots & a_{1n} \\ a_{21} & a_{22} & \cdots & a_{2n} \\ \vdots & \vdots & & \vdots \\ a_{n1} & a_{n2} & \cdots & a_{nn} \end{bmatrix}$$

By an ***elementary product from A*** we shall mean any product of n entries from A, no two of which come from the same row or same column.

Example 5

List all elementary products from the matrices

$$\text{(i)} \begin{bmatrix} a_{11} & a_{12} \\ a_{21} & a_{22} \end{bmatrix} \qquad \text{(ii)} \begin{bmatrix} a_{11} & a_{12} & a_{13} \\ a_{21} & a_{22} & a_{23} \\ a_{31} & a_{32} & a_{33} \end{bmatrix}$$

(i) Since each elementary product has two factors, and since each factor comes from a different row, an elementary product can be written in the form

$$a_{1-}a_{2-}$$

where the blanks designate column numbers. Since no two factors in the product come from the same column, the column numbers must be $\underline{1}\ \underline{2}$ or $\underline{2}\ 1$. The only elementary products are therefore $a_{11}a_{22}$ and $a_{12}a_{21}$.

(ii) Since each elementary product has three factors, each of which comes from a different row, an elementary product can be written in the form

$$a_{1-}a_{2-}a_{3-}$$

Since no two factors in the product come from the same column, the column numbers have no repetitions; consequently, they must form a permutation of the set $\{1, 2, 3\}$. These $3! = 6$ permutations yield the following list of elementary products.

$$a_{11}a_{22}a_{33} \qquad a_{12}a_{21}a_{33} \qquad a_{13}a_{21}a_{32}$$

$$a_{11}a_{23}a_{32} \qquad a_{12}a_{23}a_{31} \qquad a_{13}a_{22}a_{31}$$

As this example points out, an $n \times n$ matrix A has $n!$ elementary products. They are the products of the form $a_{1j_1}a_{2j_2} \cdots a_{nj_n}$, where $(j_1, j_2, \ldots, j_n)$ is a permutation of the set $\{1, 2, \ldots, n\}$. By a ***signed elementary product from A*** we shall mean an elementary product $a_{1j_1}a_{2j_2} \cdots a_{nj_n}$ multiplied by $+1$ or -1. We use the $+$ if $(j_1, j_2, \ldots, j_n)$ is an even permutation and the $-$ if $(j_1, j_2, \ldots, j_n)$ is an odd permutation.

Example 6

List all signed elementary products from the matrices

$$\text{(i)} \begin{bmatrix} a_{11} & a_{12} \\ a_{21} & a_{22} \end{bmatrix} \qquad \text{(ii)} \begin{bmatrix} a_{11} & a_{12} & a_{13} \\ a_{21} & a_{22} & a_{23} \\ a_{31} & a_{32} & a_{33} \end{bmatrix}$$

(i)

Elementary product	*Associated permutation*	*Even or odd*	*Signed elementary product*
$a_{11}a_{22}$	(1, 2)	even	$a_{11}a_{22}$
$a_{12}a_{21}$	(2, 1)	odd	$-a_{12}a_{21}$

(ii)

Elementary product	*Associated permutation*	*Even or odd*	*Signed elementary product*
$a_{11}a_{22}a_{33}$	(1, 2, 3)	even	$a_{11}a_{22}a_{33}$
$a_{11}a_{23}a_{32}$	(1, 3, 2)	odd	$-a_{11}a_{23}a_{32}$
$a_{12}a_{21}a_{33}$	(2, 1, 3)	odd	$-a_{12}a_{21}a_{33}$
$a_{12}a_{23}a_{31}$	(2, 3, 1)	even	$a_{12}a_{23}a_{31}$
$a_{13}a_{21}a_{32}$	(3, 1, 2)	even	$a_{13}a_{21}a_{32}$
$a_{13}a_{22}a_{31}$	(3, 2, 1)	odd	$-a_{13}a_{22}a_{31}$

We are now in a position to define the determinant function.

Definition. Let A be a square matrix. The ***determinant function*** is denoted by ***det***, and we define $\det(A)$ to be the sum of all signed elementary products from A.

Example 7

Referring to Example 6, we obtain

(i) $\det\begin{bmatrix} a_{11} & a_{12} \\ a_{21} & a_{22} \end{bmatrix} = a_{11}a_{22} - a_{12}a_{21}$

(ii) $$\det\begin{bmatrix} a_{11} & a_{12} & a_{13} \\ a_{21} & a_{22} & a_{23} \\ a_{31} & a_{32} & a_{33} \end{bmatrix} = a_{11}a_{22}a_{33} + a_{12}a_{23}a_{31} + a_{13}a_{21}a_{32} - a_{13}a_{22}a_{31} - a_{12}a_{21}a_{33} - a_{11}a_{23}a_{32}$$

It is useful to have the two formulas in this example available for ready reference. To avoid memorizing these unwieldy expressions, however, we suggest using the mnemonic devices illustrated in Figure 2.2. The first formula in Example 7 is obtained from Figure 2.2*a* by multiplying the entries on the rightward arrow and subtracting the product of the entries on the leftward arrow. The second formula in Example 7 is obtained by recopying the first and second columns as shown in Figure 2.2*b*. The determinant is then computed by summing the products on the rightward arrows and subtracting the products on the leftward arrows.

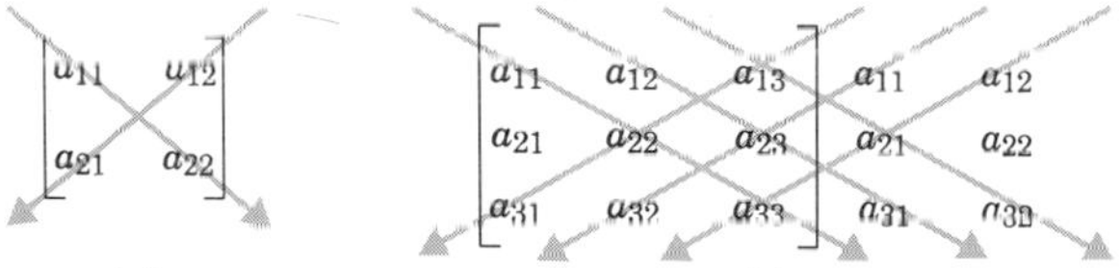

Figure 2.2 (*a*) (*b*)

Example 8

Evaluate the determinants of

$$A = \begin{bmatrix} 3 & 1 \\ 4 & -2 \end{bmatrix} \quad \text{and} \quad B = \begin{bmatrix} 1 & 2 & 3 \\ -4 & 5 & 6 \\ 7 & -8 & 9 \end{bmatrix}$$

Using the method of Figure 2.2*a* gives

$$\det(A) = (3)(-2) - (1)(4) = -10$$

Using the method of Figure 2.2*b* gives

$$\det(B) = (45) + (84) + (96) - (105) - (-48) - (-72) = 240$$

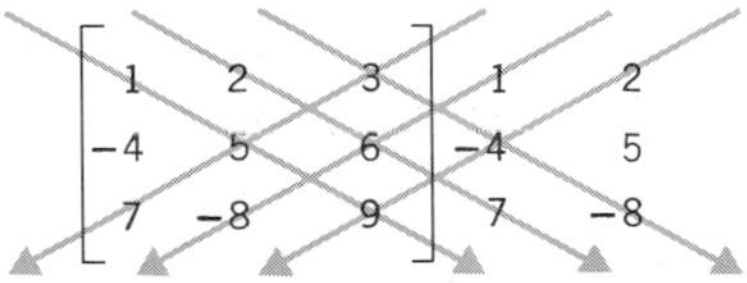

Warning. We emphasize that the methods described in Figure 2.2 *do not* work for determinants of 4×4 matrices or higher.

Evaluating determinants directly from the definition leads to computational difficulties. Indeed, the direct evaluation of a 4×4 determinant would involve computing $4! = 24$ signed elementary products and a 10×10 determinant would involve $10! = 3{,}628{,}800$ signed elementary products. Even the fastest of our current digital computers cannot handle the computation of a 25×25 determinant by this method in a practical amount of time. Much of the remainder of this chapter is devoted, therefore, to developing properties of determinants that will simplify their evaluation.

We conclude this section by noting that the determinant of A is often written symbolically as

$$\det(A) = \Sigma \pm a_{1j_1}a_{2j_2} \cdots a_{nj_n}$$

where Σ indicates that the terms are to be summed over all permutations $(j_1, j_2, \ldots, j_n)$ and the $+$ or $-$ is selected in each term according to whether the permutation is even or odd. An alternative notation for the determinant of a matrix A is $|A|$ rather than $\det(A)$. In this notation, the determinant of the matrix A in Example 8 can be written

$$\begin{vmatrix} 3 & 1 \\ 4 & -2 \end{vmatrix} = -10$$

EXERCISE SET 2.1

1. Find the number of inversions in each of the following permutations of $\{1, 2, 3, 4, 5\}$.

(a) (3 4 1 5 2) (b) (4 2 5 3 1) (c) (5 4 3 2 1)
(d) (1 2 3 4 5) (e) (1 3 5 4 2) (f) (2 3 5 4 1)

2. Classify each of the permutations in Exercise 1 as even or odd.

In Exercises 3–10 evaluate the determinant.

3. $\begin{vmatrix} 1 & 2 \\ -1 & 3 \end{vmatrix}$ **4.** $\begin{vmatrix} 6 & 4 \\ 3 & 2 \end{vmatrix}$ **5.** $\begin{vmatrix} -1 & 7 \\ -8 & -3 \end{vmatrix}$ **6.** $\begin{vmatrix} k-1 & 2 \\ 4 & k-3 \end{vmatrix}$

7. $\begin{vmatrix} 1 & -2 & 7 \\ 3 & 5 & 1 \\ 4 & 3 & 8 \end{vmatrix}$ **8.** $\begin{vmatrix} 8 & 2 & -1 \\ -3 & 4 & -6 \\ 1 & 7 & 2 \end{vmatrix}$ **9.** $\begin{vmatrix} 1 & 0 & 3 \\ 4 & 0 & -1 \\ 2 & 8 & 6 \end{vmatrix}$ **10.** $\begin{vmatrix} k & -3 & 9 \\ 2 & 4 & k+1 \\ 1 & k^2 & 3 \end{vmatrix}$

11. Find all values of λ for which $\det(A) = 0$.

(a) $A = \begin{bmatrix} \lambda - 1 & -2 \\ 1 & \lambda - 4 \end{bmatrix}$ (b) $A = \begin{bmatrix} \lambda - 6 & 0 & 0 \\ 0 & \lambda & -1 \\ 0 & 4 & \lambda - 4 \end{bmatrix}$

12. Classify each permutation of $\{1, 2, 3, 4\}$ as even or odd.

13. Use the results in Exercise 12 to construct a formula for the determinant of a 4×4 matrix.

14. Use the formula obtained in Exercise 13 to evaluate

$$\begin{bmatrix} 1 & 4 & -3 & 1 \\ 2 & 0 & 6 & 3 \\ 4 & -1 & 2 & 5 \\ 1 & 0 & -2 & 4 \end{bmatrix}$$

15. Use the determinant definition to evaluate

$$\text{(a)} \begin{vmatrix} 0 & 0 & 0 & 0 & 1 \\ 0 & 0 & 0 & 2 & 0 \\ 0 & 0 & 3 & 0 & 0 \\ 0 & 4 & 0 & 0 & 0 \\ 5 & 0 & 0 & 0 & 0 \end{vmatrix} \qquad \text{(b)} \begin{vmatrix} 0 & 4 & 0 & 0 & 0 \\ 0 & 0 & 0 & 2 & 0 \\ 0 & 0 & 3 & 0 & 0 \\ 0 & 0 & 0 & 0 & 1 \\ 5 & 0 & 0 & 0 & 0 \end{vmatrix}$$

16. Prove that if a square matrix A has a column of zeros, then $\det(A) = 0$.

2.2 EVALUATING DETERMINANTS BY ROW REDUCTION

In this section we show that the determinant of a matrix can be evaluated by reducing the matrix to row-echelon form. This method is of importance since it avoids the lengthy computations involved in directly applying the determinant definition.

We first consider two classes of matrices whose determinants can be easily evaluated, regardless of the size of the matrix.

Theorem 1. *If A is any square matrix that contains a row of zeros, then $\det(A) = 0$.*

Proof. Since a signed elementary product from A contains one factor from each row of A, every signed elementary product contains a factor from the row of zeros and consequently has value zero. Since $\det(A)$ is the sum of all signed elementary products, we obtain $\det(A) = 0$. ∎

A square matrix is called ***upper triangular*** if all the entries below the main diagonal are zeros. Similarly, a square matrix is called ***lower triangular*** if all the entries above the main diagonal are zeros. A matrix that is either upper or lower triangular is called ***triangular***.

Example 9

A general 4 × 4 upper triangular matrix has the form

$$\begin{bmatrix} a_{11} & a_{12} & a_{13} & a_{14} \\ 0 & a_{22} & a_{23} & a_{24} \\ 0 & 0 & a_{33} & a_{34} \\ 0 & 0 & 0 & a_{44} \end{bmatrix}$$

A general 4 × 4 lower triangular matrix has the form

$$\begin{bmatrix} a_{11} & 0 & 0 & 0 \\ a_{21} & a_{22} & 0 & 0 \\ a_{31} & a_{32} & a_{33} & 0 \\ a_{41} & a_{42} & a_{43} & a_{44} \end{bmatrix}$$

Example 10

Evaluate $\det(A)$, where

$$A = \begin{bmatrix} a_{11} & 0 & 0 & 0 \\ a_{21} & a_{22} & 0 & 0 \\ a_{31} & a_{32} & a_{33} & 0 \\ a_{41} & a_{42} & a_{43} & a_{44} \end{bmatrix}$$

The only elementary product from A that can be nonzero is $a_{11}a_{22}a_{33}a_{44}$. To see this, consider a typical elementary product $a_{1j_1}a_{2j_2}a_{3j_3}a_{4j_4}$. Since $a_{12} = a_{13} = a_{14} = 0$, we must have $j_1 = 1$ in order to have a nonzero elementary product. If $j_1 = 1$, we must have $j_2 \neq 1$ since no two factors come from the same column. Further, since $a_{23} = a_{24} = 0$, we must have $j_2 = 2$ in order to have a nonzero product. Continuing in this way, we obtain $j_3 = 3$ and $j_4 = 4$. Since $a_{11}a_{22}a_{33}a_{44}$ is multiplied by $+1$ in forming the signed elementary product, we obtain

$$\det(A) = a_{11}a_{22}a_{33}a_{44}$$

An argument similar to the one just presented can be applied to any triangular matrix to yield the following general result.

Theorem 2. *If A is an $n \times n$ triangular matrix, then $\det(A)$ is the product of the entries on the main diagonal; that is, $\det(A) = a_{11}a_{22} \cdots a_{nn}$.*

Example 11

$$\begin{vmatrix} 2 & 7 & -3 & 8 & 3 \\ 0 & -3 & 7 & 5 & 1 \\ 0 & 0 & 6 & 7 & 6 \\ 0 & 0 & 0 & 9 & 8 \\ 0 & 0 & 0 & 0 & 4 \end{vmatrix} = (2)(-3)(6)(9)(4) = -1296$$

The next theorem shows how an elementary row operation on a matrix affects the value of its determinant.

Theorem 3. *Let A be any $n \times n$ matrix.*

(*a*) *If A' is the matrix that results when a single row of A is multiplied by a constant k, then $\det(A') = k\det(A)$.*
(*b*) *If A' is the matrix that results when two rows of A are interchanged, then $\det(A') = -\det(A)$.*
(*c*) *If A' is the matrix that results when a multiple of one row of A is added to another row, then $\det(A') = \det(A)$.*

We shall omit the proof (see Exercise 15).

Example 12

Consider the matrices

$$A = \begin{bmatrix} 1 & 2 & 3 \\ 0 & 1 & 4 \\ 1 & 2 & 1 \end{bmatrix} \qquad A_1 = \begin{bmatrix} 4 & 8 & 12 \\ 0 & 1 & 4 \\ 1 & 2 & 1 \end{bmatrix} \qquad A_2 = \begin{bmatrix} 0 & 1 & 4 \\ 1 & 2 & 3 \\ 1 & 2 & 1 \end{bmatrix}$$

$$A_3 = \begin{bmatrix} 1 & 2 & 3 \\ -2 & -3 & 2 \\ 1 & 2 & 1 \end{bmatrix}$$

If we evaluate the determinants of these matrices by the method used in Example 8 we obtain

$$\det(A) = -2, \quad \det(A_1) = -8, \quad \det(A_2) = 2, \quad \det(A_3) = -2$$

Observe that A_1 is obtained by multiplying the first row of A by 4; A_2 by interchanging the first two rows; and A_3 by adding -2 times the third row of A to the second. As predicted by Theorem 3, we have the relationships

$$\det(A_1) = 4\det(A) \qquad \det(A_2) = -\det(A) \qquad \text{and} \qquad \det(A_3) = \det(A)$$

Example 13

Statement (*a*) in Theorem 3 has an alternate interpretation that is sometimes useful. This result allows us to bring a "common factor" from any row of a square matrix through the determinant sign. To illustrate, consider the matrices

$$A = \begin{bmatrix} a_{11} & a_{12} & a_{13} \\ a_{21} & a_{22} & a_{23} \\ a_{31} & a_{32} & a_{33} \end{bmatrix} \qquad B = \begin{bmatrix} a_{11} & a_{12} & a_{13} \\ ka_{21} & ka_{22} & ka_{23} \\ a_{31} & a_{32} & a_{33} \end{bmatrix}$$

where the second row of B has a common factor of k. Since B is the matrix that results when the second row of A is multiplied by k, statement (*a*) in Theorem 3 asserts that $\det(B) = k\det(A)$; that is,

$$\begin{vmatrix} a_{11} & a_{12} & a_{13} \\ ka_{21} & ka_{22} & ka_{23} \\ a_{31} & a_{32} & a_{33} \end{vmatrix} = k\begin{vmatrix} a_{11} & a_{12} & a_{13} \\ a_{21} & a_{22} & a_{23} \\ a_{31} & a_{32} & a_{33} \end{vmatrix}$$

We shall now formulate an alternative method for evaluating determinants that will avoid the large amount of computation involved in directly applying the determinant definition. The basic idea of this method is to apply elementary row operations to reduce the given matrix A to a matrix R that is in row-echelon form. Since a row-echelon form of a square matrix is upper triangular (Exercise 14), $\det(R)$ can be evaluated using Theorem 2. The value of $\det(A)$ can then be obtained by using Theorem 3 to relate the unknown value of $\det(A)$ to the known value of $\det(R)$. The following example illustrates this method.

Example 14

Evaluate $\det(A)$ where

$$A = \begin{bmatrix} 0 & 1 & 5 \\ 3 & -6 & 9 \\ 2 & 6 & 1 \end{bmatrix}$$

Solution. Reducing A to row-echelon form and applying Theorem 3, we obtain

$$\det(A) = \begin{vmatrix} 0 & 1 & 5 \\ 3 & -6 & 9 \\ 2 & 6 & 1 \end{vmatrix} = -\begin{vmatrix} 3 & -6 & 9 \\ 0 & 1 & 5 \\ 2 & 6 & 1 \end{vmatrix}$$

The first and second rows of A were interchanged.

$$= -3\begin{vmatrix} 1 & -2 & 3 \\ 0 & 1 & 5 \\ 2 & 6 & 1 \end{vmatrix}$$

A common factor of 3 from the first row of the preceding matrix was taken through the det sign (see Example 13).

$$= -3\begin{vmatrix} 1 & -2 & 3 \\ 0 & 1 & 5 \\ 0 & 10 & -5 \end{vmatrix}$$

-2 times the first row of the preceding matrix was added to the third row.

$$= -3\begin{vmatrix} 1 & -2 & 3 \\ 0 & 1 & 5 \\ 0 & 0 & -55 \end{vmatrix}$$

-10 times the second row of the preceding matrix was added to the third row.

$$= (-3)(-55)\begin{vmatrix} 1 & -2 & 3 \\ 0 & 1 & 5 \\ 0 & 0 & 1 \end{vmatrix}$$

A common factor of -55 from the last row of the preceding matrix was taken through the det sign.

$$= (-3)(-55)(1) = 165$$

Example 15

Evaluate det(A), where

$$A = \begin{bmatrix} 1 & 3 & -2 & 4 \\ 2 & 6 & -4 & 8 \\ 3 & 9 & 1 & 5 \\ 1 & 1 & 4 & 8 \end{bmatrix}$$

$$\det(A) = \begin{vmatrix} 1 & 3 & -2 & 4 \\ 0 & 0 & 0 & 0 \\ 3 & 9 & 1 & 5 \\ 1 & 1 & 4 & 8 \end{vmatrix}$$

-2 times the first row of A was added to the second row.

No further reduction is needed since it follows from Theorem 1 that det(A) = 0.

It should be evident from this example that whenever a square matrix has two proportional rows (like the first and second rows of A), it is possible to introduce a row of zeros by adding a suitable multiple of one of these rows to the other. Thus, *if a square matrix has two proportional rows, its determinant is zero.*

Example 16

Each of the following matrices has two proportional rows; thus, by inspection, each has a zero determinant.

$$\begin{bmatrix} -1 & 4 \\ -2 & 8 \end{bmatrix} \qquad \begin{bmatrix} 2 & 7 & 8 \\ 3 & 2 & 4 \\ 2 & 7 & 8 \end{bmatrix} \qquad \begin{bmatrix} 3 & -1 & 4 & -5 \\ 6 & -2 & 5 & 2 \\ 5 & 8 & 1 & 4 \\ -9 & 3 & -12 & 15 \end{bmatrix}$$

EXERCISE SET 2.2

1. Evaluate the following by inspection.

(a) $\begin{vmatrix} 2 & -40 & 17 \\ 0 & 1 & 11 \\ 0 & 0 & 3 \end{vmatrix}$ (b) $\begin{vmatrix} 1 & 0 & 0 & 0 \\ -9 & -1 & 0 & 0 \\ 12 & 7 & 8 & 0 \\ 4 & 5 & 7 & 2 \end{vmatrix}$

(c) $\begin{vmatrix} 1 & 2 & 3 \\ 3 & 7 & 6 \\ 1 & 2 & 3 \end{vmatrix}$ (d) $\begin{vmatrix} 3 & -1 & 2 \\ 6 & -2 & 4 \\ 1 & 7 & 3 \end{vmatrix}$

In Exercises 2–9 evaluate the determinants of the given matrices by reducing the matrix to row-echelon form.

2. $\begin{bmatrix} 2 & 3 & 4 \\ 0 & 0 & -3 \\ -1 & 2 & 7 \end{bmatrix}$ **3.** $\begin{bmatrix} 2 & 1 & 1 \\ 4 & 2 & 3 \\ 1 & 3 & 0 \end{bmatrix}$

4. $\begin{bmatrix} 6 & 6 & 2 \\ -3 & 3 & 1 \\ 3 & 9 & 2 \end{bmatrix}$ **5.** $\begin{bmatrix} 3 & 1 & 4 \\ 1 & 7 & 3 \\ 5 & -12 & 5 \end{bmatrix}$

6. $\begin{bmatrix} -1 & 2 & 1 & 2 \\ 1 & 2 & 4 & 1 \\ 2 & 0 & -1 & 3 \\ 3 & 2 & -1 & 0 \end{bmatrix}$ **7.** $\begin{bmatrix} 4 & 6 & 8 & -6 \\ 0 & -3 & 0 & -1 \\ 3 & 3 & -4 & -2 \\ -2 & 3 & 4 & 2 \end{bmatrix}$

8. $\begin{bmatrix} \frac{1}{2} & \frac{1}{2} & 1 & \frac{1}{2} \\ -\frac{1}{2} & \frac{1}{2} & 0 & \frac{1}{2} \\ \frac{2}{3} & \frac{1}{3} & \frac{1}{3} & 0 \\ \frac{1}{3} & 1 & \frac{1}{3} & 0 \end{bmatrix}$ **9.** $\begin{bmatrix} 0 & 0 & 2 & 2 & 2 \\ 3 & -3 & 3 & 3 & 3 \\ -4 & 4 & 4 & 4 & 4 \\ 4 & 3 & 1 & 9 & 2 \\ 1 & 2 & 1 & 3 & 1 \end{bmatrix}$

10. Assume $\det \begin{bmatrix} a & b & c \\ d & e & f \\ g & h & i \end{bmatrix} = 5$. Find

(a) $\det \begin{bmatrix} d & e & f \\ g & h & i \\ a & b & c \end{bmatrix}$ (b) $\det \begin{bmatrix} -a & -b & -c \\ 2d & 2e & 2f \\ -g & -h & -i \end{bmatrix}$

(c) $\det \begin{bmatrix} a+d & b+e & c+f \\ d & e & f \\ g & h & i \end{bmatrix}$ (d) $\det \begin{bmatrix} a & b & c \\ d-3a & e-3b & f-3c \\ 2g & 2h & 2i \end{bmatrix}$

11. Show that

$$\begin{vmatrix} 1 & 1 & 1 \\ a & b & c \\ a^2 & b^2 & c^2 \end{vmatrix} = (b-a)(c-a)(c-b)$$

12. Use an argument like that given in Example 10 to show

(a) $\det \begin{bmatrix} 0 & 0 & a_{13} \\ 0 & a_{22} & a_{23} \\ a_{31} & a_{32} & a_{33} \end{bmatrix} = -a_{13}a_{22}a_{31}$

(b) $\det \begin{bmatrix} 0 & 0 & 0 & a_{14} \\ 0 & 0 & a_{23} & a_{24} \\ 0 & a_{32} & a_{33} & a_{34} \\ a_{41} & a_{42} & a_{43} & a_{44} \end{bmatrix} = a_{14}a_{23}a_{32}a_{41}$

13. Prove that Theorem 1 is true when the word "row" is replaced by "column."

14. Prove that a row-echelon form of a square matrix is upper triangular.

15. Prove the following special cases of Theorem 3.

(a) $$\begin{vmatrix} ka_{11} & ka_{12} & ka_{13} \\ a_{21} & a_{22} & a_{23} \\ a_{31} & a_{32} & a_{33} \end{vmatrix} = k\begin{vmatrix} a_{11} & a_{12} & a_{13} \\ a_{21} & a_{22} & a_{23} \\ a_{31} & a_{32} & a_{33} \end{vmatrix}$$

(b) $$\begin{vmatrix} a_{21} & a_{22} & a_{23} \\ a_{11} & a_{12} & a_{13} \\ a_{31} & a_{32} & a_{33} \end{vmatrix} = -\begin{vmatrix} a_{11} & a_{12} & a_{13} \\ a_{21} & a_{22} & a_{23} \\ a_{31} & a_{32} & a_{33} \end{vmatrix}$$

(c) $$\begin{vmatrix} a_{11} + ka_{21} & a_{12} + ka_{22} & a_{13} + ka_{23} \\ a_{21} & a_{22} & a_{23} \\ a_{31} & a_{32} & a_{33} \end{vmatrix} = \begin{vmatrix} a_{11} & a_{12} & a_{13} \\ a_{21} & a_{22} & a_{23} \\ a_{31} & a_{32} & a_{33} \end{vmatrix}$$

2.3 PROPERTIES OF THE DETERMINANT FUNCTION

In this section we develop some of the fundamental properties of the determinant function. Our work here will give us some further insight into the relationship between a square matrix and its determinant. One of the immediate consequences of this material will be an important determinant test for the invertibility of a matrix.

If A is any $m \times n$ matrix, then the ***transpose of A*** is denoted by A^t and is defined to be the $n \times m$ matrix whose first column is the first row of A, whose second column is the second row of A, whose third column is the third row of A, etc.

Example 17

The transposes of the matrices

$$A = \begin{bmatrix} a_{11} & a_{12} & a_{13} & a_{14} \\ a_{21} & a_{22} & a_{23} & a_{24} \\ a_{31} & a_{32} & a_{33} & a_{34} \end{bmatrix}$$

$$B = \begin{bmatrix} 2 & 3 \\ 1 & 4 \\ 5 & 6 \end{bmatrix} \qquad C = [1 \quad 3 \quad 5] \qquad D = \begin{bmatrix} 3 & 5 & -2 \\ 5 & 4 & 1 \\ -2 & 1 & 7 \end{bmatrix}$$

are

$$A^t = \begin{bmatrix} a_{11} & a_{21} & a_{31} \\ a_{12} & a_{22} & a_{32} \\ a_{13} & a_{23} & a_{33} \\ a_{14} & a_{24} & a_{34} \end{bmatrix}$$

$$B^t = \begin{bmatrix} 2 & 1 & 5 \\ 3 & 4 & 6 \end{bmatrix} \qquad C^t = \begin{bmatrix} 1 \\ 3 \\ 5 \end{bmatrix} \qquad D^t = \begin{bmatrix} 3 & 5 & -2 \\ 5 & 4 & 1 \\ -2 & 1 & 7 \end{bmatrix}$$

Assuming that the sizes of the matrices are such that the operations can be performed, the transpose operation has the following properties (see Exercise 13):

Properties of the Transpose Operation.

(i) $(A^t)^t = A$
(ii) $(A + B)^t = A^t + B^t$
(iii) $(kA)^t = kA^t$, *where k is any scalar*
(iv) $(AB)^t = B^t A^t$

Recall that the determinant of an $n \times n$ matrix A is defined to be the sum of all signed elementary products from A. Since an elementary product has one factor from each row and one factor from each column, it is obvious that A and A^t have precisely the same set of elementary products. Although we shall omit the details, it can be shown that A and A^t actually have the same *signed* elementary products; this yields the following theorem.

Theorem 4. *If A is any square matrix, then* $\det(A) = \det(A^t)$.

Because of this result, nearly every theorem about determinants that contains the word "row" in its statement is also true when the word "column" is substituted for "row." To prove a column statement one need only transpose the matrix in question to convert the column statement to a row statement, and then apply the corresponding known result for rows. To illustrate the idea, suppose we want to prove that interchanging two columns of a square matrix A changes the sign of $\det(A)$. We can proceed as follows: Let A' be the matrix that results when column r and column s of A are interchanged. Thus $(A')^t$ is the matrix that results when *row* r and *row* s of A^t are interchanged. Therefore

$$\begin{aligned} \det(A') &= \det(A')^t && \text{(Theorem 4)} \\ &= -\det(A^t) && \text{(Theorem 3}b\text{)} \\ &= -\det(A) && \text{(Theorem 4)} \end{aligned}$$

which proves the result.

The following examples illustrate several points about determinants that depend on column properties of the matrix.

Example 18

By inspection, the matrix

$$\begin{bmatrix} 1 & -2 & 7 \\ -4 & 8 & 5 \\ 2 & -4 & 3 \end{bmatrix}$$

has a zero determinant since the first and second columns are proportional.

Example 19

Compute the determinant of

$$A = \begin{bmatrix} 1 & 0 & 0 & 3 \\ 2 & 7 & 0 & 6 \\ 0 & 6 & 3 & 0 \\ 7 & 3 & 1 & -5 \end{bmatrix}$$

This determinant could be computed as before by using elementary row operations to reduce A to row-echelon form. On the other hand, we can put A in lower triangular form in one step by adding -3 times the first column to the fourth to obtain

$$\det(A) = \det \begin{bmatrix} 1 & 0 & 0 & 0 \\ 2 & 7 & 0 & 0 \\ 0 & 6 & 3 & 0 \\ 7 & 3 & 1 & -26 \end{bmatrix} = (1)(7)(3)(-26) = -546$$

This example points out that it is always wise to keep an eye open for a clever column operation that will shorten the computations.

Suppose that A and B are $n \times n$ matrices and k is any scalar. We now consider possible relationships between $\det(A)$, $\det(B)$, and

$$\det(kA), \quad \det(A + B), \quad \text{and } \det(AB)$$

Since a common factor of any row of a matrix can be moved through the det sign, and since each of the n rows in kA has a common factor of k, we obtain

$$\det(kA) = k^n \det(A) \tag{2.1}$$

Example 20

Consider the matrices

$$A = \begin{bmatrix} 3 & 1 \\ 2 & 2 \end{bmatrix} \quad \text{and} \quad 5A = \begin{bmatrix} 15 & 5 \\ 10 & 10 \end{bmatrix}$$

By direct calculation $\det(A) = 4$ and $\det(5A) = 100$. This agrees with relationship (2.1), which asserts that $\det(5A) = 5^2\det(A)$.

Unfortunately, no simple relationship exists between $\det(A)$, $\det(B)$, and $\det(A + B)$ in general. In particular, we emphasize that $\det(A + B)$ is usually *not* equal to $\det(A) + \det(B)$. The following example illustrates this point.

Example 21

Consider

$$A = \begin{bmatrix} 1 & 2 \\ 2 & 5 \end{bmatrix} \quad B = \begin{bmatrix} 3 & 1 \\ 1 & 3 \end{bmatrix} \quad A + B = \begin{bmatrix} 4 & 3 \\ 3 & 8 \end{bmatrix}$$

We have $\det(A) = 1$, $\det(B) = 8$, and $\det(A + B) = 23$; thus $\det(A + B) \neq \det(A) + \det(B)$.

In spite of this negative result, there is one important relationship concerning sums of determinants that is often useful. To illustrate, consider two 2×2 matrices

$$A = \begin{bmatrix} a_{11} & a_{12} \\ a_{21} & a_{22} \end{bmatrix} \quad \text{and} \quad A' = \begin{bmatrix} a_{11} & a_{12} \\ a'_{21} & a'_{22} \end{bmatrix}$$

that differ only in the second row. From the formula in Example 7, we obtain

$$\begin{aligned} \det(A) + \det(A') &= (a_{11}a_{22} - a_{12}a_{21}) + (a_{11}a'_{22} - a_{12}a'_{21}) \\ &= a_{11}(a_{22} + a'_{22}) - a_{12}(a_{21} + a'_{21}) \\ &= \det \begin{bmatrix} a_{11} & a_{12} \\ a_{21} + a'_{21} & a_{22} + a'_{22} \end{bmatrix} \end{aligned}$$

Thus

$$\det \begin{bmatrix} a_{11} & a_{12} \\ a_{21} & a_{22} \end{bmatrix} + \det \begin{bmatrix} a_{11} & a_{12} \\ a'_{21} & a'_{22} \end{bmatrix} = \det \begin{bmatrix} a_{11} & a_{12} \\ a_{21} + a'_{21} & a_{22} + a'_{22} \end{bmatrix}$$

This example is a special case of the following general result:

Let A, A', and A'' be $n \times n$ matrices that differ only in a single row, say the rth, and assume that the rth row of A'' can be obtained by adding corresponding entries in the rth rows of A and A'. Then

$$\det(A'') = \det(A) + \det(A')$$

A similar result holds for columns.

Example 22

By evaluating the determinants, the reader can check that

$$\det \begin{bmatrix} 1 & 7 & 5 \\ 2 & 0 & 3 \\ 1 + 0 & 4 + 1 & 7 + (-1) \end{bmatrix} = \det \begin{bmatrix} 1 & 7 & 5 \\ 2 & 0 & 3 \\ 1 & 4 & 7 \end{bmatrix} + \det \begin{bmatrix} 1 & 7 & 5 \\ 2 & 0 & 3 \\ 0 & 1 & -1 \end{bmatrix}$$

Theorem 5. *If A and B are square matrices of the same size, then $\det(AB) = \det(A)\det(B)$.*

The elegant simplicity of this result contrasted with the complex nature of both matrix multiplication and the determinant definition is both refreshing and surprising. We shall omit the proof.

Example 23

Consider the matrices

$$A = \begin{bmatrix} 3 & 1 \\ 2 & 1 \end{bmatrix} \qquad B = \begin{bmatrix} -1 & 3 \\ 5 & 8 \end{bmatrix} \qquad AB = \begin{bmatrix} 2 & 17 \\ 3 & 14 \end{bmatrix}$$

We have $\det(A)\det(B) = (1)(-23) = -23$. On the other hand, by direct computation $\det(AB) = -23$, so that $\det(AB) = \det(A)\det(B)$.

In Theorem 13 of Chapter 1, we listed three important statements that are equivalent to the invertibility of a matrix. The next example will help us to add another result to that list.

Example 24

The purpose of this example is to show that if the reduced row-echelon form R of a *square* matrix has no rows consisting entirely of zeros, then R must be the identity matrix. This can be illustrated by considering the following 3×3 matrix, which we assume to be in reduced row-echelon form

$$R = \begin{bmatrix} r_{11} & r_{12} & r_{13} \\ r_{21} & r_{22} & r_{23} \\ r_{31} & r_{32} & r_{33} \end{bmatrix}$$

Either the last row in this matrix consists entirely of zeros or it does not. If not, the matrix contains no zero rows, and consequently each of the three rows has a leading entry of 1. Since these leading 1's occur progressively further to the right as we move down the matrix, each of these 1's must occur on the main diagonal. Since the other entries in the same column as one of these 1's are zero, R must be I. Thus either R has a row of zeros or $R = I$.

Theorem 6. *A square matrix A is invertible if and only if $\det(A) \neq 0$.*

Proof. If A is invertible, then $I = AA^{-1}$ so that $1 = \det(I) = \det(A)\det(A^{-1})$. Thus $\det(A) \neq 0$. Conversely, assume that $\det(A) \neq 0$. We shall show that A is row equivalent to I, and thus conclude from Theorem 10 in Chapter 1 that A is invertible. Let R be the reduced row-echelon form of A. Since R can be obtained from A by a finite sequence of elementary row operations, we can find elementary matrices $E_1, E_2, \ldots, E_k$ such that $E_k \cdots E_2E_1A = R$ or $A = E_1^{-1}E_2^{-1} \cdots E_k^{-1}R$. Thus

$$\det(A) = \det(E_1^{-1})\det(E_2^{-1}) \cdots \det(E_k^{-1})\det(R)$$

Since we are assuming that $\det(A) \neq 0$ it follows from this equation that $\det(R) \neq 0$. Therefore R does not have any zero rows, so that $R = I$ (see Example 24). ▮

Corollary. If A is invertible then

$$\det(A^{-1}) = \frac{1}{\det(A)}$$

Proof. Since $A^{-1}A = I$, $\det(A^{-1}A) = \det(I)$; that is, $\det(A^{-1})\det(A) = 1$. Since $\det(A) \neq 0$, the proof can be completed by dividing through by $\det(A)$. ▮

Example 25

Since the first and third rows of

$$A = \begin{bmatrix} 1 & 2 & 3 \\ 1 & 0 & 1 \\ 2 & 4 & 6 \end{bmatrix}$$

are proportional, $\det(A) = 0$. Thus A is not invertible.

EXERCISE SET 2.3

1. Find the transposes of

(a) $\begin{bmatrix} 2 & 1 \\ -3 & 1 \\ 0 & 2 \end{bmatrix}$ (b) $\begin{bmatrix} 6 & 1 & 1 \\ -8 & 4 & 3 \\ 0 & 1 & 3 \end{bmatrix}$ (c) $[7 \quad 0 \quad 2]$ (d) $\begin{bmatrix} a_{11} & a_{12} & a_{13} \\ a_{21} & a_{22} & a_{23} \end{bmatrix}$

2. Verify Theorem 4 for

$$A = \begin{bmatrix} 1 & 2 & 7 \\ -1 & 0 & 6 \\ 3 & 2 & 8 \end{bmatrix}$$

3. Verify that $\det(AB) = \det(A)\det(B)$ when

$$A = \begin{bmatrix} 2 & 1 & 0 \\ 3 & 4 & 0 \\ 0 & 0 & 2 \end{bmatrix} \quad \text{and} \quad B = \begin{bmatrix} 1 & -1 & 3 \\ 7 & 1 & 2 \\ 5 & 0 & 1 \end{bmatrix}$$

4. Use Theorem 6 to determine which of the following matrices are invertible.

(a) $\begin{bmatrix} 1 & 0 & 0 \\ 3 & 6 & 7 \\ 0 & 8 & -1 \end{bmatrix}$ (b) $\begin{bmatrix} -2 & 1 & -4 \\ 1 & 1 & 2 \\ 3 & 1 & 6 \end{bmatrix}$ (c) $\begin{bmatrix} 7 & 2 & 1 \\ 7 & 2 & 1 \\ 3 & 6 & 6 \end{bmatrix}$ (d) $\begin{bmatrix} 0 & 7 & 5 \\ 0 & 1 & -1 \\ 0 & 3 & 2 \end{bmatrix}$

5. Assume $\det(A) = 5$, where

$$A = \begin{bmatrix} a & b & c \\ d & e & f \\ g & h & i \end{bmatrix}$$

Find

(a) $\det(3A)$ (b) $\det(2A^{-1})$ (c) $\det((2A)^{-1})$ (d) $\det \begin{bmatrix} a & g & d \\ b & h & e \\ c & i & f \end{bmatrix}$

6. Without directly evaluating, show that $x = 0$ and $x = 2$ satisfy

$$\begin{vmatrix} x^2 & x & 2 \\ 2 & 1 & 1 \\ 0 & 0 & -5 \end{vmatrix} = 0$$

7. Without directly evaluating, show that

$$\det \begin{bmatrix} b + c & c + a & b + a \\ a & b & c \\ 1 & 1 & 1 \end{bmatrix} = 0$$

8. For which value(s) of k does A fail to be invertible?

(a) $A = \begin{bmatrix} k - 3 & -2 \\ -2 & k - 2 \end{bmatrix}$ (b) $A = \begin{bmatrix} 1 & 2 & 4 \\ 3 & 1 & 6 \\ k & 3 & 2 \end{bmatrix}$

9. Let A and B be $n \times n$ matrices. Show that if A is invertible, then $\det(B) = \det(A^{-1}BA)$.

10. (a) Find a nonzero 3×3 matrix A such that $A = A^t$.
(b) Find a nonzero 3×3 matrix A such that $A = -A^t$.

11. Let a_{ij} be the entry in the ith row and jth column of A. In which row and column of A^t will a_{ij} appear?

12. Let $AX = 0$ be a system of n linear equations in n unknowns. Show that the system has a nontrivial solution if and only if $\det(A) = 0$.

13. Let

$$A = \begin{bmatrix} a_{11} & a_{12} & a_{13} \\ a_{21} & a_{22} & a_{23} \\ a_{31} & a_{32} & a_{33} \end{bmatrix} \quad \text{and} \quad B = \begin{bmatrix} b_{11} & b_{12} & b_{13} \\ b_{21} & b_{22} & b_{23} \\ b_{31} & b_{32} & b_{33} \end{bmatrix}$$

Show that
(a) $(A^t)^t = A$ (b) $(A + B)^t = A^t + B^t$ (c) $(AB)^t = B^tA^t$ (d) $(kA)^t = kA^t$

14. Prove that $(A^tB^t)^t = BA$.

15. A square matrix A is called ***symmetric*** if $A^t = A$ and ***skew-symmetric*** if $A^t = -A$. Show that if B is a square matrix, then
(a) BB^t and $B + B^t$ are symmetric (b) $B - B^t$ is skew-symmetric

2.4 COFACTOR EXPANSION; CRAMER'S RULE

In this section we consider another method for evaluating determinants. As a consequence of our work here, we shall obtain a formula for the inverse of an invertible matrix as well as a formula for the solution to certain systems of linear equations in terms of determinants.

Definition. If A is a square matrix, then the ***minor of entry a_{ij}*** is denoted by M_{ij} and is defined to be the determinant of the submatrix that remains after the ith row and jth column are deleted from A. The number $(-1)^{i+j}M_{ij}$ is denoted by C_{ij} and is called the ***cofactor of entry a_{ij}***.

Example 26

Let

$$A = \begin{bmatrix} 3 & 1 & -4 \\ 2 & 5 & 6 \\ 1 & 4 & 8 \end{bmatrix}$$

The minor of entry a_{11} is

$$M_{11} = \begin{vmatrix} 3 & 1 & -4 \\ 2 & 5 & 6 \\ 1 & 4 & 8 \end{vmatrix} = \begin{vmatrix} 5 & 6 \\ 4 & 8 \end{vmatrix} = 16$$

The cofactor of a_{11} is

$$C_{11} = (-1)^{1+1}M_{11} = M_{11} = 16$$

Similarly, the minor of entry a_{32} is

$$M_{32} = \begin{vmatrix} 3 & 1 & -4 \\ 2 & 5 & 6 \\ 1 & 4 & 8 \end{vmatrix} = \begin{vmatrix} 3 & -4 \\ 2 & 6 \end{vmatrix} = 26$$

The cofactor of a_{32} is

$$C_{32} = (-1)^{3+2}M_{32} = -M_{32} = -26$$

Notice that the cofactor and the minor of an element a_{ij} differ only in sign, that is, $C_{ij} = \pm M_{ij}$. A quick way for determining whether to use the + or − is to use the fact that the sign relating C_{ij} and M_{ij} is in the ith row and jth column of the array

$$\begin{bmatrix} + & - & + & - & + & \cdots \\ - & + & - & + & - & \cdots \\ + & - & + & - & + & \cdots \\ - & + & - & + & - & \cdots \\ \vdots & \vdots & \vdots & \vdots & \vdots & \end{bmatrix}$$

For example, $C_{11} = M_{11}$, $C_{21} = -M_{21}$, $C_{12} = -M_{12}$, $C_{22} = M_{22}$, etc.

Consider the general 3×3 matrix

$$A = \begin{bmatrix} a_{11} & a_{12} & a_{13} \\ a_{21} & a_{22} & a_{23} \\ a_{31} & a_{32} & a_{33} \end{bmatrix}$$

In Example 7 we showed

$$\det(A) = a_{11}a_{22}a_{33} + a_{12}a_{23}a_{31} + a_{13}a_{21}a_{32} \\ - a_{13}a_{22}a_{31} - a_{12}a_{21}a_{33} - a_{11}a_{23}a_{32} \quad (2.2)$$

which can be rewritten as

$$\det(A) = a_{11}(a_{22}a_{33} - a_{23}a_{32}) + a_{21}(a_{13}a_{32} - a_{12}a_{33}) + a_{31}(a_{12}a_{23} - a_{13}a_{22})$$

Since the expressions in parentheses are just the cofactors C_{11}, C_{21}, and C_{31} (verify) we have

$$\det(A) = a_{11}C_{11} + a_{21}C_{21} + a_{31}C_{31} \quad (2.3)$$

Equation (2.3) shows that the determinant of A can be computed by multiplying the entries in the first column of A by their cofactors and adding the resulting products. This method of evaluating $\det(A)$ is called ***cofactor expansion*** along the first column of A.

Example 27

Let

$$A = \begin{bmatrix} 3 & 1 & 0 \\ -2 & -4 & 3 \\ 5 & 4 & -2 \end{bmatrix}$$

Evaluate $\det(A)$ by cofactor expansion along the first column of A.

Solution. From (2.3)

$$\det(A) = 3\begin{vmatrix} -4 & 3 \\ 4 & -2 \end{vmatrix} - (-2)\begin{vmatrix} 1 & 0 \\ 4 & -2 \end{vmatrix} + 5\begin{vmatrix} 1 & 0 \\ -4 & 3 \end{vmatrix}$$
$$= 3(-4) - (-2)(-2) + 5(3) = -1$$

By rearranging the terms in (2.2) in various ways, it is possible to obtain other formulas like (2.3). There should be no trouble checking that all of the following are correct (see Exercise 23):

$$\begin{aligned} \det(A) &= a_{11}C_{11} + a_{12}C_{12} + a_{13}C_{13} \\ &= a_{11}C_{11} + a_{21}C_{21} + a_{31}C_{31} \\ &= a_{21}C_{21} + a_{22}C_{22} + a_{23}C_{23} \\ &= a_{12}C_{12} + a_{22}C_{22} + a_{32}C_{32} \\ &= a_{31}C_{31} + a_{32}C_{32} + a_{33}C_{33} \\ &= a_{13}C_{13} + a_{23}C_{23} + a_{33}C_{33} \end{aligned} \quad (2.4)$$

Notice that in each equation the entries and cofactors all come from the same row or column. These equations are called the ***cofactor expansions*** of $\det(A)$.

The results we have just given for 3×3 matrices form a special case of the following general theorem, which we state without proof.

Theorem 7. *The determinant of an $n \times n$ matrix A can be computed by multiplying the entries in any row (or column) by their cofactors and adding the resulting products; that is, for each $1 \leq i \leq n$ and $1 \leq j \leq n$,*

$$det(A) = a_{1j}C_{1j} + a_{2j}C_{2j} + \cdots + a_{nj}C_{nj}$$
(*cofactor expansion along the jth column*)

and

$$det(A) = a_{i1}C_{i1} + a_{i2}C_{i2} + \cdots + a_{in}C_{in}$$
(*cofactor expansion along the ith row*)

Example 28

Let A be the matrix in Example 27. Evaluate $\det(A)$ by cofactor expansion along the first row.

Solution.

$$\det(A) = \begin{vmatrix} 3 & 1 & 0 \\ -2 & -4 & 3 \\ 5 & 4 & -2 \end{vmatrix} = 3\begin{vmatrix} -4 & 3 \\ 4 & -2 \end{vmatrix} - (1)\begin{vmatrix} -2 & 3 \\ 5 & -2 \end{vmatrix} + 0\begin{vmatrix} -2 & -4 \\ 5 & 4 \end{vmatrix}$$

$$= 3(-4) - (1)(-11) = -1$$

This agrees with the result obtained in Example 27.

REMARK. In this example it was unnecessary to compute the last cofactor, since it was multiplied by zero. In general, the best strategy for evaluating a determinant by cofactor expansion is to expand along a row or column having the largest number of zeros.

Although cofactor expansion is not usually as efficient as reduction to triangular form for the evaluation of determinants, the two methods can sometimes be combined in particular situations to yield an effective computational technique. The following example illustrates this idea.

Example 29

Evaluate $\det(A)$ where

$$A = \begin{bmatrix} 3 & 5 & -2 & 6 \\ 1 & 2 & -1 & 1 \\ 2 & 4 & 1 & 5 \\ 3 & 7 & 5 & 3 \end{bmatrix}$$

Solution. By adding suitable multiples of the second row to the remaining rows, we obtain

$$\det(A) = \begin{vmatrix} 0 & -1 & 1 & 3 \\ 1 & 2 & -1 & 1 \\ 0 & 0 & 3 & 3 \\ 0 & 1 & 8 & 0 \end{vmatrix}$$

$$= -\begin{vmatrix} -1 & 1 & 3 \\ 0 & 3 & 3 \\ 1 & 8 & 0 \end{vmatrix} \qquad \boxed{\text{Cofactor expansion along the first column.}}$$

$$= -\begin{vmatrix} -1 & 1 & 3 \\ 0 & 3 & 3 \\ 0 & 9 & 3 \end{vmatrix} \qquad \boxed{\text{We added the first row to the third row.}}$$

$$= -(-1)\begin{vmatrix} 3 & 3 \\ 9 & 3 \end{vmatrix} \qquad \boxed{\text{Cofactor expansion along the first column.}}$$

$$= -18$$

In a cofactor expansion we compute $\det(A)$ by multiplying the entries in a row or column by their cofactors and adding the resulting products. It turns out that if one multiplies the entries in any row by the corresponding cofactors from a *different* row, the sum of these products is always zero. (This result holds also for columns.) Although we omit the general proof, the next example illustrates the idea of the proof in a special case.

Example 30

Let

$$A = \begin{bmatrix} a_{11} & a_{12} & a_{13} \\ a_{21} & a_{22} & a_{23} \\ a_{31} & a_{32} & a_{33} \end{bmatrix}$$

Consider the quantity

$$a_{11}C_{31} + a_{12}C_{32} + a_{13}C_{33}$$

which is formed by multiplying the entries in the first row by the cofactors of the corresponding entries in the third row and adding the resulting products. We now show this quantity is equal to zero by the following trick. Construct a new matrix A' by replacing the third row of A with another copy of the first row. Thus

$$A' = \begin{bmatrix} a_{11} & a_{12} & a_{13} \\ a_{21} & a_{22} & a_{23} \\ a_{11} & a_{12} & a_{13} \end{bmatrix}$$

Let C'_{31}, C'_{32} and C'_{33} be the cofactors of the entries in the third row of A'. Since the first two rows of A and A' are the same, and since the computation of C_{31}, C_{32},

C_{33}, C'_{31}, C'_{32}, and C'_{33} involve only entries from the first two rows of A and A', it follows that

$$C_{31} = C'_{31}, \quad C_{32} = C'_{32}, \quad C_{33} = C'_{33}$$

Since A' has two identical rows

$$\det(A') = 0 \tag{2.5}$$

On the other hand, evaluating $\det(A')$ by cofactor expansion along the third row gives

$$\begin{aligned}\det(A') &= a_{11}C'_{31} + a_{12}C'_{32} + a_{13}C'_{33} \\ &= a_{11}C_{31} + a_{12}C_{32} + a_{13}C_{33}\end{aligned} \tag{2.6}$$

From (2.5) and (2.6) we obtain

$$a_{11}C_{31} + a_{12}C_{32} + a_{13}C_{33} = 0$$

Definition. If A is any $n \times n$ matrix and C_{ij} is the cofactor of a_{ij}, then the matrix

$$\begin{bmatrix} C_{11} & C_{12} & \cdots & C_{1n} \\ C_{21} & C_{22} & \cdots & C_{2n} \\ \vdots & \vdots & & \vdots \\ C_{n1} & C_{n2} & \cdots & C_{nn} \end{bmatrix}$$

is called the ***matrix of cofactors from A***. The transpose of this matrix is called the ***adjoint*** of A and is denoted by $\mathrm{adj}(A)$.

Example 31

Let

$$A = \begin{bmatrix} 3 & 2 & -1 \\ 1 & 6 & 3 \\ 2 & -4 & 0 \end{bmatrix}$$

The cofactors of A are

$$\begin{array}{lll} C_{11} = 12 & C_{12} = 6 & C_{13} = -16 \\ C_{21} = 4 & C_{22} = 2 & C_{23} = 16 \\ C_{31} = 12 & C_{32} = -10 & C_{33} = 16 \end{array}$$

so that the matrix of cofactors is

$$\begin{bmatrix} 12 & 6 & -16 \\ 4 & 2 & 16 \\ 12 & -10 & 16 \end{bmatrix}$$

and the adjoint of A is

$$\mathrm{adj}(A) = \begin{bmatrix} 12 & 4 & 12 \\ 6 & 2 & -10 \\ -16 & 16 & 16 \end{bmatrix}$$

We are now in a position to establish a formula for the inverse of an invertible matrix.

Theorem 8. *If A is an invertible matrix, then*

$$A^{-1} = \frac{1}{det(A)} adj(A)$$

Proof. We show first that

$$A \operatorname{adj}(A) = \det(A)I$$

Consider the product

$$A \operatorname{adj}(A) = \begin{bmatrix} a_{11} & a_{12} & \cdots & a_{1n} \\ a_{21} & a_{22} & \cdots & a_{2n} \\ \vdots & \vdots & & \vdots \\ a_{i1} & a_{i2} & \cdots & a_{in} \\ \vdots & \vdots & & \vdots \\ a_{n1} & a_{n2} & \cdots & a_{nn} \end{bmatrix} \begin{bmatrix} C_{11} & C_{21} & \cdots & C_{j1} & \cdots & C_{n1} \\ C_{12} & C_{22} & \cdots & C_{j2} & \cdots & C_{n2} \\ \vdots & \vdots & & \vdots & & \vdots \\ C_{1n} & C_{2n} & \cdots & C_{jn} & \cdots & C_{nn} \end{bmatrix}$$

The entry in the ith row and jth column of $A \operatorname{adj}(A)$ is

$$a_{i1}C_{j1} + a_{i2}C_{j2} + \cdots + a_{in}C_{jn} \tag{2.7}$$

(see the shaded lines above).

If $i = j$, then (2.7) is the cofactor expansion of $\det(A)$ along the ith row of A (Theorem 7). On the other hand, if $i \neq j$, then the a's and the cofactors come from different rows of A, so the value of (2.7) is zero. Therefore

$$A \operatorname{adj}(A) = \begin{bmatrix} \det(A) & 0 & \cdots & 0 \\ 0 & \det(A) & \cdots & 0 \\ \vdots & \vdots & & \vdots \\ 0 & 0 & \cdots & \det(A) \end{bmatrix} = \det(A)I \tag{2.8}$$

Since A is invertible, $\det(A) \neq 0$. Therefore, equation (2.8) can be rewritten as

$$\frac{1}{\det(A)}[A \operatorname{adj}(A)] = I$$

or

$$A\left[\frac{1}{\det(A)} \operatorname{adj}(A)\right] = I$$

Multiplying both sides on the left by A^{-1} yields

$$A^{-1} = \frac{1}{\det(A)} \operatorname{adj}(A) \quad ▮ \tag{2.9}$$

Example 32

Use (2.9) to find the inverse of the matrix A in Example 31.

Solution. The reader can check that $\det(A) = 64$. Thus

$$A^{-1} = \frac{1}{\det(A)}\operatorname{adj}(A) = \frac{1}{64}\begin{bmatrix} 12 & 4 & 12 \\ 6 & 2 & -10 \\ -16 & 16 & 16 \end{bmatrix}$$

$$= \begin{bmatrix} \frac{12}{64} & \frac{4}{64} & \frac{12}{64} \\ \frac{6}{64} & \frac{2}{64} & -\frac{10}{64} \\ -\frac{16}{64} & \frac{16}{64} & \frac{16}{64} \end{bmatrix}$$

We note that for matrices larger than 3×3 the matrix inversion method in this example is computationally inferior to the technique given in Section 1.6. On the other hand, the inversion method in Section 1.6 is just a computational procedure or algorithm for computing A^{-1} and is not very useful for studying properties of the inverse. Formula 2.9 can often be used to obtain properties of the inverse without actually computing it (Exercise 20).

In a similar vein, it is often useful to have a formula for the solution of a system of equations that can be used to study properties of the solution without solving the system. The next theorem establishes such a formula for systems of n equations in n unknowns. The formula is known as *Cramer's rule.*

Theorem 9 (***Cramer's Rule***). *If $AX = B$ is a system of n linear equations in n unknowns such that $\det(A) \neq 0$, then the system has a unique solution. This solution is*

$$x_1 = \frac{\det(A_1)}{\det(A)}, \quad x_2 = \frac{\det(A_2)}{\det(A)}, \ldots, x_n = \frac{\det(A_n)}{\det(A)}$$

where A_j is the matrix obtained by replacing the entries of the jth column of A by the entries in the matrix

$$B = \begin{bmatrix} b_1 \\ b_2 \\ \vdots \\ b_n \end{bmatrix}$$

Proof. If $\det(A) \neq 0$, then A is invertible and, by Theorem 11 in Section 1.7, $X = A^{-1}B$ is the unique solution of $AX = B$. Therefore, by Theorem 8 we have

$$X = A^{-1}B = \frac{1}{\det(A)}\operatorname{adj}(A)B = \frac{1}{\det(A)}\begin{bmatrix} C_{11} & C_{21} & \cdots & C_{n1} \\ C_{12} & C_{22} & \cdots & C_{n2} \\ \vdots & \vdots & & \vdots \\ C_{1n} & C_{2n} & \cdots & C_{nn} \end{bmatrix}\begin{bmatrix} b_1 \\ b_2 \\ \vdots \\ b_n \end{bmatrix}$$

Multiplying the matrices out gives

$$X = \frac{1}{\det(A)}\begin{bmatrix} b_1C_{11} + b_2C_{21} + \cdots + b_nC_{n1} \\ b_1C_{12} + b_2C_{22} + \cdots + b_nC_{n2} \\ \vdots \qquad\qquad \vdots \qquad\qquad\quad \vdots \\ b_1C_{1n} + b_2C_{2n} + \cdots + b_nC_{nn} \end{bmatrix}$$

The entry in the jth row of X is therefore

$$x_j = \frac{b_1C_{1j} + b_2C_{2j} + \cdots + b_nC_{nj}}{\det(A)} \tag{2.10}$$

Now let

$$A_j = \begin{bmatrix} a_{11} & a_{12} & \cdots & a_{1j-1} & b_1 & a_{1j+1} & \cdots & a_{1n} \\ a_{21} & a_{22} & \cdots & a_{2j-1} & b_2 & a_{2j+1} & \cdots & a_{2n} \\ \vdots & \vdots & & \vdots & \vdots & \vdots & & \vdots \\ a_{n1} & a_{n2} & \cdots & a_{nj-1} & b_n & a_{nj+1} & \cdots & a_{nn} \end{bmatrix}$$

Since A_j differs from A only in the jth column, the cofactors of entries $b_1, b_2, \ldots, b_n$ in A_j are the same as the cofactors of the corresponding entries in the jth column of A. The cofactor expansion of $\det(A_j)$ along the jth column is therefore

$$\det(A_j) = b_1C_{1j} + b_2C_{2j} + \cdots + b_nC_{nj}$$

Substituting this result in (2.10) gives

$$x_j = \frac{\det(A_j)}{\det(A)} \qquad \blacksquare$$

Example 33

Use Cramer's Rule to solve

$$\begin{aligned} x_1 + \qquad\; + 2x_3 &= 6 \\ -3x_1 + 4x_2 + 6x_3 &= 30 \\ -x_1 - 2x_2 + 3x_3 &= 8 \end{aligned}$$

Solution.

$$A = \begin{bmatrix} 1 & 0 & 2 \\ -3 & 4 & 6 \\ -1 & -2 & 3 \end{bmatrix} \qquad A_1 = \begin{bmatrix} 6 & 0 & 2 \\ 30 & 4 & 6 \\ 8 & -2 & 3 \end{bmatrix}$$

$$A_2 = \begin{bmatrix} 1 & 6 & 2 \\ -3 & 30 & 6 \\ -1 & 8 & 3 \end{bmatrix} \qquad A_3 = \begin{bmatrix} 1 & 0 & 6 \\ -3 & 4 & 30 \\ -1 & -2 & 8 \end{bmatrix}$$

Therefore

$$x_1 = \frac{\det(A_1)}{\det(A)} = \frac{-40}{44} = \frac{-10}{11}, \qquad x_2 = \frac{\det(A_2)}{\det(A)} = \frac{72}{44} = \frac{18}{11},$$

$$x_3 = \frac{\det(A_3)}{\det(A)} = \frac{152}{44} = \frac{38}{11}$$

To solve a system of n equations in n unknowns by Cramer's Rule, it is necessary to evaluate $n + 1$ determinants of $n \times n$ matrices. For systems with more than three equations, Gaussian elimination is superior computationally since it is only necessary to reduce one n by $n + 1$ augmented matrix. Cramer's rule, however, gives a formula for the solution.

EXERCISE SET 2.4

1. Let

$$A = \begin{bmatrix} 1 & 6 & -3 \\ -2 & 7 & 1 \\ 3 & -1 & 4 \end{bmatrix}$$

(a) Find all the minors.
(b) Find all the cofactors.

2. Let

$$A = \begin{bmatrix} 4 & 0 & 4 & 4 \\ -1 & 0 & 1 & 1 \\ 1 & -3 & 0 & 3 \\ 6 & 3 & 14 & 2 \end{bmatrix}$$

Find
(a) M_{13} and C_{13} (b) M_{23} and C_{23} (c) M_{22} and C_{22} (d) M_{21} and C_{21}

3. Evaluate the determinant of the matrix in Exercise 1 by a cofactor expansion along
(a) the first row (b) the first column (c) the second row
(d) the second column (e) the third row (f) the third column

4. For the matrix in Exercise 1, find
(a) adj(A) (b) A^{-1} using the method of Example 32.

In Exercises 5–10 evaluate det(A) by a cofactor expansion along a row or column of your choice.

5. $A = \begin{bmatrix} 0 & 6 & 0 \\ 8 & 6 & 8 \\ 3 & 2 & 2 \end{bmatrix}$ **6.** $A = \begin{bmatrix} 1 & 3 & 7 \\ 2 & 0 & -8 \\ -1 & -3 & 4 \end{bmatrix}$

7. $\begin{bmatrix} 1 & 1 & 1 \\ k & k & k \\ k^2 & k^2 & k^2 \end{bmatrix}$ **8.** $\begin{bmatrix} k-1 & 2 & 3 \\ 2 & k-3 & 4 \\ 3 & 4 & k-4 \end{bmatrix}$

9. $A = \begin{bmatrix} 4 & 4 & 0 & 4 \\ 1 & 1 & 0 & -1 \\ 3 & 0 & -3 & 1 \\ 6 & 14 & 3 & 6 \end{bmatrix}$ **10.** $A = \begin{bmatrix} 4 & 3 & 1 & 9 & 2 \\ 0 & 3 & 2 & 4 & 2 \\ 0 & 3 & 4 & 6 & 4 \\ 1 & -1 & 2 & 2 & 2 \\ 0 & 0 & 3 & 3 & 3 \end{bmatrix}$

11. Let

$$A = \begin{bmatrix} 1 & 3 & 1 & 1 \\ 2 & 5 & 2 & 2 \\ 1 & 3 & 8 & 9 \\ 1 & 3 & 2 & 2 \end{bmatrix}$$

(a) Evaluate A^{-1} using the method of Example 32.
(b) Evaluate A^{-1} using the method of Example 29 in Section 1.6.
(c) Which method involves less computation?

In Exercises 12–17 solve by Cramer's rule, where it applies.

12. $\begin{aligned} 3x_1 - 4x_2 &= -5 \\ 2x_1 + x_2 &= 4 \end{aligned}$

13. $\begin{aligned} 4x + 5y \phantom{{}+2z} &= 2 \\ 11x + y + 2z &= 3 \\ x + 5y + 2z &= 1 \end{aligned}$

14. $\begin{aligned} x + y - 2z &= 1 \\ 2x - y + z &= 2 \\ x - 2y - 4z &= -4 \end{aligned}$

15. $\begin{aligned} x_1 - 3x_2 + x_3 &= 4 \\ 2x_1 - x_2 \phantom{{}+x_3} &= -2 \\ 4x_1 \phantom{{}-3x_2} - 3x_3 &= 0 \end{aligned}$

16. $\begin{aligned} 2x_1 - x_2 + x_3 - 4x_4 &= -32 \\ 7x_1 + 2x_2 + 9x_3 - x_4 &= 14 \\ 3x_1 - x_2 + x_3 + x_4 &= 11 \\ x_1 + x_2 - 4x_3 - 2x_4 &= -4 \end{aligned}$

17. $\begin{aligned} 2x_1 - x_2 + x_3 &= 8 \\ 4x_1 + 3x_2 + x_3 &= 7 \\ 6x_1 + 2x_2 + 2x_3 &= 15 \end{aligned}$

18. Use Cramer's rule to solve for z without solving for x, y, and w.

$$\begin{aligned} 4x + y + z + w &= 6 \\ 3x + 7y - z + w &= 1 \\ 7x + 3y - 5z + 8w &= -3 \\ x + y + z + 2w &= 3 \end{aligned}$$

19. Let $AX = B$ be the system in Exercise 18.
(a) Solve by Cramer's Rule.
(b) Solve by Gauss-Jordan elimination.
(c) Which method involves the least amount of computation?

20. Prove that if $\det(A) = 1$ and all the entries in A are integers, then all the entries in A^{-1} are integers.

21. Let $AX = B$ be a system of n linear equations in n unknowns with integer coefficients and integer constants. Prove that if $\det(A) = 1$, then the solution X has integer entries.

22. Prove that if A is an invertible upper triangular matrix, then A^{-1} is upper triangular.

23. Derive the first and last cofactor expansions listed in (2.4).

24. Prove: The equation of the line through the distinct points (a_1, b_1) and (a_2, b_2) can be written

$$\begin{vmatrix} x & y & 1 \\ a_1 & b_1 & 1 \\ a_2 & b_2 & 1 \end{vmatrix} = 0$$

25. Use the result in Exercise 24 to prove: (x_1, y_1), (x_2, y_2), and (x_3, y_3) are collinear points if and only if

$$\begin{vmatrix} x_1 & y_1 & 1 \\ x_2 & y_2 & 1 \\ x_3 & y_3 & 1 \end{vmatrix} = 0$$

26. Prove: The equation of the plane through the noncollinear points (a_1, b_1, c_1), (a_2, b_2, c_2), and (a_3, b_3, c_3) can be written:

$$\begin{vmatrix} x & y & z & 1 \\ a_1 & b_1 & c_1 & 1 \\ a_2 & b_2 & c_2 & 1 \\ a_3 & b_3 & c_3 & 1 \end{vmatrix} = 0$$

3 Vectors in 2-Space and 3-Space

Readers familiar with the contents of this chapter can go to Chapter 4 with no loss of continuity.

3.1 INTRODUCTION TO VECTORS (GEOMETRIC)

In this section vectors in 2-space and 3-space are introduced geometrically. Arithmetic operations on vectors are defined and some basic properties of these operations are established.

Many physical quantities like area, length, and mass are completely described once a real number representing the magnitude of the quantity is given. Other physical quantities, called ***vectors***, are not completely determined until both a magnitude and a direction are specified. Force, displacement, and velocity are examples of vectors.

Vectors can be represented geometrically as directed line segments or arrows in 2-space or 3-space; the direction of the arrow specifies the direction of the vector and the length of the arrow describes its magnitude. The tail of the arrow is called the ***initial point*** of the vector, and the tip of the arrow the ***terminal point***. We shall denote vectors by lower case boldface type like **a**, **k**, **v**, **w** and **x**. When discussing vectors, we shall refer to numbers as ***scalars***. All our scalars will be real numbers and will be denoted by ordinary lowercase type like a, k, v, w and x.

If as in Figure 3.1a the initial point of a vector **v** is A and the terminal point is B, we write

$$\mathbf{v} = \overrightarrow{AB}$$

Vectors having the same length and same direction, like those in Figure 3.1b, are called ***equivalent***. Since we want a vector to be determined solely by its length

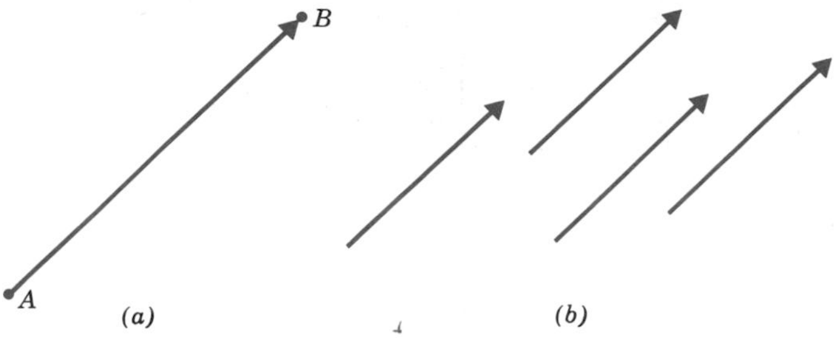

Figure 3.1 (*a*) The vector $\overrightarrow{AB}$. (*b*) Equivalent vectors.

and direction, equivalent vectors are regarded as ***equal*** even though they may be located in different positions. If $\mathbf{v}$ and $\mathbf{w}$ are equivalent we write

$$\mathbf{v} = \mathbf{w}$$

Definition. If $\mathbf{v}$ and $\mathbf{w}$ are any two vectors, then the ***sum*** $\mathbf{v} + \mathbf{w}$ is the vector determined as follows. Position the vector $\mathbf{w}$ so that its initial point coincides with the terminal point of $\mathbf{v}$. The vector $\mathbf{v} + \mathbf{w}$ is represented by the arrow from the initial point of $\mathbf{v}$ to the terminal point of $\mathbf{w}$ (Figure 3.2*a*).

In Figure 3.3 we have constructed two sums, $\mathbf{v} + \mathbf{w}$ (blue arrows) and $\mathbf{w} + \mathbf{v}$ (white arrows). It is evident that

$$\mathbf{v} + \mathbf{w} = \mathbf{w} + \mathbf{v}$$

and that the sum coincides with the diagonal of the parallelogram determined by $\mathbf{v}$ and $\mathbf{w}$ when these vectors are located so they have the same initial point.

A vector of length zero is called a ***zero vector*** and is denoted by $\mathbf{0}$. We define

$$\mathbf{0} + \mathbf{v} = \mathbf{v} + \mathbf{0} = \mathbf{v}$$

for every vector $\mathbf{v}$. Since there is no natural direction for a zero vector, we shall agree that it can be assigned any direction that is convenient for the problem

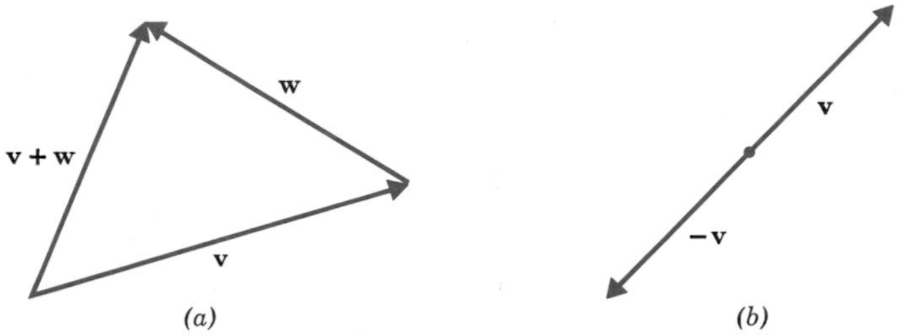

Figure 3.2

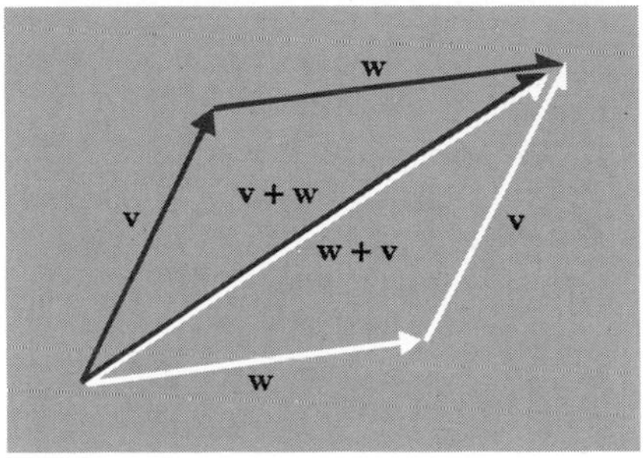

Figure 3.3

being considered. If $\mathbf{v}$ is any nonzero vector, it is obvious that the only vector $\mathbf{w}$ satisfying $\mathbf{v} + \mathbf{w} = \mathbf{0}$ is the vector having the same magnitude as $\mathbf{v}$ but oppositely directed (Figure 3.2*b*). This vector is called the ***negative*** (or ***additive inverse***) of $\mathbf{v}$ and we write

$$\mathbf{w} = -\mathbf{v}$$

In addition, we define $-\mathbf{0} = \mathbf{0}$.

Definition. If $\mathbf{v}$ and $\mathbf{w}$ are any two vectors, then subtraction is defined by

$$\mathbf{v} - \mathbf{w} = \mathbf{v} + (-\mathbf{w})$$

(Figure 3.4*a*)

To obtain the difference $\mathbf{v} - \mathbf{w}$ without constructing $-\mathbf{w}$, position $\mathbf{v}$ and $\mathbf{w}$ so their initial points coincide; the vector from the terminal point of $\mathbf{w}$ to the terminal point of $\mathbf{v}$ is then the vector $\mathbf{v} - \mathbf{w}$ (Figure 3.4*b*).

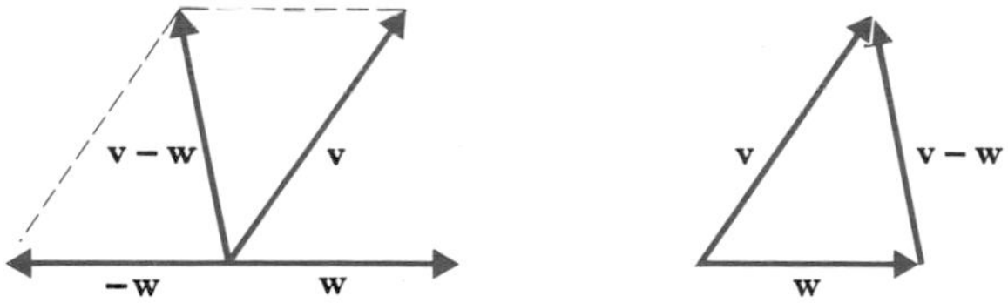

Figure 3.4 (*a*) (*b*)

Definition. If $\mathbf{v}$ is a vector and k is a real number (scalar), then the ***product*** $k\mathbf{v}$ is defined to be the vector whose length is $|k|$ times the length of $\mathbf{v}$ and whose direction is the same as that of $\mathbf{v}$ if $k > 0$ and opposite to that of $\mathbf{v}$ if $k < 0$. We define $k\mathbf{v} = \mathbf{0}$ if $k = 0$ or $\mathbf{v} = \mathbf{0}$.

Figure 3.5 illustrates the relation between a vector $\mathbf{v}$ and the vectors $\frac{1}{2}\mathbf{v}$, $(-1)\mathbf{v}$, $2\mathbf{v}$, and $(-3)\mathbf{v}$.

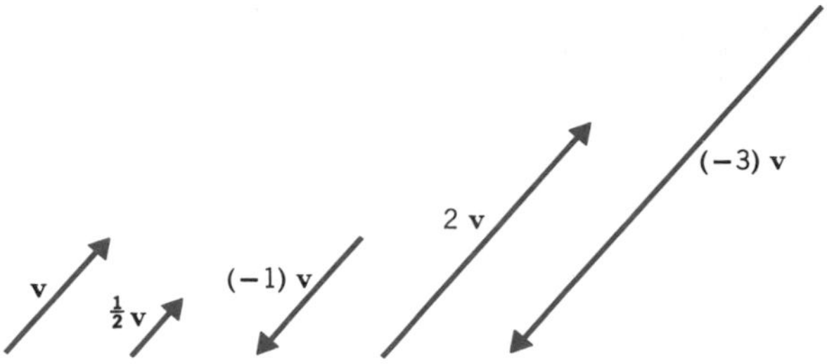

Figure 3.5

Note that the vector $(-1)\mathbf{v}$ has the same length as $\mathbf{v}$ but is oppositely directed. Thus $(-1)\mathbf{v}$ is just the negative of $\mathbf{v}$; that is,

$$(-1)\mathbf{v} = -\mathbf{v}$$

Problems involving vectors can often be simplified by introducing a rectangular coordinate system. For the moment we shall restrict the discussion to vectors in 2-space (the plane). Let $\mathbf{v}$ be any vector in the plane, and assume, as in Figure 3.6, that $\mathbf{v}$ has been positioned so its initial point is at the origin of a rectangular coordinate system. The coordinates (v_1, v_2) of the terminal point of $\mathbf{v}$ are called the ***components of*** $\mathbf{v}$, and we write

$$\mathbf{v} = (v_1, v_2)$$

If equivalent vectors, $\mathbf{v}$ and $\mathbf{w}$, are located so their initial points fall at the origin, then it is obvious that their terminal points must coincide (since the vectors have the same length and direction). The vectors thus have the same components. It is equally obvious that vectors with the same components must have the same length and same direction, and consequently be equivalent. In summary, two vectors

$$\mathbf{v} = (v_1, v_2) \quad \text{and} \quad \mathbf{w} = (w_1, w_2)$$

are equivalent if and only if

$$v_1 = w_1 \quad \text{and} \quad v_2 = w_2$$

The operations of vector addition and multiplication by scalars are very easy to carry out in terms of components. As illustrated in Figure 3.7, if

$$\mathbf{v} = (v_1, v_2) \quad \text{and} \quad \mathbf{w} = (w_1, w_2)$$

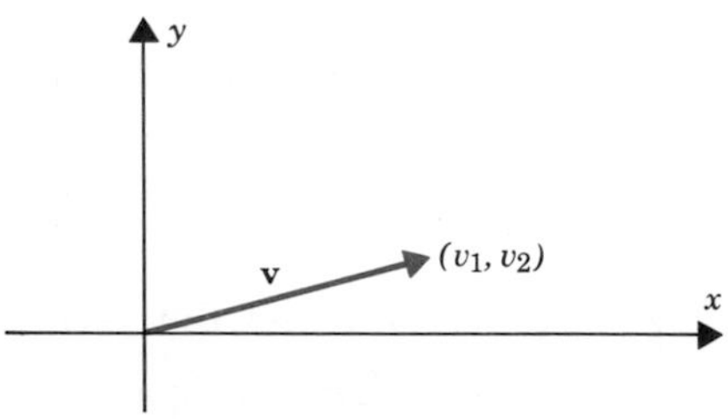

Figure 3.6

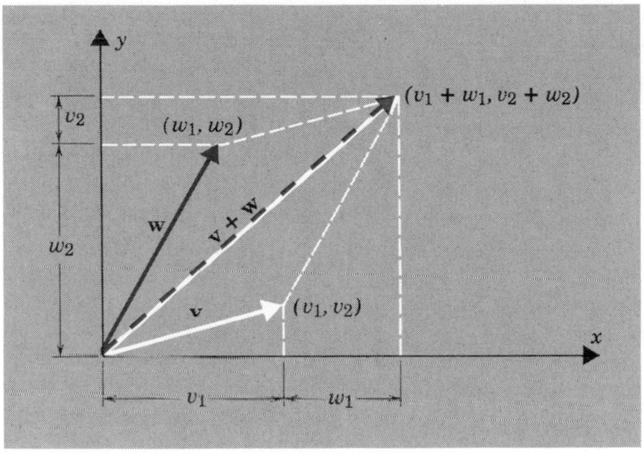

Figure 3.7

then

$$\mathbf{v} + \mathbf{w} = (v_1 + w_1, v_2 + w_2)$$

If $\mathbf{v} = (v_1, v_2)$ and k is any scalar, then by using a geometric argument involving similar triangles, it can be shown (Exercise 14) that

$$k\mathbf{v} = (kv_1, kv_2)$$

(Figure 3.8).

Thus, for example, if $\mathbf{v} = (1, -2)$ and $\mathbf{w} = (7, 6)$, then

$$\mathbf{v} + \mathbf{w} = (1, -2) + (7, 6) = (1 + 7, -2 + 6) = (8, 4)$$

and

$$4\mathbf{v} = 4(1, -2) = (4(1), 4(-2)) = (4, -8)$$

Just as vectors in the plane can be described by pairs of real numbers, vectors in 3-space can be described by triples of real numbers, by introducing a ***rectangular***

Figure 3.8

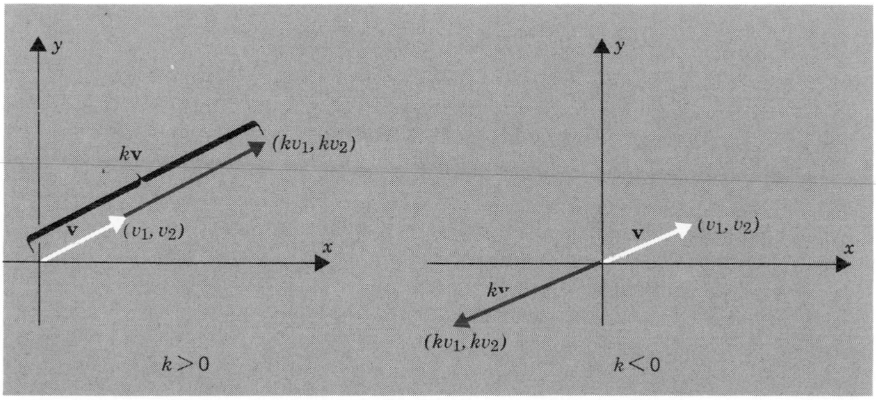

coordinate system. To construct such a coordinate system, select a point O, called the ***origin***, and choose three mutually perpendicular lines, called ***coordinate axes***, passing through the origin. Label these axes x, y, and z and select a positive direction for each coordinate axis as well as a unit of length for measuring distances. (Figure 3.9*a*). Each pair of coordinate axes determines a plane called a ***coordinate plane***. These are referred to as the ***xy-plane***, the ***xz-plane*** and the ***yz-plane***. To each point P in 3-space we assign a triple of numbers (x, y, z) called the ***coordinates of P*** as follows. Pass three planes through P parallel to the coordinate planes, and denote the points of intersections of these planes with the three coordinate axes by X, Y, and Z (Figure 3.9*b*). The coordinates of P are defined to be the signed lengths

$$x = OX \qquad y = OY \qquad z = OZ$$

In Figure 3.10 we have constructed the points whose coordinates are (4, 5, 6) and $(-3, 2, -4)$.

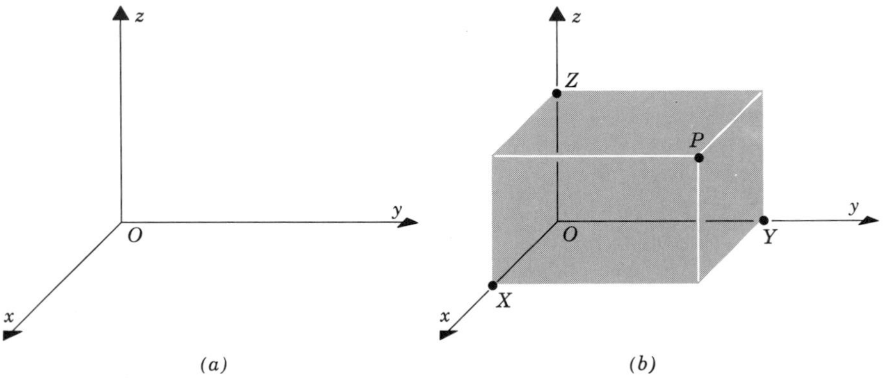

Figure 3.9

Figure 3.10

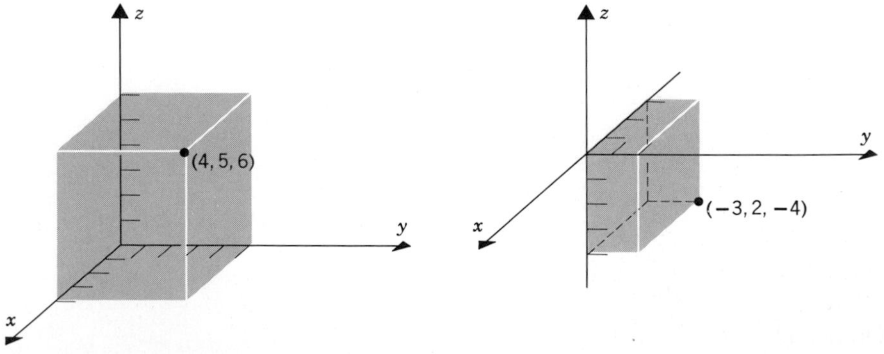

Rectangular coordinate systems in 3-space fall into two categories, ***left-handed*** and ***right-handed***. A right-handed system has the property that an ordinary screw pointed in the positive direction on the z-axis would be advanced if the positive x-axis is rotated 90° toward the positive y-axis (Figure 3.11a). The system is left-handed if the screw would retract (Figure 3.11b).

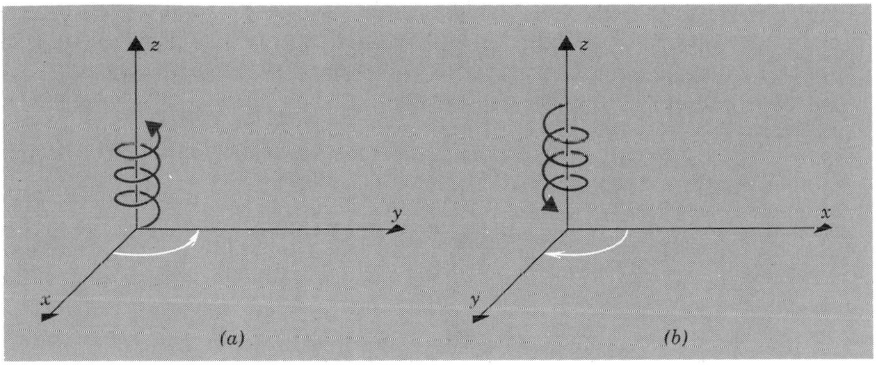

Figure 3.11 (a) Right-handed. (b) Left-handed.

In this book we shall use only right-handed coordinate systems.

If, as in Figure 3.12, a vector $\mathbf{v}$ in 3-space is located so its initial point is at the origin of a rectangular coordinate system, then the coordinates of the terminal point are called the ***components*** of $\mathbf{v}$ and we write

$$\mathbf{v} = (v_1, v_2, v_3)$$

If $\mathbf{v} = (v_1, v_2, v_3)$ and $\mathbf{w} = (w_1, w_2, w_3)$ are two vectors in 3-space, then arguments similar to those used for vectors in a plane can be used to establish the following results.

(i) $\mathbf{v}$ and $\mathbf{w}$ are equivalent if and only if $v_1 = w_1$, $v_2 = w_2$ and $v_3 = w_3$
(ii) $\mathbf{v} + \mathbf{w} = (v_1 + w_1, v_2 + w_2, v_3 + w_3)$
(iii) $k\mathbf{v} = (kv_1, kv_2, kv_3)$ where k is any scalar

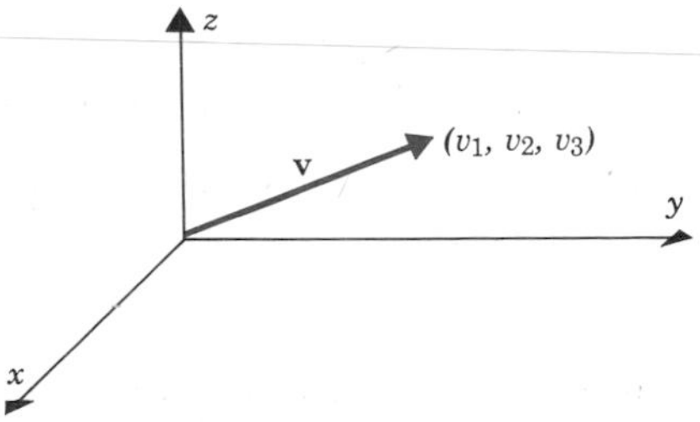

Figure 3.12

Example 1

If $\mathbf{v} = (1, -3, 2)$ and $\mathbf{w} = (4, 2, 1)$, then

$$\mathbf{v} + \mathbf{w} = (5, -1, 3), \quad 2\mathbf{v} = (2, -6, 4), \quad -\mathbf{w} = (-4, -2, -1),$$
$$\mathbf{v} - \mathbf{w} = \mathbf{v} + (-\mathbf{w}) = (-3, -5, 1).$$

Sometimes vectors arise that do not have their initial points at the origin. To find the components of a vector $\mathbf{v}$ having initial point $P_1(x_1, y_1, z_1)$ and terminal point $P_2(x_2, y_2, z_2)$, we construct an equivalent vector whose initial point is at the origin. In Figure 3.13 $\overrightarrow{OQ}$ is such a vector. Therefore, the components of $\mathbf{v} = \overrightarrow{P_1P_2}$ are the coordinates (a, b, c) of the point Q.

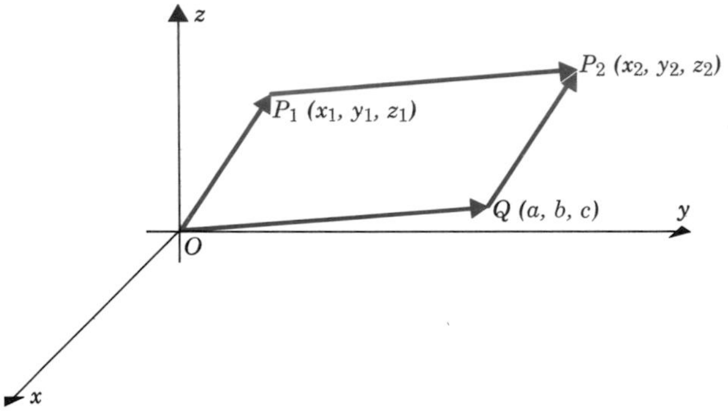

Figure 3.13

From Figure 3.13, $\overrightarrow{OQ} + \overrightarrow{OP_1} = \overrightarrow{OP_2}$, or in terms of components

$$(a, b, c) + (x_1, y_1, z_1) = (x_2, y_2, z_2)$$

or

$$(a + x_1, b + y_1, c + z_1) = (x_2, y_2, z_2)$$

By equating corresponding components and solving for a, b, and c, we find that the components (a, b, c) of $\mathbf{v} = \overrightarrow{P_1P_2}$ are given by

$$a = x_2 - x_1 \qquad b = y_2 - y_1 \qquad c = z_2 - z_1$$

Example 2

The components of the vector $\mathbf{v} = \overrightarrow{P_1P_2}$ with initial point $P_1(2, -1, 4)$ and terminal point $P_2(7, 5, -8)$ are

$$\mathbf{v} = (7 - 2, 5 - (-1), (-8) - 4) = (5, 6, -12)$$

Analogously, in 2-space the vector $\mathbf{v}$ with initial point $P_1(x_1, y_1)$ and terminal point $P_2(x_2, y_2)$ is $\mathbf{v} = (x_2 - x_1, y_2 - y_1)$.

Example 3

The solutions to many problems can be simplified by translating the coordinate axes to obtain new axes parallel to the original ones.

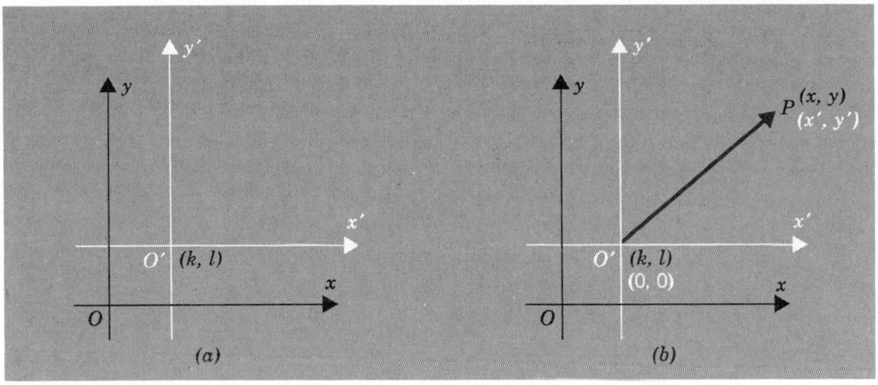

Figure 3.14

In Figure 3.14a we have translated the xy-coordinate axes to obtain an $x'y'$-coordinate system whose origin O' is at the point $(x, y) = (k, l)$. A point P in 2-space now has both (x, y) coordinates and (x', y') coordinates. To see how the two are related, consider the vector $\overrightarrow{O'P}$ (Figure 3.14b). In the xy-system its initial point is at (k, l) and its terminal point is at (x, y); thus $\overrightarrow{O'P} = (x - k, y - l)$. In the $x'y'$-system its initial point is at (0, 0) and its terminal point is at (x', y'); thus $\overrightarrow{O'P} = (x', y')$. Therefore

$$x' = x - k \qquad y' = y - l$$

These are called the ***translation equations***.

To illustrate, if the new origin is at $(k, l) = (4, 1)$ and the xy-coordinates of a point P are (2, 0), then $x'y'$-coordinates of P are $x' = 2 - 4 = -2$ and $y' = 0 - 1 = -1$.

In 3-space the translation equations are

$$x' = x - k \qquad y' = y - l \qquad z' = z - m$$

where (k, l, m) are the xyz-coordinates of the new origin.

EXERCISE SET 3.1

1. Draw a right-handed coordinate system, and locate the points whose coordinates are:
(a) (2, 3, 4) (b) (−2, 3, 4) (c) (2, −3, 4) (d) (2, 3, −4)
(e) (−2, −3, 4) (f) (−2, 3, −4) (g) (2, −3, −4) (h) (−2, −3, −4)
(i) (0, 2, 0) (j) (0, 0, −2) (k) (2, 0, 2) (l) (−2, 0, 0)

2. Sketch the following vectors with the initial points located at the origin:
(a) $\mathbf{v}_1 = (2, 5)$ (b) $\mathbf{v}_2 = (-3, 7)$ (c) $\mathbf{v}_3 = (-5, -4)$
(d) $\mathbf{v}_4 = (6, -2)$ (e) $\mathbf{v}_5 = (2, 0)$ (f) $\mathbf{v}_6 = (0, -8)$
(g) $\mathbf{v}_7 = (2, 3, 4)$ (h) $\mathbf{v}_8 = (2, 0, 2)$ (i) $\mathbf{v}_9 = (0, 0, -2)$

3. Find the components of the vector having initial point P_1 and terminal point P_2.
(a) $P_1(3, 5), P_2(2, 8)$ (b) $P_1(7, -2), P_2(0, 0)$
(c) $P_1(6, 5, 8), P_2(8, -7, -3)$ (d) $P_1(0, 0, 0), P_2(-8, 7, 4)$

4. Find a vector with initial point $P(2, -1, 4)$ that has the same direction as $\mathbf{v} = (7, 6, -3)$.

5. Find a vector, oppositely directed to $\mathbf{v} = (-2, 4, -1)$, having terminal point $Q(2, 0, -7)$.

6. Let $\mathbf{u} = (1, 2, 3)$, $\mathbf{v} = (2, -3, 1)$, and $\mathbf{w} = (3, 2, -1)$. Find the components of
(a) $\mathbf{u} - \mathbf{w}$ (b) $7\mathbf{v} + 3\mathbf{w}$ (c) $-\mathbf{w} + \mathbf{v}$
(d) $3(\mathbf{u} - 7\mathbf{v})$ (e) $-3\mathbf{v} - 8\mathbf{w}$ (f) $2\mathbf{v} - (\mathbf{u} + \mathbf{w})$

7. Let **u**, **v**, and **w** be the vectors in Exercise 6. Find the components of the vector **x** that satisfies $2\mathbf{u} - \mathbf{v} + \mathbf{x} = 7\mathbf{x} + \mathbf{w}$.

8. Let **u**, **v**, and **w** be the vectors in Exercise 6. Find scalars c_1, c_2, and c_3 such that $c_1\mathbf{u} + c_2\mathbf{v} + c_3\mathbf{w} = (6, 14, -2)$.

9. Show that there do not exist scalars c_1, c_2, and c_3 such that

$$c_1(1, 2, -3) + c_2(5, 7, 1) + c_3(6, 9, -2) = (4, 5, 0).$$

10. Find all scalars c_1, c_2, and c_3 such that

$$c_1(2, 7, 8) + c_2(1, -1, 3) + c_3(3, 6, 11) = (0, 0, 0).$$

11. Let P be the point $(2, 3, -2)$ and Q the point $(7, -4, 1)$.
(a) Find the midpoint of the line segment connecting P and Q.
(b) Find the point on the line segment connecting P and Q that is 3/4 of the way from P to Q.

12. Suppose an xy-coordinate system is translated to obtain an $x'y'$-coordinate system whose origin O' has xy-coordinates $(2, -3)$.
(a) Find the $x'y'$-coordinates of the point P whose xy-coordinates are $(7, 5)$.
(b) Find the xy-coordinates of the point Q whose $x'y'$-coordinates are $(-3, 6)$.
(c) Draw the xy and $x'y'$-coordinate axes and locate the points P and Q.

13. Suppose an xyz-coordinate system is translated to obtain an $x'y'z'$-coordinate system. Let **v** be a vector whose components are $\mathbf{v} = (v_1, v_2, v_3)$ in the xyz-system. Show that **v** has the same components in the $x'y'z'$-system.

14. Prove geometrically that if $\mathbf{v} = (v_1, v_2)$, then $k\mathbf{v} = (kv_1, kv_2)$. (Restrict the proof to the case $k > 0$ illustrated in Figure 3.8. The complete proof would involve many cases depending on the quadrant in which the vector falls and the sign of k.)

3.2 NORM OF A VECTOR; VECTOR ARITHMETIC

In this section we establish the basic rules of vector arithmetic.

Theorem 1. *If* **u**, **v** *and* **w** *are vectors in 2- or 3-space and k and l are scalars, then the following relationships hold.*

(a) $\mathbf{u} + \mathbf{v} = \mathbf{v} + \mathbf{u}$
(b) $(\mathbf{u} + \mathbf{v}) + \mathbf{w} = \mathbf{u} + (\mathbf{v} + \mathbf{w})$

(*c*) $\mathbf{u} + \mathbf{0} = \mathbf{0} + \mathbf{u} = \mathbf{u}$
(*d*) $\mathbf{u} + (-\mathbf{u}) = \mathbf{0}$
(*e*) $k(l\mathbf{u}) = (kl)\mathbf{u}$
(*f*) $k(\mathbf{u} + \mathbf{v}) = k\mathbf{u} + k\mathbf{v}$
(*g*) $(k + l)\mathbf{u} = k\mathbf{u} + l\mathbf{u}$
(*h*) $1\mathbf{u} = \mathbf{u}$

Before discussing the proof, we note that we have developed two approaches to vectors: *geometric*, in which vectors are represented by arrows or directed line segments, and *analytic*, in which vectors are represented by pairs or triples of numbers called components. As a consequence, the results in Theorem 1 can be established either geometrically or analytically. To illustrate, we shall prove part (b) both ways. The remaining proofs are left as exercises.

Proof of part (*b*) (*analytic*). We shall give the proof for vectors in 3-space. The proof for 2-space is similar. If $\mathbf{u} = (u_1, u_2, u_3)$, $\mathbf{v} = (v_1, v_2, v_3)$, and $\mathbf{w} = (w_1, w_2, w_3)$ then

$$\begin{aligned}(\mathbf{u} + \mathbf{v}) + \mathbf{w} &= [(u_1, u_2, u_3) + (v_1, v_2, v_3)] + (w_1, w_2, w_3)\\ &= (u_1 + v_1, u_2 + v_2, u_3 + v_3) + (w_1, w_2, w_3)\\ &= ([u_1 + v_1] + w_1, [u_2 + v_2] + w_2, [u_3 + v_3] + w_3)\\ &= (u_1 + [v_1 + w_1], u_2 + [v_2 + w_2], u_3 + [v_3 + w_3])\\ &= (u_1, u_2, u_3) + (v_1 + w_1, v_2 + w_2, v_3 + w_3)\\ &= \mathbf{u} + (\mathbf{v} + \mathbf{w}) \quad \blacksquare\end{aligned}$$

Proof of part (*b*) (*geometric*). Let **u**, **v**, and **w** be represented by $\overrightarrow{PQ}$, $\overrightarrow{QR}$, and $\overrightarrow{RS}$ as shown in Figure 3.15. Then

$$\mathbf{v} + \mathbf{w} = \overrightarrow{QS} \text{ and } \mathbf{u} + (\mathbf{v} + \mathbf{w}) = \overrightarrow{PS}$$

Also

$$\mathbf{u} + \mathbf{v} = \overrightarrow{PR} \text{ and } (\mathbf{u} + \mathbf{v}) + \mathbf{w} = \overrightarrow{PS}$$

Therefore

$$\mathbf{u} + (\mathbf{v} + \mathbf{w}) = (\mathbf{u} + \mathbf{v}) + \mathbf{w} \quad \blacksquare$$

The length of a vector **v** is often called the ***norm*** of **v** and is denoted by $\|\mathbf{v}\|$. It follows from the theorem of Pythagoras that the norm of a vector $\mathbf{v} = (v_1, v_2)$ in

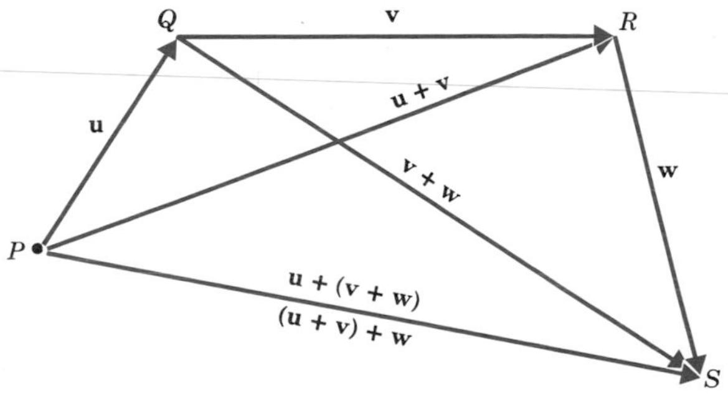

Figure 3.15

2-space is

$$\|\mathbf{v}\| = \sqrt{v_1^2 + v_2^2}$$

(Figure 3.16*a*). Let $\mathbf{v} = (v_1, v_2, v_3)$ be a vector in 3-space. Using Figure 3.16*b* and two applications of the Theorem of Pythagoras, we obtain

$$\begin{aligned}\|\mathbf{v}\|^2 &= (OR)^2 + (RP)^2 \\ &= (OQ)^2 + (OS)^2 + (RP)^2 \\ &= v_1^2 + v_2^2 + v_3^2\end{aligned}$$

Thus

$$\|\mathbf{v}\| = \sqrt{v_1^2 + v_2^2 + v_3^2} \tag{3.1}$$

If $P_1(x_1, y_1, z_1)$ and $P_2(x_2, y_2, z_2)$ are two points in 3-space, then the distance d between them is the norm of the vector $\overrightarrow{P_1P_2}$ (Figure 3.17). Since

$$\overrightarrow{P_1P_2} = (x_2 - x_1, y_2 - y_1, z_2 - z_1)$$

it follows from (3.1) that

$$d = \sqrt{(x_2 - x_1)^2 + (y_2 - y_1)^2 + (z_2 - z_1)^2}$$

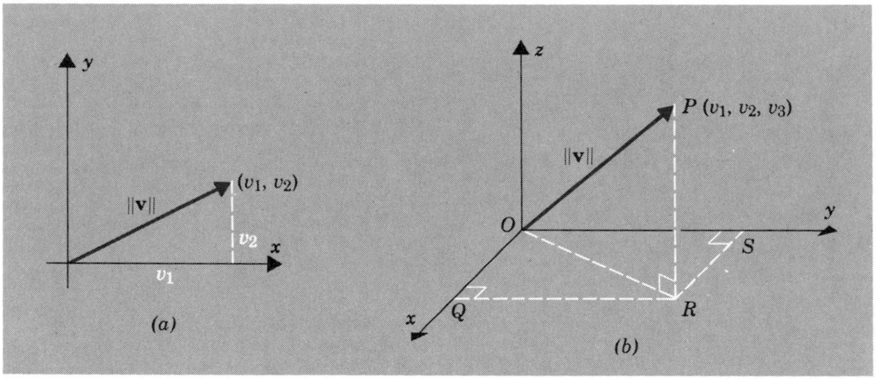

Figure 3.16

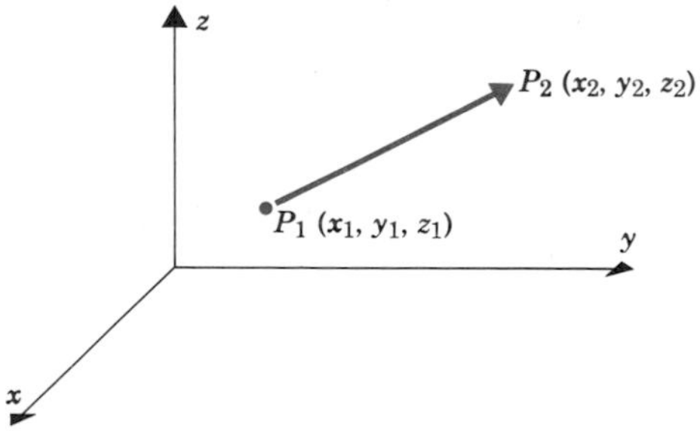

Figure 3.17

Similarly, if $P_1(x_1, y_1)$ and $P_2(x_2, y_2)$ are points in 2-space, then the distance between them is given by

$$d = \sqrt{(x_2 - x_1)^2 + (y_2 - y_1)^2}$$

Example 4

The norm of the vector $\mathbf{v} = (-3, 2, 1)$ is

$$\|\mathbf{v}\| = \sqrt{(-3)^2 + (2)^2 + (1)^2} = \sqrt{14}$$

The distance d between the points $P_1(2, -1, -5)$ and $P_2(4, -3, 1)$ is

$$d = \sqrt{(4-2)^2 + (-3+1)^2 + (1+5)^2} = \sqrt{44} = 2\sqrt{11}$$

EXERCISE SET 3.2

1. Compute the norm of $\mathbf{v}$ when
(a) $\mathbf{v} = (3, 4)$ (b) $\mathbf{v} = (-1, 7)$ (c) $\mathbf{v} = (0, -3)$
(d) $\mathbf{v} = (1, 1, 1)$ (e) $\mathbf{v} = (-8, 7, 4)$ (f) $\mathbf{v} = (9, 0, 0)$

2. Compute the distance between P_1 and P_2.
(a) $P_1(2, 3)$, $P_2(4, 6)$ (b) $P_1(-2, 7)$, $P_2(0, -3)$
(c) $P_1(8, -4, 2)$, $P_2(-6, -1, 0)$ (d) $P_1(1, 1, 1)$, $P_2(6, -7, 3)$

3. Let $\mathbf{u} = (1, -3, 2)$, $\mathbf{v} = (1, 1, 0)$, and $\mathbf{w} = (2, 2, -4)$. Find:
(a) $\|\mathbf{u} + \mathbf{v}\|$ (b) $\|\mathbf{u}\| + \|\mathbf{v}\|$ (c) $\|-2\mathbf{u}\| + 2\|\mathbf{u}\|$
(d) $\|3\mathbf{u} - 5\mathbf{v} + \mathbf{w}\|$ (e) $\dfrac{1}{\|\mathbf{w}\|}\mathbf{w}$ (f) $\left\|\dfrac{1}{\|\mathbf{w}\|}\mathbf{w}\right\|$

4. Find all scalars k such that $\|k\mathbf{v}\| = 3$, where $\mathbf{v} = (1, 2, 4)$.

5. Verify parts (b), (e), (f), and (g) of Theorem 1 for $\mathbf{u} = (1, -3, 7)$, $\mathbf{v} = (6, 6, 9)$, $\mathbf{w} = (-8, 1, 2)$, $k = -3$, and $l = 6$.

6. Show that if $\mathbf{v}$ is nonzero, then $\dfrac{1}{\|\mathbf{v}\|}\mathbf{v}$ has norm 1.

7. Use Exercise 6 to find a vector of norm 1 having the same direction as $\mathbf{v} = (1, 1, 1)$.

8. Let $\mathbf{p}_0 = (x_0, y_0, z_0)$ and $\mathbf{p} = (x, y, z)$. Describe the set of all points (x, y, z) for which $\|\mathbf{p} - \mathbf{p}_0\| = 1$.

9. Prove geometrically that if $\mathbf{u}$ and $\mathbf{v}$ are vectors in 2 or 3-space then $\|\mathbf{u} + \mathbf{v}\| \leq \|\mathbf{u}\| + \|\mathbf{v}\|$.

10. Prove parts (a), (c), and (e) of Theorem 1 analytically.

11. Prove parts (d), (g), and (h) of Theorem 1 analytically.

12. Prove part (f) of Theorem 1 geometrically.

3.3 DOT PRODUCT; PROJECTIONS

In this section we introduce a kind of multiplication of vectors in 2-space and 3-space. Arithmetic properties of this multiplication are established and some applications are given.

Let **u** and **v** be two nonzero vectors in 2-space or 3-space, and assume these vectors have been positioned so their initial points coincide. By the ***angle between* u *and* v**, we shall mean the angle θ determined by **u** and **v** that satisfies $0 \leq \theta \leq \pi$ (Figure 3.18).

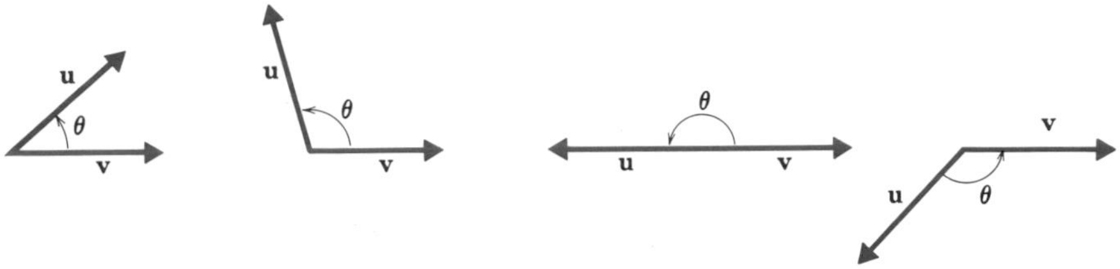

Figure 3.18

Definition. If **u** and **v** are vectors in 2-space or 3-space and θ is the angle between **u** and **v**, then the ***dot product*** or ***Euclidean inner product*** $\mathbf{u} \cdot \mathbf{v}$ is defined by

$$\mathbf{u} \cdot \mathbf{v} = \begin{cases} \|\mathbf{u}\| \, \|\mathbf{v}\| \cos \theta, & \text{if } \mathbf{u} \neq \mathbf{0} \text{ and } \mathbf{v} \neq \mathbf{0} \\ 0 \quad , & \text{if } \mathbf{u} = \mathbf{0} \text{ or } \mathbf{v} = \mathbf{0} \end{cases}$$

Example 5

As shown in Figure 3.19, the angle between the vectors $\mathbf{u} = (0, 0, 1)$ and $\mathbf{v} = (0, 2, 2)$ is 45°. Thus

$$\mathbf{u} \cdot \mathbf{v} = \|\mathbf{u}\| \, \|\mathbf{v}\| \cos \theta = (\sqrt{0^2 + 0^2 + 1^2})(\sqrt{0^2 + 2^2 + 2^2})\left(\frac{1}{\sqrt{2}}\right) = 2$$

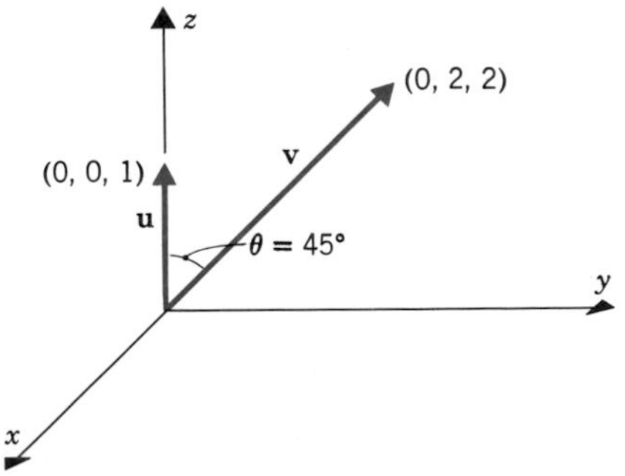

Figure 3.19

Let $\mathbf{u} = (u_1, u_2, u_3)$ and $\mathbf{v} = (v_1, v_2, v_3)$ be two nonzero vectors. If, as in Figure 3.20, θ is the angle between $\mathbf{u}$ and $\mathbf{v}$, then the law of cosines yields

$$\|\overrightarrow{PQ}\|^2 = \|\mathbf{u}\|^2 + \|\mathbf{v}\|^2 - 2\|\mathbf{u}\|\,\|\mathbf{v}\| \cos\theta \tag{3.2}$$

Since $\overrightarrow{PQ} = \mathbf{v} - \mathbf{u}$, we can rewrite (3.2) as

$$\|\mathbf{u}\|\,\|\mathbf{v}\| \cos\theta = \tfrac{1}{2}(\|\mathbf{u}\|^2 + \|\mathbf{v}\|^2 - \|\mathbf{v} - \mathbf{u}\|^2)$$

or

$$\mathbf{u} \cdot \mathbf{v} = \tfrac{1}{2}(\|\mathbf{u}\|^2 + \|\mathbf{v}\|^2 - \|\mathbf{v} - \mathbf{u}\|^2)$$

Substituting

$$\|\mathbf{u}\|^2 = u_1^2 + u_2^2 + u_3^2 \qquad \|\mathbf{v}\|^2 = v_1^2 + v_2^2 + v_3^2$$

and

$$\|\mathbf{v} - \mathbf{u}\|^2 = (v_1 - u_1)^2 + (v_2 - u_2)^2 + (v_3 - u_3)^2$$

we obtain after simplying

$$\mathbf{u} \cdot \mathbf{v} = u_1v_1 + u_2v_2 + u_3v_3 \tag{3.3}$$

If $\mathbf{u} = (u_1, u_2)$ and $\mathbf{v} = (v_1, v_2)$ are two vectors in 2-space, then the formula corresponding to (3.3) is

$$\mathbf{u} \cdot \mathbf{v} = u_1v_1 + u_2v_2$$

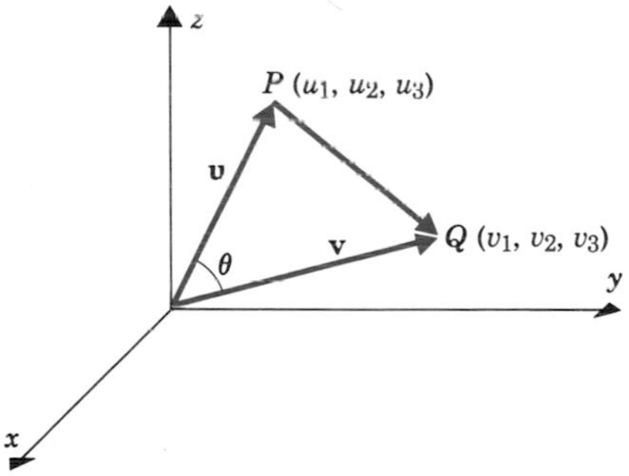

Figure 3.20

Example 6

Consider the vectors

$$\mathbf{u} = (2, -1, 1) \qquad \text{and} \qquad \mathbf{v} = (1, 1, 2)$$

Find $\mathbf{u} \cdot \mathbf{v}$ and determine the angle θ between $\mathbf{u}$ and $\mathbf{v}$.

Solution.

$$\mathbf{u} \cdot \mathbf{v} = u_1v_1 + u_2v_2 + u_3v_3 = (2)(1) + (-1)(1) + (1)(2) = 3$$

Also $\|\mathbf{u}\| = \|\mathbf{v}\| = \sqrt{6}$, so that

$$\cos\theta = \frac{\mathbf{u}\cdot\mathbf{v}}{\|\mathbf{u}\|\,\|\mathbf{v}\|} = \frac{3}{\sqrt{6}\sqrt{6}} = \frac{1}{2}$$

Thus $\theta = 60°$

Example 7

Find the angle between a diagonal of a cube and one of its edges.

Solution. Let k be the length of an edge and introduce a coordinate system as shown in Figure 3.21.

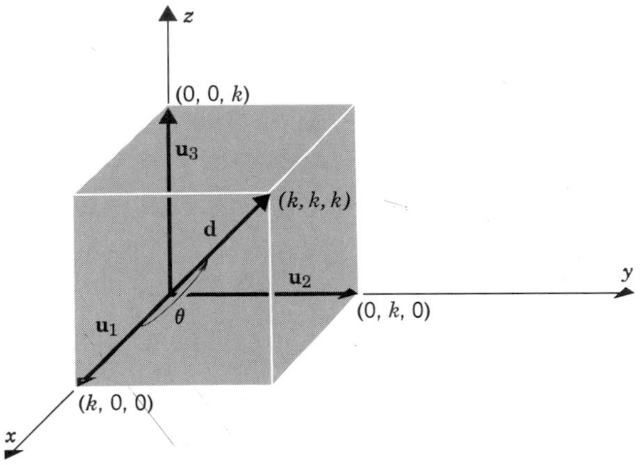

Figure 3.21

If we let $\mathbf{u}_1 = (k, 0, 0)$, $\mathbf{u}_2 = (0, k, 0)$, and $\mathbf{u}_3 = (0, 0, k)$, then the vector

$$\mathbf{d} = (k, k, k) = \mathbf{u}_1 + \mathbf{u}_2 + \mathbf{u}_3$$

is a diagonal of the cube. The angle θ between $\mathbf{d}$ and the edge $\mathbf{u}_1$ satisfies

$$\cos\theta = \frac{\mathbf{u}_1\cdot\mathbf{d}}{\|\mathbf{u}_1\|\,\|\mathbf{d}\|} = \frac{k^2}{(k)(\sqrt{3k^2})} = \frac{1}{\sqrt{3}}$$

Thus

$$\theta = \cos^{-1}\left(\frac{1}{\sqrt{3}}\right) \approx 54°44'.$$

The next theorem shows how the dot product can be used to obtain information about the angle between two vectors; it also establishes an important relationship between the norm and the dot product.

Theorem 2. *Let* **u** *and* **v** *be vectors in 2- or 3-space.*

(*a*) $\mathbf{v} \cdot \mathbf{v} = \|\mathbf{v}\|^2$; *that is,* $\|\mathbf{v}\| = (\mathbf{v} \cdot \mathbf{v})^{1/2}$
(*b*) *If* **u** *and* **v** *are nonzero vectors and* θ *is the angle between them, then*

$$\theta \text{ is acute} \quad \text{if and only if} \quad \mathbf{u} \cdot \mathbf{v} > 0$$

$$\theta \text{ is obtuse} \quad \text{if and only if} \quad \mathbf{u} \cdot \mathbf{v} < 0$$

$$\theta = \frac{\pi}{2} \quad \text{if and only if} \quad \mathbf{u} \cdot \mathbf{v} = 0$$

Proof.

(*a*) Since the angle θ between **v** and **v** is 0 we have

$$\mathbf{v} \cdot \mathbf{v} = \|\mathbf{v}\| \, \|\mathbf{v}\| \cos \theta = \|\mathbf{v}\|^2 \cos 0 = \|\mathbf{v}\|^2$$

(*b*) Since $\|\mathbf{u}\| > 0$, $\|\mathbf{v}\| > 0$, and $\mathbf{u} \cdot \mathbf{v} = \|\mathbf{u}\| \, \|\mathbf{v}\| \cos \theta$, $\mathbf{u} \cdot \mathbf{v}$ has the same sign as $\cos \theta$. Since θ satisfies $0 \leq \theta \leq \pi$, the angle θ is acute if and only if $\cos \theta > 0$, θ is obtuse if and only if $\cos \theta < 0$, and $\theta = \pi/2$ if and only if $\cos \theta = 0$. ▮

Example 8

If $\mathbf{u} = (1, -2, 3)$, $\mathbf{v} = (-3, 4, 2)$, and $\mathbf{w} = (3, 6, 3)$, then

$$\begin{aligned} \mathbf{u} \cdot \mathbf{v} &= (1)(-3) + (-2)(4) + (3)(2) = -5 \\ \mathbf{v} \cdot \mathbf{w} &= (-3)(3) + (4)(6) + (2)(3) = 21 \\ \mathbf{u} \cdot \mathbf{w} &= (1)(3) + (-2)(6) + (3)(3) = 0 \end{aligned}$$

Therefore **u** and **v** make an obtuse angle, **v** and **w** make an acute angle, and **u** and **w** are perpendicular.

The main arithmetic properties of the dot product are listed in the next theorem.

Theorem 3. If **u**, **v** *and* **w** *are vectors in 2- or 3-space and k is a scalar, then*

(*a*) $\mathbf{u} \cdot \mathbf{v} = \mathbf{v} \cdot \mathbf{u}$
(*b*) $\mathbf{u} \cdot (\mathbf{v} + \mathbf{w}) = \mathbf{u} \cdot \mathbf{v} + \mathbf{u} \cdot \mathbf{w}$
(*c*) $k(\mathbf{u} \cdot \mathbf{v}) = (k\mathbf{u}) \cdot \mathbf{v} = \mathbf{u} \cdot (k\mathbf{v})$
(*d*) $\mathbf{v} \cdot \mathbf{v} > 0$ *if* $\mathbf{v} \neq \mathbf{0}$ *and* $\mathbf{v} \cdot \mathbf{v} = 0$ *if* $\mathbf{v} = \mathbf{0}$

Proof. We shall prove (*c*) for vectors in 3-space and leave the remaining proofs as exercises. Let $\mathbf{u} = (u_1, u_2, u_3)$ and $\mathbf{v} = (v_1, v_2, v_3)$; then

$$\begin{aligned} k(\mathbf{u} \cdot \mathbf{v}) &= k(u_1 v_1 + u_2 v_2 + u_3 v_3) \\ &= (ku_1)v_1 + (ku_2)v_2 + (ku_3)v_3 \\ &= (k\mathbf{u}) \cdot \mathbf{v} \end{aligned}$$

Similarly $$k(\mathbf{u} \cdot \mathbf{v}) = \mathbf{u} \cdot (k\mathbf{v})$$ ▮

Based on part (*b*) of Theorem 2, we define two vectors $\mathbf{u}$ and $\mathbf{v}$ to be ***orthogonal*** (written $\mathbf{u} \perp \mathbf{v}$) if $\mathbf{u} \cdot \mathbf{v} = 0$. If we agree that the zero vector makes an angle of $\pi/2$ with every vector, then two vectors are orthogonal if and only if they are geometrically perpendicular.

The dot product is useful in problems where it is of interest to "decompose" a vector into a sum of two perpendicular vectors. If $\mathbf{u}$ and $\mathbf{v}$ are nonzero vectors in 2- or 3-space, then it is always possible to write $\mathbf{u}$ as

$$\mathbf{u} = \mathbf{w}_1 + \mathbf{w}_2$$

where $\mathbf{w}_1$ is a scalar multiple of $\mathbf{v}$, and $\mathbf{w}_2$ is perpendicular to $\mathbf{v}$ (Figure 3.22). The vector $\mathbf{w}_1$ is called the ***orthogonal projection of* u *on* v** and the vector $\mathbf{w}_2$ is called the ***component of* u *orthogonal to* v**.

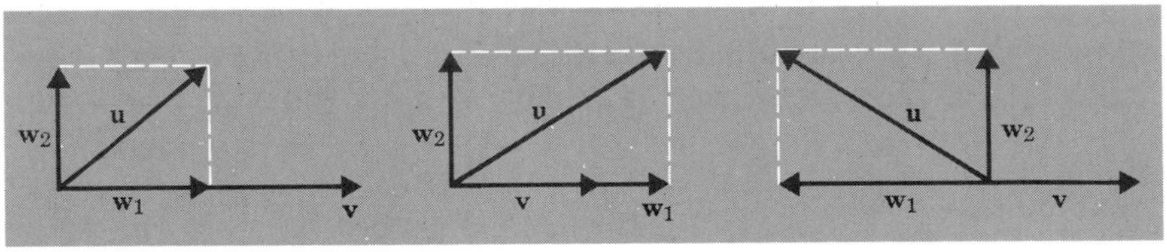

Figure 3.22

The vectors $\mathbf{w}_1$ and $\mathbf{w}_2$ can be obtained as follows. Since $\mathbf{w}_1$ is a scalar multiple of $\mathbf{v}$, we can write it in the form $\mathbf{w}_1 = k\mathbf{v}$. Thus

$$\mathbf{u} = \mathbf{w}_1 + \mathbf{w}_2 = k\mathbf{v} + \mathbf{w}_2 \tag{3.4}$$

Taking the dot product of both sides of (3.4) with $\mathbf{v}$, and using Theorems 2 and 3 we obtain

$$\mathbf{u} \cdot \mathbf{v} = (k\mathbf{v} + \mathbf{w}_2) \cdot \mathbf{v} = k\|\mathbf{v}\|^2 + \mathbf{w}_2 \cdot \mathbf{v}$$

Since $\mathbf{w}_2$ is perpendicular to $\mathbf{v}$, we have $\mathbf{w}_2 \cdot \mathbf{v} = 0$ so that this equation yields

$$k = \frac{\mathbf{u} \cdot \mathbf{v}}{\|\mathbf{v}\|^2}$$

and since $\mathbf{w}_1 = k\mathbf{v}$ we obtain

$$\mathbf{w}_1 = \frac{\mathbf{u} \cdot \mathbf{v}}{\|\mathbf{v}\|^2}\mathbf{v} \qquad \text{orthogonal projection of } \mathbf{u} \text{ on } \mathbf{v}$$

Solving $\mathbf{u} = \mathbf{w}_1 + \mathbf{w}_2$ for $\mathbf{w}_2$ gives

$$\mathbf{w}_2 = \mathbf{u} - \frac{\mathbf{u} \cdot \mathbf{v}}{\|\mathbf{v}\|^2}\mathbf{v} \qquad \text{component of } \mathbf{u} \text{ orthogonal to } \mathbf{v}$$

Example 9

Consider the vectors

$$\mathbf{u} = (2, -1, 3) \qquad \text{and} \qquad \mathbf{v} = (4, -1, 2)$$

Since

$$\mathbf{u} \cdot \mathbf{v} = (2)(4) + (-1)(-1) + (3)(2) = 15$$

and

$$\|\mathbf{v}\|^2 = 4^2 + (-1)^2 + 2^2 = 21$$

the orthogonal projection of **u** on **v** is

$$\mathbf{w}_1 = \frac{\mathbf{u} \cdot \mathbf{v}}{\|\mathbf{v}\|^2}\mathbf{v} = \frac{15}{21}(4, -1, 2) = \left(\frac{20}{7}, \frac{-5}{7}, \frac{10}{7}\right)$$

The component of **u** orthogonal to **v** is

$$\mathbf{w}_2 = \mathbf{u} - \mathbf{w}_1 = (2, -1, 3) - \left(\frac{20}{7}, \frac{-5}{7}, \frac{10}{7}\right) = \left(\frac{-6}{7}, \frac{-2}{7}, \frac{11}{7}\right)$$

As a check, the reader may wish to verify that $\mathbf{w}_2$ is perpendicular to **v** by showing that $\mathbf{w}_2 \cdot \mathbf{v} = 0$.

EXERCISE SET 3.3

1. Find $\mathbf{u} \cdot \mathbf{v}$ for:
   (a) $\mathbf{u} = (1, 2)$, $\mathbf{v} = (6, -8)$ (b) $\mathbf{u} = (-7, -3)$, $\mathbf{v} = (0, 1)$
   (c) $\mathbf{u} = (1, -3, 7)$, $\mathbf{v} = (8, -2, -2)$ (d) $\mathbf{u} = (-3, 1, 2)$, $\mathbf{v} = (4, 2, -5)$

2. In each part of Exercise 1, find the cosine of the angle θ between **u** and **v**.

3. Determine if **u** and **v** make an actue angle, make an obtuse angle, or are orthogonal.
   (a) $\mathbf{u} = (7, 3, 5)$, $\mathbf{v} = (-8, 4, 2)$ (b) $\mathbf{u} = (6, 1, 3)$, $\mathbf{v} = (4, 0, -6)$
   (c) $\mathbf{u} = (1, 1, 1)$, $\mathbf{v} = (-1, 0, 0)$ (d) $\mathbf{u} = (4, 1, 6)$, $\mathbf{v} = (-3, 0, 2)$

4. Find the orthogonal projection of **u** on **v** if:
   (a) $\mathbf{u} = (2, 1)$, $\mathbf{v} = (-3, 2)$ (b) $\mathbf{u} = (2, 6)$, $\mathbf{v} = (-9, 3)$
   (c) $\mathbf{v} = (-7, 1, 3)$, $\mathbf{v} = (5, 0, 1)$ (d) $\mathbf{u} = (0, 0, 1)$, $\mathbf{v} = (8, 3, 4)$

5. In each part of Exercise 4, find the component of **u** orthogonal to **v**.

6. Verify Theorem 3 for $\mathbf{u} = (6, -1, 2)$, $\mathbf{v} = (2, 7, 4)$, and $k = -5$.

7. Find two vectors of norm 1 that are orthogonal to $(3, -2)$.

8. Let $\mathbf{u} = (1, 2)$, $\mathbf{v} = (4, -2)$, and $\mathbf{w} = (6, 0)$. Find:
   (a) $\mathbf{u} \cdot (7\mathbf{v} + \mathbf{w})$ (b) $\|(\mathbf{u} \cdot \mathbf{w})\,\mathbf{w}\|$ (c) $\|\mathbf{u}\|\,(\mathbf{v} \cdot \mathbf{w})$ (d) $(\|\mathbf{u}\|\,\mathbf{v}) \cdot \mathbf{w}$

9. Explain why each of the following expressions makes no sense.
   (a) $\mathbf{u} \cdot (\mathbf{v} \cdot \mathbf{w})$ (b) $(\mathbf{u} \cdot \mathbf{v}) + \mathbf{w}$ (c) $\|\mathbf{u} \cdot \mathbf{v}\|$ (d) $k \cdot (\mathbf{u} + \mathbf{v})$

10. Use vectors to find the cosines of the interior angles of the triangle with vertices $(-1, 0)$, $(-2, 1)$, and $(1, 4)$.

11. Establish the identity

$$\|\mathbf{u} + \mathbf{v}\|^2 + \|\mathbf{u} - \mathbf{v}\|^2 = 2\|\mathbf{u}\|^2 + 2\|\mathbf{v}\|^2$$

12. Establish the identity

$$\mathbf{u} \cdot \mathbf{v} = \tfrac{1}{4}\|\mathbf{u} + \mathbf{v}\|^2 - \tfrac{1}{4}\|\mathbf{u} - \mathbf{v}\|^2$$

13. Find the angle between a diagonal of a cube and one of its faces.

14. The ***direction cosines*** of a vector $\mathbf{v}$ in 3-space are the numbers $\cos \alpha$, $\cos \beta$, and $\cos \gamma$ where α, β, and γ are the angles between $\mathbf{v}$ and the positive x, y, and z aexs. Show that if $\mathbf{v} = (a, b, c)$, then $\cos \alpha = a/\sqrt{a^2 + b^2 + c^2}$. Find $\cos \beta$ and $\cos \gamma$.

15. Show that if $\mathbf{v}$ is orthogonal to $\mathbf{w}_1$ and $\mathbf{w}_2$ then $\mathbf{v}$ is orthogonal to $k_1\mathbf{w}_1 + k_2\mathbf{w}_2$ for all scalars k_1 and k_2.

16. Let $\mathbf{u}$ and $\mathbf{v}$ be nonzero vectors in 2 or 3-space. If $k = \|\mathbf{u}\|$ and $l = \|\mathbf{v}\|$, show that the vector

$$\mathbf{w} = \frac{1}{k + l}(k\mathbf{v} + l\mathbf{u})$$

bisects the angle between $\mathbf{u}$ and $\mathbf{v}$.

3.4 CROSS PRODUCT

In many applications of vectors to problems in geometry, physics, and engineering, it is of interest to construct a vector in 3-space that is perpendicular to two given vectors. In this section we introduce a type of vector multiplication that facilitates this construction.

Definition. If $\mathbf{u} = (u_1, u_2, u_3)$ and $\mathbf{v} = (v_1, v_2, v_3)$ are vectors in 3-space, then the ***cross product*** $\mathbf{u} \times \mathbf{v}$ is the vector defined by

$$\mathbf{u} \times \mathbf{v} = (u_2v_3 - u_3v_2, u_3v_1 - u_1v_3, u_1v_2 - u_2v_1)$$

or in determinant notation

$$\mathbf{u} \times \mathbf{v} = \left(\begin{vmatrix} u_2 & u_3 \\ v_2 & v_3 \end{vmatrix}, -\begin{vmatrix} u_1 & u_3 \\ v_1 & v_3 \end{vmatrix}, \begin{vmatrix} u_1 & u_2 \\ v_1 & v_2 \end{vmatrix} \right) \tag{3.5}$$

REMARK. There is a pattern in Formula 3.5 that is useful to keep in mind. If we form the 2×3 matrix

$$\begin{bmatrix} u_1 & u_2 & u_3 \\ v_1 & v_2 & v_3 \end{bmatrix}$$

where the entries in the first row are the components of the first factor **u** and those in the second row are the components of the second factor **v**, then the determinant in the first component of $\mathbf{u} \times \mathbf{v}$ is obtained by deleting the first column of the matrix, the determinant in the second component by deleting the second column of the matrix, and the determinant in the third component by deleting the third column of the matrix.

Example 10

Find $\mathbf{u} \times \mathbf{v}$, where $\mathbf{u} = (1, 2, -2)$ and $\mathbf{v} = (3, 0, 1)$.

Solution.

$$\begin{bmatrix} 1 & 2 & -2 \\ 3 & 0 & 1 \end{bmatrix}$$

$$\begin{aligned} \mathbf{u} \times \mathbf{v} &= \left(\begin{vmatrix} 2 & -2 \\ 0 & 1 \end{vmatrix}, -\begin{vmatrix} 1 & -2 \\ 3 & 1 \end{vmatrix}, \begin{vmatrix} 1 & 2 \\ 3 & 0 \end{vmatrix} \right) \\ &= (2, -7, -6) \end{aligned}$$

Whereas the dot product of two vectors is a scalar, the cross product is another vector. The following theorem gives an important relationship between dot product and cross product and also shows that $\mathbf{u} \times \mathbf{v}$ is orthogonal to both **u** and **v**.

Theorem 4. *If* **u** *and* **v** *are vectors in 3-space, then:*

(*a*) $\mathbf{u} \cdot (\mathbf{u} \times \mathbf{v}) = 0$ ($\mathbf{u} \times \mathbf{v}$ *is orthogonal to* **u**)
(*b*) $\mathbf{v} \cdot (\mathbf{u} \times \mathbf{v}) = 0$ ($\mathbf{u} \times \mathbf{v}$ *is orthogonal to* **v**)
(*c*) $\|\mathbf{u} \times \mathbf{v}\|^2 = \|\mathbf{u}\|^2 \|\mathbf{v}\|^2 - (\mathbf{u} \cdot \mathbf{v})^2$ (*Lagrange's identity*)

Proof. Let $\mathbf{u} = (u_1, u_2, u_3)$ and $\mathbf{v} = (v_1, v_2, v_3)$.

(*a*)
$$\begin{aligned} \mathbf{u} \cdot (\mathbf{u} \times \mathbf{v}) &= (u_1, u_2, u_3) \cdot (u_2v_3 - u_3v_2, u_3v_1 - u_1v_3, u_1v_2 - u_2v_1) \\ &= u_1(u_2v_3 - u_3v_2) + u_2(u_3v_1 - u_1v_3) + u_3(u_1v_2 - u_2v_1) \\ &= 0 \end{aligned}$$

(*b*) Similar to (*a*).
(*c*) Since

$$\|\mathbf{u} \times \mathbf{v}\|^2 = (u_2v_3 - u_3v_2)^2 + (u_3v_1 - u_1v_3)^2 + (u_1v_2 - u_2v_1)^2 \tag{3.6}$$

and

$$\begin{aligned} &\|\mathbf{u}\|^2 \|\mathbf{v}\|^2 - (\mathbf{u} \cdot \mathbf{v})^2 \\ &\quad = (u_1^2 + u_2^2 + u_3^2)(v_1^2 + v_2^2 + v_3^2) - (u_1v_1 + u_2v_2 + u_3v_3)^2 \end{aligned} \tag{3.7}$$

Lagrange's identity can be established by "multiplying out" the right sides of (3.6) and (3.7) and verifying their equality. ∎

Example 11

Consider the vectors

$$\mathbf{u} = (1, 2, -2) \qquad \text{and} \qquad \mathbf{v} = (3, 0, 1)$$

In Example 10 we showed that

$$\mathbf{u} \times \mathbf{v} = (2, -7, -6)$$

Since

$$\mathbf{u} \cdot (\mathbf{u} \times \mathbf{v}) = (1)(2) + (2)(-7) + (-2)(-6) = 0$$

and

$$\mathbf{v} \cdot (\mathbf{u} \times \mathbf{v}) = (3)(2) + (0)(-7) + (1)(-6) = 0$$

$\mathbf{u} \times \mathbf{v}$ is orthogonal to both $\mathbf{u}$ and $\mathbf{v}$ as guaranteed by Theorem 4.

The main arithmetic properties of the cross product are listed in the next theorem.

Theorem 5. *If* $\mathbf{u}$, $\mathbf{v}$ *and* $\mathbf{w}$ *are any vectors in 3-space and* k *is any scalar, then*:

(*a*) $\mathbf{u} \times \mathbf{v} = -(\mathbf{v} \times \mathbf{u})$
(*b*) $\mathbf{u} \times (\mathbf{v} + \mathbf{w}) = (\mathbf{u} \times \mathbf{v}) + (\mathbf{u} \times \mathbf{w})$
(*c*) $(\mathbf{u} + \mathbf{v}) \times \mathbf{w} = (\mathbf{u} \times \mathbf{w}) + (\mathbf{v} \times \mathbf{w})$
(*d*) $k(\mathbf{u} \times \mathbf{v}) = (k\mathbf{u}) \times \mathbf{v} = \mathbf{u} \times (k\mathbf{v})$
(*e*) $\mathbf{u} \times \mathbf{0} = \mathbf{0} \times \mathbf{u} = \mathbf{0}$
(*f*) $\mathbf{u} \times \mathbf{u} = \mathbf{0}$

The proofs follow immediately from Formula 3.5 and properties of determinants; for example, (*a*) can be proved as follows:

Proof. (*a*) Interchanging $\mathbf{u}$ and $\mathbf{v}$ in (3.5) interchanges the rows of the three determinants on the right side of (3.5) and thereby changes the sign of each component in the cross product. Thus $\mathbf{u} \times \mathbf{v} = -(\mathbf{v} \times \mathbf{u})$. ▮

The proofs of the remaining parts are left as exercises.

Example 12

Consider the vectors

$$\mathbf{i} = (1, 0, 0) \qquad \mathbf{j} = (0, 1, 0) \qquad \mathbf{k} = (0, 0, 1)$$

These vectors each have length 1 and lie along the coordinate axes (Figu. ? 23). They are called the ***standard unit vectors*** in 3-space. Every vector $\mathbf{v} = (v_1, v_2, v_3)$ in 3-space is expressible in terms of $\mathbf{i}$, $\mathbf{j}$, and $\mathbf{k}$ since we can write

$$\mathbf{v} = (v_1, v_2, v_3) = v_1(1, 0, 0) + v_2(0, 1, 0) + v_3(0, 0, 1) = v_1\mathbf{i} + v_2\mathbf{j} + v_3\mathbf{k}$$

For example,

$$(2, -3, 4) = 2\mathbf{i} - 3\mathbf{j} + 4\mathbf{k}$$

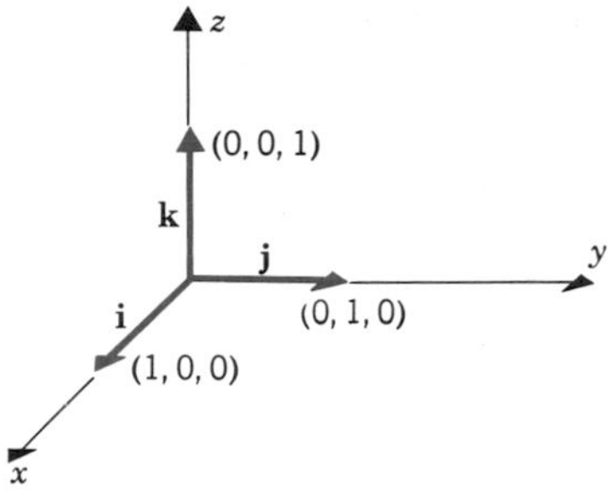

Figure 3.23

From (3.5) we obtain

$$\mathbf{i} \times \mathbf{j} = \left(\begin{vmatrix} 0 & 0 \\ 1 & 0 \end{vmatrix}, -\begin{vmatrix} 1 & 0 \\ 0 & 0 \end{vmatrix}, \begin{vmatrix} 1 & 0 \\ 0 & 1 \end{vmatrix} \right) = (0, 0, 1) = \mathbf{k}$$

The reader should have no trouble obtaining the following results:

$$\mathbf{i} \times \mathbf{i} = \mathbf{j} \times \mathbf{j} = \mathbf{k} \times \mathbf{k} = \mathbf{0}$$
$$\mathbf{i} \times \mathbf{j} = \mathbf{k}, \quad \mathbf{j} \times \mathbf{k} = \mathbf{i}, \quad \mathbf{k} \times \mathbf{i} = \mathbf{j}$$
$$\mathbf{j} \times \mathbf{i} = -\mathbf{k}, \quad \mathbf{k} \times \mathbf{j} = -\mathbf{i}, \quad \mathbf{i} \times \mathbf{k} = -\mathbf{j}$$

The following diagram is helpful for remembering these results.

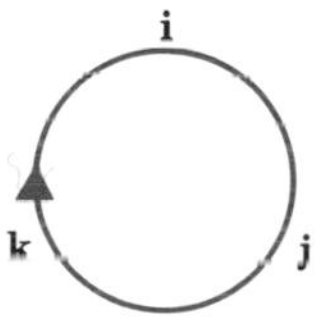

Referring to this diagram, the cross product of two consecutive vectors going clockwise is the next vector around, and the product of two consecutive vectors going counterclockwise is the negative of the next vector around.

It is also worth noting that a cross product can be represented symbolically in the form of a 3 × 3 determinant,

$$\mathbf{u} \times \mathbf{v} = \begin{vmatrix} \mathbf{i} & \mathbf{j} & \mathbf{k} \\ u_1 & u_2 & u_3 \\ v_1 & v_2 & v_3 \end{vmatrix} = \begin{vmatrix} u_2 & u_3 \\ v_2 & v_3 \end{vmatrix} \mathbf{i} - \begin{vmatrix} u_1 & u_3 \\ v_1 & v_3 \end{vmatrix} \mathbf{j} + \begin{vmatrix} u_1 & u_2 \\ v_1 & v_2 \end{vmatrix} \mathbf{k}$$

For example, if $\mathbf{u} = (1, 2, -2)$ and $\mathbf{v} = (3, 0, 1)$ then

$$\mathbf{u} \times \mathbf{v} = \begin{vmatrix} \mathbf{i} & \mathbf{j} & \mathbf{k} \\ 1 & 2 & -2 \\ 3 & 0 & 1 \end{vmatrix} = 2\mathbf{i} - 7\mathbf{j} - 6\mathbf{k}$$

which agrees with result obtained in Example 10.

Warning. It is *not* true in general that $\mathbf{u} \times (\mathbf{v} \times \mathbf{w}) = (\mathbf{u} \times \mathbf{v}) \times \mathbf{w}$. For example,

$$\mathbf{i} \times (\mathbf{j} \times \mathbf{j}) = \mathbf{i} \times \mathbf{0} = \mathbf{0}$$

and

$$(\mathbf{i} \times \mathbf{j}) \times \mathbf{j} = \mathbf{k} \times \mathbf{j} = -(\mathbf{j} \times \mathbf{k}) = -\mathbf{i}$$

so that

$$\mathbf{i} \times (\mathbf{j} \times \mathbf{j}) \neq (\mathbf{i} \times \mathbf{j}) \times \mathbf{j}$$

We know from Theorem 4 that $\mathbf{u} \times \mathbf{v}$ is orthogonal to both $\mathbf{u}$ and $\mathbf{v}$. If $\mathbf{u}$ and $\mathbf{v}$ are nonzero vectors, it can be shown that the direction of $\mathbf{u} \times \mathbf{v}$ can be determined using the following "right-hand rule"* (Figure 3.24). Let θ be the angle between $\mathbf{u}$ and $\mathbf{v}$, and suppose $\mathbf{u}$ is rotated through the angle θ until it coincides with $\mathbf{v}$. If the fingers of the right hand are cupped so they point in the direction of rotation, then the thumb indicates (roughly) the direction of $\mathbf{u} \times \mathbf{v}$.

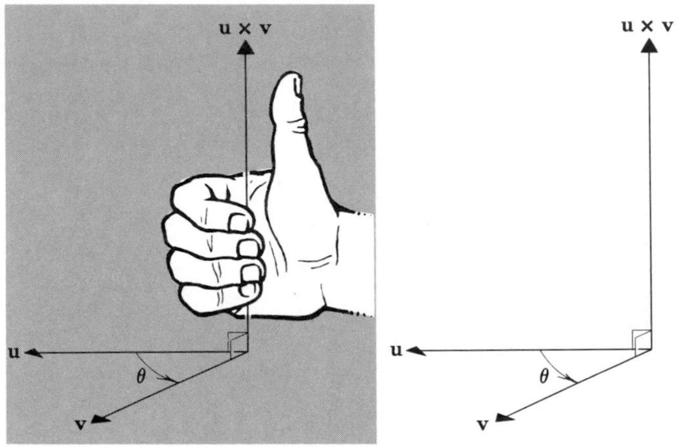

Figure 3.24

The reader may find it instructive to practice this rule with the products

$$\mathbf{i} \times \mathbf{j} = \mathbf{k} \qquad \mathbf{j} \times \mathbf{k} = \mathbf{i} \qquad \mathbf{k} \times \mathbf{i} = \mathbf{j}$$

discussed in Example 12.

If $\mathbf{u}$ and $\mathbf{v}$ are nonzero vectors in 3-space, then the norm of $\mathbf{u} \times \mathbf{v}$ has a useful geometric interpretation. Lagrange's identity, given in Theorem 4, states that

$$\|\mathbf{u} \times \mathbf{v}\|^2 = \|\mathbf{u}\|^2 \|\mathbf{v}\|^2 - (\mathbf{u} \cdot \mathbf{v})^2 \tag{3.8}$$

If θ denotes the angle between $\mathbf{u}$ and $\mathbf{v}$, then $\mathbf{u} \cdot \mathbf{v} = \|\mathbf{u}\| \|\mathbf{v}\| \cos \theta$, so that (3.8) can

* Recall that we agreed to consider only right-handed coordinate systems in this text. Had we used left-handed systems instead, a "left-hand rule" would apply here.

be rewritten as

$$\begin{aligned}\|\mathbf{u} \times \mathbf{v}\|^2 &= \|\mathbf{u}\|^2 \|\mathbf{v}\|^2 - \|\mathbf{u}\|^2 \|\mathbf{v}\|^2 \cos^2 \theta \\ &= \|\mathbf{u}\|^2 \|\mathbf{v}\|^2 (1 - \cos^2 \theta) \\ &= \|\mathbf{u}\|^2 \|\mathbf{v}\|^2 \sin^2 \theta\end{aligned}$$

Thus

$$\|\mathbf{u} \times \mathbf{v}\| = \|\mathbf{u}\| \|\mathbf{v}\| \sin \theta \tag{3.9}$$

But $\|\mathbf{v}\| \sin \theta$ is the altitude of the parallelogram determined by **u** and **v** (Figure 3.25). Thus from (3.9), the area A of this parallelogram is given by

$$A = \text{(base)(altitude)} = \|\mathbf{u}\| \|\mathbf{v}\| \sin \theta = \|\mathbf{u} \times \mathbf{v}\|$$

In other words, *the norm of* $\mathbf{u} \times \mathbf{v}$ *is equal to the area of the parallelogram determined by* **u** *and* **v**.

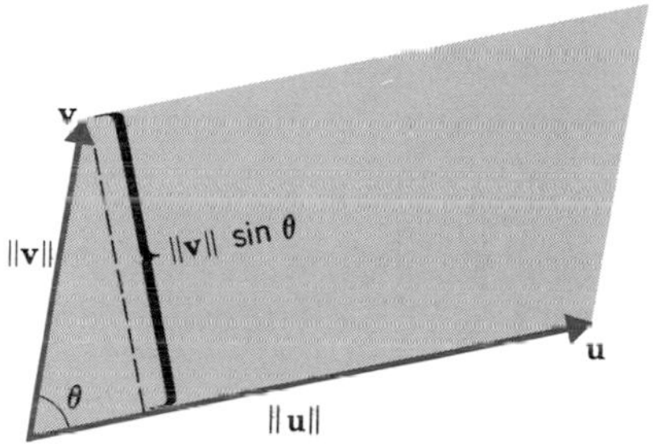

Figure 3.25

Example 13

Find the area of the triangle determined by the points $P_1(2, 2, 0)$, $P_2(-1, 0, 2)$, and $P_3(0, 4, 3)$.

Solution. The area A of the triangle is 1/2 the area of the parallelogram determined by the vectors $\overrightarrow{P_1P_2}$ and $\overrightarrow{P_1P_3}$ (Figure 3.26).

Using the method discussed in Example 2 of Section 3.1, $\overrightarrow{P_1P_2} = (-3, -2, 2)$ and $\overrightarrow{P_1P_3} = (-2, 2, 3)$. It follows that

$$\overrightarrow{P_1P_2} \times \overrightarrow{P_1P_3} = (-10, 5, -10)$$

and consequently

$$A = \tfrac{1}{2}\|\overrightarrow{P_1P_2} \times \overrightarrow{P_1P_3}\| = \tfrac{1}{2}(15) = \tfrac{15}{2}$$

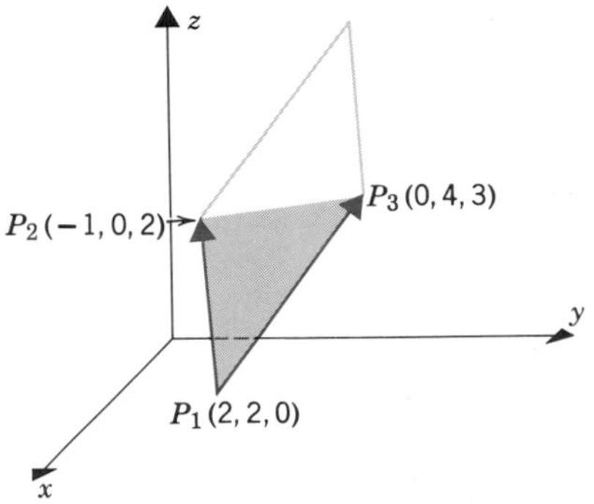

Figure 3.26

Initially, we defined a vector to be a directed line segment or arrow in 2-space or 3-space; coordinate systems and components were introduced later in order to simplify computations with vectors. Thus a vector has a "mathematical existence" regardless of whether a coordinate system has been introduced. Further, the components of a vector are not determined by the vector alone; they depend as well on the coordinate system chosen. For example, in Figure 3.27 we have indicated a fixed-plane vector **v** and two different coordinate systems. In the xy-coordinate system the components of **v** are (1, 1), and in the $x'y'$-system they are $(\sqrt{2}, 0)$.

This raises an important question about our definition of cross product. Since we defined the cross product $\mathbf{u} \times \mathbf{v}$ in terms of the components of **u** and **v**, and since these components depend on the coordinate system chosen, it seems possible that two *fixed* vectors **u** and **v** might have different cross products in different coordinate systems. Fortunately, this is not the case. To see this, we need only recall that:

(i) $\mathbf{u} \times \mathbf{v}$ is perpendicular to both **u** and **v**.
(ii) The orientation of $\mathbf{u} \times \mathbf{v}$ is determined by the right-hand rule.
(iii) $\|\mathbf{u} \times \mathbf{v}\| = \|\mathbf{u}\| \|\mathbf{v}\| \sin \theta$.

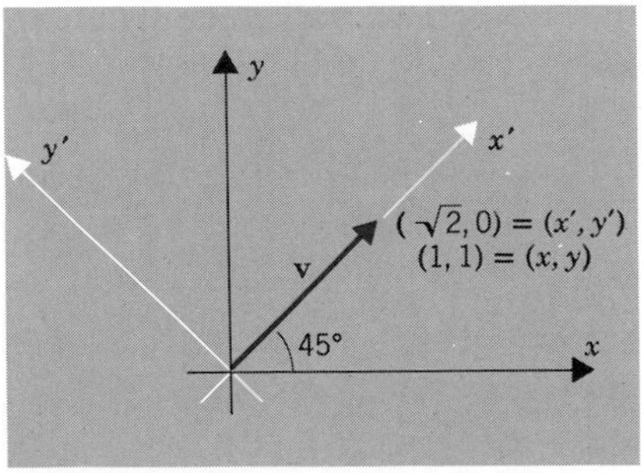

Figure 3.27

These three properties completely determine the vector $\mathbf{u} \times \mathbf{v}$; properties (i) and (ii) determine the direction, and property (iii) determines the length. Since these properties depend only on the lengths and relative positions of **u** and **v** and not on the particular right-hand coordinate system being used, the vector $\mathbf{u} \times \mathbf{v}$ will remain unchanged if a different right-hand coordinate system is introduced. This fact is described by stating that the definition of $\mathbf{u} \times \mathbf{v}$ is ***coordinate free***. This result is of importance to physicists and engineers who often work with many coordinate systems in the same problem.

Example 14

Consider two perpendicular vectors **u** and **v**, each of length 1 (as shown in Figure 3.28*a*.) If we introduce an *xyz*-coordinate system as shown in Figure 3.28*b* then

$$\mathbf{u} = (1, 0, 0) = \mathbf{i} \quad \text{and} \quad \mathbf{v} = (0, 1, 0) = \mathbf{j}$$

so that

$$\mathbf{u} \times \mathbf{v} = \mathbf{i} \times \mathbf{j} = \mathbf{k} = (0, 0, 1)$$

On the other hand, if we introduce an $x'y'z'$-coordinate system as shown in Figure 3.28*c* then

$$\mathbf{u} = (0, 0, 1) = \mathbf{k} \quad \text{and} \quad \mathbf{v} = (1, 0, 0) = \mathbf{i}$$

so that

$$\mathbf{u} \times \mathbf{v} = \mathbf{k} \times \mathbf{i} = \mathbf{j} = (0, 1, 0)$$

But it is clear from Figures 3.28*b* and *c* that the vector (0, 0, 1) in the *xyz*-system is the same as the vector (0, 1, 0) in the $x'y'z'$-system. Thus we obtain the same vector $\mathbf{u} \times \mathbf{v}$ whether we compute with coordinates from the *xyz*-system or with the coordinates from the $x'y'z'$-system.

Figure 3.28

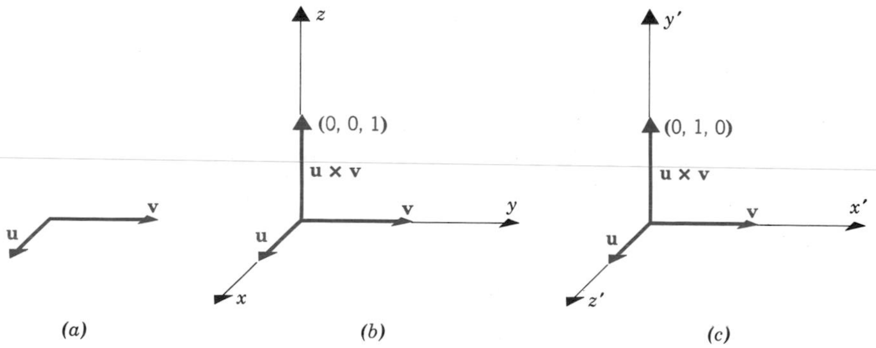

EXERCISE SET 3.4

1. Let $\mathbf{u} = (2, -1, 3)$, $\mathbf{v} = (0, 1, 7)$, and $\mathbf{w} = (1, 4, 5)$. Compute:
(a) $\mathbf{v} \times \mathbf{w}$ (b) $\mathbf{u} \times (\mathbf{v} \times \mathbf{w})$ (c) $(\mathbf{u} \times \mathbf{v}) \times \mathbf{w}$
(d) $(\mathbf{u} \times \mathbf{v}) \times (\mathbf{v} \times \mathbf{w})$ (e) $\mathbf{u} \times (\mathbf{v} - 2\mathbf{w})$ (f) $(\mathbf{u} \times \mathbf{v}) - 2\mathbf{w}$

2. In each part find a vector orthogonal to both $\mathbf{u}$ and $\mathbf{v}$.
(a) $\mathbf{u} = (-7, 3, 1)$ $\mathbf{v} = (2, 0, 4)$
(b) $\mathbf{u} = (-1, -1, -1)$ $\mathbf{v} = (2, 0, 2)$

3. In each part find the area of the triangle having vertices P, Q, and R.
(a) $P(1, 5, -2)$ $Q(0, 0, 0)$ $R(3, 5, 1)$
(b) $P(2, 0, -3)$ $Q(1, 4, 5)$ $R(7, 2, 9)$

4. Verify Theorem 4 for the vectors $\mathbf{u} = (1, -5, 6)$ and $\mathbf{v} = (2, 1, 2)$.

5. Verify Theorem 5 for $\mathbf{u} = (2, 0, -1)$, $\mathbf{v} = (6, 7, 4)$, $\mathbf{w} = (1, 1, 1)$ and $k = -3$.

6. What is wrong with the expression $\mathbf{u} \times \mathbf{v} \times \mathbf{w}$?

7. Let $\mathbf{u} = (-1, 3, 2)$ and $\mathbf{w} = (1, 1, -1)$. Find all $\mathbf{x}$ vectors that satisfy $\mathbf{u} \times \mathbf{x} = \mathbf{w}$.

8. Let $\mathbf{u} = (u_1, u_2, u_3)$, $\mathbf{v} = (v_1, v_2, v_3)$, and $\mathbf{w} = (w_1, w_2, w_3)$. Show that

$$\mathbf{u} \cdot (\mathbf{v} \times \mathbf{w}) = \begin{vmatrix} u_1 & u_2 & u_3 \\ v_1 & v_2 & v_3 \\ w_1 & w_2 & w_3 \end{vmatrix}$$

9. Use the result of Exercise 8 to compute $\mathbf{u} \cdot (\mathbf{v} \times \mathbf{w})$ when $\mathbf{u} = (-1, 4, 7)$, $\mathbf{v} = (6, -7, 3)$, and $\mathbf{w} = (4, 0, 1)$.

10. Let $\mathbf{m}$ and $\mathbf{n}$ be vectors whose components in the xyz-system of Figure 3.28 are $\mathbf{m} = (0, 0, 1)$ and $\mathbf{n} = (0, 1, 0)$.
(a) Find the components of $\mathbf{m}$ and $\mathbf{n}$ in the $x'y'z'$-system of Figure 3.28.
(b) Compute $\mathbf{m} \times \mathbf{n}$ using the components in the xyz-system.
(c) Compute $\mathbf{m} \times \mathbf{n}$ using the components in the $x'y'z'$-system.
(d) Show that the vectors obtained in (b) and (c) are the same.

11. Prove the following identities.
(a) $(\mathbf{u} + k\mathbf{v}) \times \mathbf{v} = \mathbf{u} \times \mathbf{v}$
(b) $(\mathbf{u} \times \mathbf{v}) \cdot \mathbf{z} = \mathbf{u} \cdot (\mathbf{v} \times \mathbf{z})$

12. Let $\mathbf{u}$, $\mathbf{v}$, and $\mathbf{w}$ be nonzero vectors in 3-space, no two of which are collinear. Show:
(a) $\mathbf{u} \times (\mathbf{v} \times \mathbf{w})$ lies in the plane determined by $\mathbf{v}$ and $\mathbf{w}$ (assuming that the vectors are positioned so they have the same initial points)
(b) $(\mathbf{u} \times \mathbf{v}) \times \mathbf{w}$ lies in the plane determined by $\mathbf{u}$ and $\mathbf{v}$.

13. Prove that $\mathbf{x} \times (\mathbf{y} \times \mathbf{z}) = (\mathbf{x} \cdot \mathbf{z})\mathbf{y} - (\mathbf{x} \cdot \mathbf{y})\mathbf{z}$. [***Hint***. First prove the result in the case where $\mathbf{z} = \mathbf{i} = (1, 0, 0)$, then when $\mathbf{z} = \mathbf{j} = (0, 1, 0)$, and then when $\mathbf{z} = \mathbf{k} = (0, 0, 1)$. Finally prove it for an arbitrary vector $\mathbf{z} = (z_1, z_2, z_3)$ by writing $\mathbf{z} = z_1\mathbf{i} + z_2\mathbf{j} + z_3\mathbf{k}$.]

14. Prove parts (a) and (b) of Theorem 5.

15. Prove parts (c) and (d) of Theorem 5.

16. Prove parts (e) and (f) of Theorem 5.

3.5 LINES AND PLANES IN 3-SPACE

In this section we shall use vectors to derive equations of lines and planes in 3-space. We shall also use these equations to solve some basic geometric problems.

In plane analytic geometry a line can be specified by giving its slope and one of its points. Similarly, a plane in 3-space can be determined by giving its inclination and specifying one of its points. A convenient method for describing the inclination is to specify a vector (called a ***normal***) that is perpendicular to the plane.

Suppose we want the equation of the plane passing through the point $P_0(x_0, y_0, z_0)$ and having the nonzero vector $\mathbf{n} = (a, b, c)$ as a normal. It is evident from Figure 3.29 that the plane consists precisely of those points $P(x, y, z)$ for which the vector $\overrightarrow{P_0P}$ is orthogonal to $\mathbf{n}$; that is, for which

$$\mathbf{n} \cdot \overrightarrow{P_0P} = 0 \tag{3.10}$$

Since $\overrightarrow{P_0P} = (x - x_0, y - y_0, z - z_0)$, (3.9) can be rewritten as

$$a(x - x_0) + b(y - y_0) + c(z - z_0) = 0 \tag{3.11}$$

We shall call this a ***point-normal*** form of the equation of a plane.

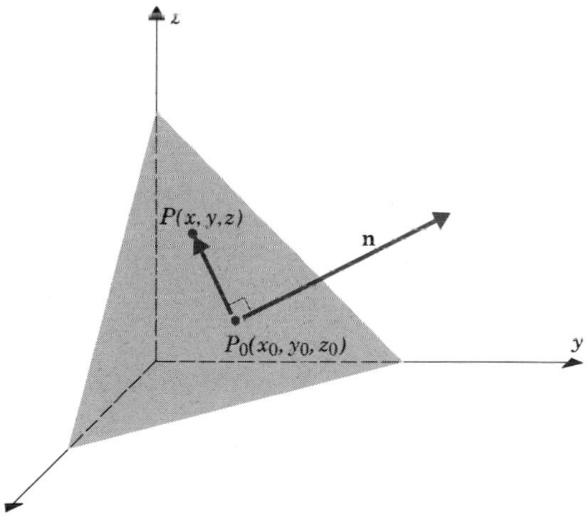

Figure 3.29

Example 15

Find the equation of the plane passing through the point $(3, -1, 7)$ and perpendicular to the vector $\mathbf{n} = (4, 2, -5)$.

Solution. From (3.11) a point-normal form is

$$4(x - 3) + 2(y + 1) - 5(z - 7) = 0$$

By multiplying out and collecting terms, (3.11) can be rewritten in the form

$$ax + by + cz + d = 0 \tag{3.12}$$

where a, b, c, and d are constants, and a, b, and c are not all zero. To illustrate, the equation in Example 15 can be rewritten as

$$4x + 2y - 5z + 25 = 0$$

As our next theorem shows, every equation having the form of (3.12) represents a plane in 3-space.

Theorem 6. *If a, b, c, and d are constants and a, b, and c are not all zero, then the graph of the equation*

$$ax + by + cz + d = 0$$

is a plane having the vector $\mathbf{n} = (a, b, c)$ as a normal.

Proof. By hypothesis a, b, and c are not all zero. Assume, for the moment, that $a \neq 0$. The equation $ax + by + cz + d = 0$ can be rewritten as $a(x + (d/a)) + by + cz = 0$. But this is a point-normal form of the plane passing through the point $(-d/a, 0, 0)$ and having $\mathbf{n} = (a, b, c)$ as a normal.

If $a = 0$, then either $b \neq 0$ or $c \neq 0$. A straightforward modification of the above argument will handle these other cases. ∎

Equation (3.12) is a linear equation in x, y and z; it is called the ***general form*** of the equation of a plane.

Just as the solutions to a system of linear equations

$$\begin{aligned} ax + by &= k_1 \\ cx + dy &= k_2 \end{aligned}$$

correspond to points of intersection of the lines $ax + by = k_1$ and $cx + dy = k_2$ in the xy-plane, so the solutions of a system

$$\begin{aligned} ax + by + cz &= k_1 \\ dx + ey + fz &= k_2 \\ gx + hy + iz &= k_3 \end{aligned} \tag{3.13}$$

correspond to points of intersection of the planes $ax + by + cz = k_1$, $dx + ey + fz = k_2$ and $gx + hy + iz = k_3$.

In Figure 3.30 we have illustrated some of the geometric possibilities when (3.13) has zero, one, or infinitely many solutions.

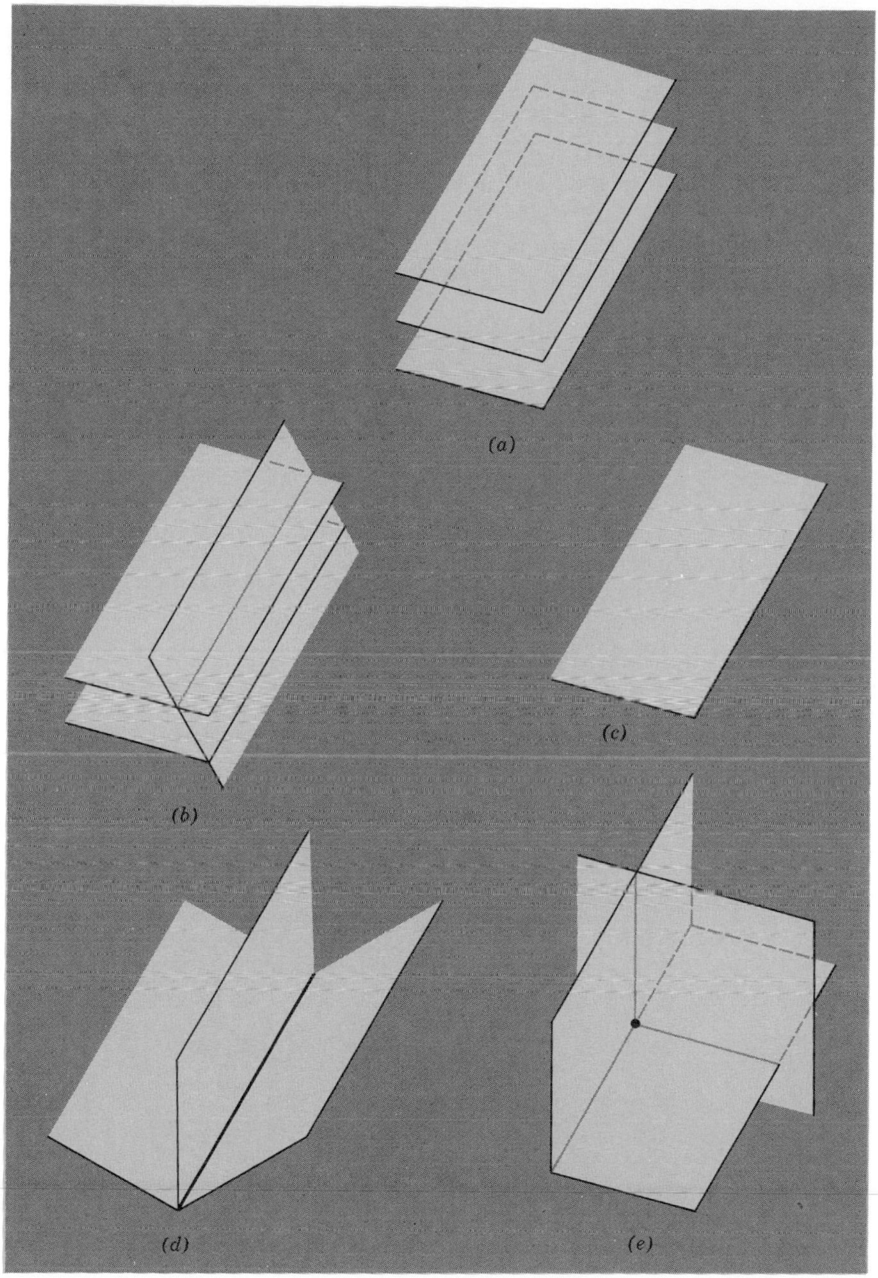

Figure 3.30 (*a*) No solutions (3 parallel planes). (*b*) No solutions (2 planes parallel). (*c*) Infinitely many solutions (3 coincident planes). (*d*) Infinitely many solutions (3 planes intersecting in a line). (*e*) One solution (3 planes intersecting at a point).

Example 16

Find the equation of the plane through the points $P_1(1, 2, -1)$, $P_2(2, 3, 1)$ and $P_3(3, -1, 2)$.

Solution. Since the three points lie in the plane, their coordinates must satisfy the general equation $ax + by + cz + d = 0$ of the plane. Thus

$$\begin{aligned} a + 2b - c + d &= 0 \\ 2a + 3b + c + d &= 0 \\ 3a - b + 2c + d &= 0 \end{aligned}$$

Solving this system gives

$$a = -\tfrac{9}{16}t \qquad b = -\tfrac{1}{16}t \qquad c = \tfrac{5}{16}t \qquad d = t$$

Letting $t = -16$, for example, yields the desired equation

$$9x + y - 5z - 16 = 0$$

We note that any other choice of t gives a multiple of this equation so that any value of $t \neq 0$ would do equally well.

Alternate solution. Since $P_1(1, 2, -1)$, $P_2(2, 3, 1)$, and $P_3(3, -1, 2)$ lie in the plane, the vectors $\overrightarrow{P_1P_2} = (1, 1, 2)$ and $\overrightarrow{P_1P_3} = (2, -3, 3)$ are parallel to the plane. Therefore $\overrightarrow{P_1P_2} \times \overrightarrow{P_1P_3} = (9, 1, -5)$ is normal to the plane, since it is perpendicular to both $\overrightarrow{P_1P_2}$ and $\overrightarrow{P_1P_3}$. From this and the fact that P_1 lies in the plane, a point normal form for the equation of the plane is

$$9(x - 1) + (y - 2) - 5(z + 1) = 0$$

or

$$9x + y - 5z - 16 = 0$$

We shall now show how to obtain equations for lines in 3-space. Suppose l is the line in 3-space through the point $P_0(x_0, y_0, z_0)$ and parallel to the nonzero vector $\mathbf{v} = (a, b, c)$. It is clear (Figure 3.31) that l consists precisely of those points

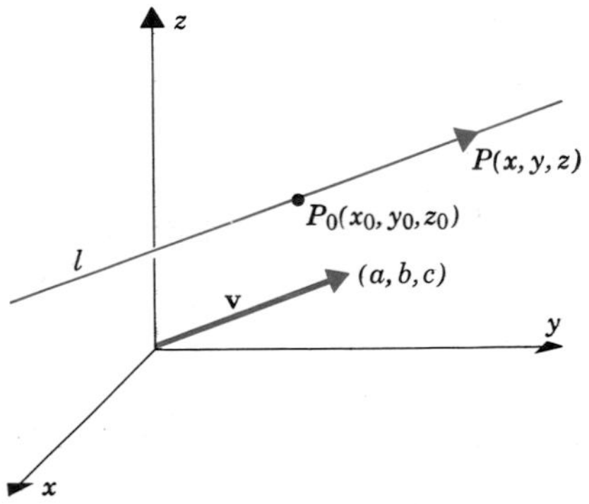

Figure 3.31

$P(x, y, z)$ for which the vector $\overrightarrow{P_0P}$ is parallel to $\mathbf{v}$, that is, for which there is a scalar t such that

$$\overrightarrow{P_0P} = t\mathbf{v} \tag{3.14}$$

In terms of components (3.14) can be written

$$(x - x_0, y - y_0, z - z_0) = (ta, tb, tc)$$

from which it follows that

$$\begin{aligned} x &= x_0 + ta \\ y &= y_0 + tb \qquad \text{where } -\infty < t < +\infty \\ z &= z_0 + tc \end{aligned}$$

These equations are called ***parametric equations*** for l since the line l is traced out by $P(x, y, z)$ as the parameter t varies from $-\infty$ to $+\infty$.

Example 17

The line through the point $(1, 2, -3)$ and parallel to the vector $\mathbf{v} = (4, 5, -7)$ has parametric equations

$$\begin{aligned} x &= 1 + 4t \\ y &= 2 + 5t \qquad \text{where } -\infty < t < +\infty \\ z &= -3 - 7t \end{aligned}$$

Example 18

(*a*) Find parametric equations for the line l passing through the points $P_1(2, 4, -1)$ and $P_2(5, 0, 7)$.
(*b*) Where does the line intersect the xy-plane?

Solution. (*a*) Since the vector $\overrightarrow{P_1P_2} = (3, -4, 8)$ is parallel to l and $P_1(2, 4, -1)$ lies on l, the line l is given by

$$\begin{aligned} x &= 2 + 3t \\ y &= 4 - 4t \qquad \text{where } -\infty < t < +\infty \\ z &= -1 + 8t \end{aligned}$$

(*b*) The line intersects the xy-plane at the point where $z = -1 + 8t = 0$; that is, when $t = 1/8$. Substituting this value of t in the parametric equations for l yields as the point of intersection

$$(x, y, z) = (\tfrac{19}{8}, \tfrac{7}{2}, 0)$$

Example 19

Find parametric equations for the line of intersection of the planes

$$3x + 2y - 4z - 6 = 0 \qquad \text{and} \qquad x - 3y - 2z - 4 = 0$$

Solution. The line of intersection consists of all points (x, y, z) that satisfy the two equations in the system

$$3x + 2y - 4z = 6$$
$$x - 3y - 2z = 4$$

Solving this system gives

$$x = \tfrac{26}{11} + \tfrac{16}{11}t \qquad y = -\tfrac{6}{11} - \tfrac{2}{11}t \qquad z = t$$

The parametric equations for l are therefore

$$\begin{aligned} x &= \tfrac{26}{11} + \tfrac{16}{11}t \\ y &= -\tfrac{6}{11} - \tfrac{2}{11}t \qquad \text{where } -\infty < t < +\infty \\ z &= t \end{aligned}$$

In some problems, a line

$$\begin{aligned} x &= x_0 + at \\ y &= y_0 + bt \qquad \text{where } -\infty < t < +\infty \\ z &= z_0 + ct \end{aligned} \tag{3.15}$$

is given, and it is of interest to find two planes whose intersection is the given line. Since there are infinitely many planes through the line, there are always infinitely many such pairs of planes. To find two such planes when a, b, and c are all different from zero, we can rewrite each equation in (3.15) in the form

$$\frac{x - x_0}{a} = t \qquad \frac{y - y_0}{b} = t \qquad \frac{z - z_0}{c} = t$$

Eliminating the parameter t shows that the line consists of all points (x, y, z) that satisfy the equations

$$\frac{x - x_0}{a} = \frac{y - y_0}{b} = \frac{z - z_0}{c}$$

called ***symmetric equations*** for the line. Therefore, the line can be viewed as the intersection of the planes

$$\frac{x - x_0}{a} = \frac{y - y_0}{b} \qquad \text{and} \qquad \frac{y - y_0}{b} = \frac{z - z_0}{c}$$

or as the intersection of

$$\frac{x - x_0}{a} = \frac{z - z_0}{c} \qquad \text{and} \qquad \frac{y - y_0}{b} = \frac{z - z_0}{c}$$

and so forth.

Example 20

Find two planes whose intersection is the line

$$x = 3 + 2t$$
$$y = -4 + 7t \qquad \text{where } -\infty < t < +\infty$$
$$z = 1 + 3t$$

Solution. Since the symmetric equations for this line are

$$\frac{x-3}{2} = \frac{y+4}{7} = \frac{z-1}{3} \tag{3.16}$$

it is the intersection of the planes

$$\frac{x-3}{2} = \frac{y+4}{7} \qquad \text{and} \qquad \frac{y+4}{7} = \frac{z-1}{3}$$

or equivalently

$$7x - 2y - 29 = 0 \qquad \text{and} \qquad 3y - 7z + 19 = 0$$

Other solutions can be obtained by choosing different pairs of equations from (3.16).

EXERCISE SET 3.5

1. In each part find a point-normal form of the equation of the plane passing through P and having $\mathbf{n}$ as a normal.
   (a) $P(2, 6, 1)$; $\mathbf{n} = (1, 4, 2)$ (b) $P(-1, -1, 2)$; $\mathbf{n} = (-1, 7, 6)$
   (c) $P(1, 0, 0)$; $\mathbf{n} = (0, 0, 1)$ (d) $P(0, 0, 0)$; $\mathbf{n} = (2, 3, 4)$
2. Write the equations of the planes in Exercise 1 in general form.
3. Find a point-normal form of:
   (a) $2x - 3y + 7z - 10 = 0$ (b) $x + 3z = 0$
4. In each part find an equation for the plane passing through the given points.
   (a) $(-2, 1, 1)$ $(0, 2, 3)$ $(1, 0, -1)$
   (b) $(3, 2, 1)$ $(2, 1, -1)$ $(-1, 3, 2)$
5. In each part find parametric equations for the line passing through P and parallel to $\mathbf{n}$.
   (a) $P(2, 4, 6)$; $\mathbf{n} = (1, 2, 5)$ (b) $P(-3, 2, -4)$; $\mathbf{n} = (5, -7, -3)$
   (c) $P(1, 1, 5)$; $\mathbf{n} = (0, 0, 1)$ (d) $P(0, 0, 0)$; $\mathbf{n} = (1, 1, 1)$
6. Find symmetric equations for lines in parts (a) and (b) of Exercise 5.
7. In each part find parametric equations for the line passing through the given points.
   (a) $(6, -1, 5)$, $(7, 2, -4)$ (b) $(0, 0, 0)$, $(-1, -1, -1)$
8. In each part find parametric equations for the line of intersection of the given planes.
   (a) $-2x + 3y + 7z + 2 = 0$ and $x + 2y - 3z + 5 = 0$
   (b) $3x - 5y + 2z = 0$ and $z = 0$

9. In each part find equations for two planes whose intersection is the given line.

(a) $\begin{aligned} x &= 3 + 4t \\ y &= -7 + 2t \\ z &= 6 - t \end{aligned} \quad -\infty < t < +\infty$ (b) $\begin{aligned} x &= 5t \\ y &= 3t \\ z &= 6t \end{aligned} \quad -\infty < t < +\infty$

10. Find equations for the xy-plane, the xz-plane and the yz-plane.

11. Show that the line

$$\begin{aligned} x &= 0 \\ y &= t \qquad -\infty < t < +\infty \\ z &= t \end{aligned}$$

(a) lies in the plane $6x + 4y - 4z = 0$
(b) is parallel to and below the plane $5x - 3y + 3z = 1$
(c) is parallel to and above the plane $6x + 2y - 2z = 3$.

12. Find the point of intersection of the line

$$\begin{aligned} x - 4 &= 5t \\ y + 2 &= t \\ z - 4 &= -t \end{aligned}$$

and the plane $3x - y + 7z + 8 = 0$.

13. Find a plane passing through the point $(2, -7, 6)$ and parallel to the plane

$$5x - 2y + z - 9 = 0.$$

14. Show that the line

$$\begin{aligned} x - 4 &= 2t \\ y &= -t \qquad -\infty < t < +\infty \\ z + 1 &= -4t \end{aligned}$$

is parallel to the plane $3x + 2y + z - 7 = 0$.

15. Show that the lines

$$\begin{aligned} x + 1 &= 4t \\ y - 3 &= t \\ z - 1 &= 0 \end{aligned} \quad \text{and} \quad \begin{aligned} x + 13 &= 12t \\ y - 1 &= 6t \\ z - 2 &= 3t \end{aligned}$$

intersect. Find the point of intersection.

16. Find an equation for the plane determined by the lines in Exercise 15.

4 Vector Spaces

4.1 EUCLIDEAN n-SPACE

The idea of using pairs of numbers to locate points in the plane and triples of numbers to locate points in 3-space was first clearly spelled out in the mid-seventeenth century. By the latter part of the nineteenth century mathematicians and physicists began to realize that there was no need to stop with triples. It was recognized that quadruples of numbers (a_1, a_2, a_3, a_4) could be regarded as points in "4-dimensional" space, quintuples $(a_1, a_2, \ldots, a_5)$ as points in "5-dimensional" space, etc. Although our geometric visualization does not extend beyond 3-space, it is nevertheless possible to extend many familiar ideas beyond 3-space by working with analytic or numerical properties of points and vectors rather than the geometric properties. In this section we shall make these ideas more precise.

Definition. If n is a positive integer, then an ***ordered n-tuple*** is a sequence of n real numbers $(a_1, a_2, \ldots, a_n)$. The set of all ordered n-tuples is called ***n-space*** and is denoted by R^n.

(When $n = 2$ or 3, it is usual to use the terms "ordered pair" and "ordered triple" rather than ordered 2-tuple and 3-tuple.)

It might have occurred to the reader in his study of 3-space that the symbol (a_1, a_2, a_3) has two different geometric interpretations. It can be interpreted as a point, in which case a_1, a_2, and a_3 are the coordinates (Figure 4.1a) or it can be interpreted as a vector, in which case a_1, a_2, and a_3 are the components (Figure 4.1b). It follows, therefore, that an ordered n-tuple $(a_1, a_2, \ldots, a_n)$ can be viewed either as a "generalized point" or a "generalized vector"—the distinction is mathematically unimportant. Thus we are free to describe the 5-tuple $(-2, 4, 0, 1, 6)$ either as a point in R^5 or a vector in R^5. We will use both descriptions.

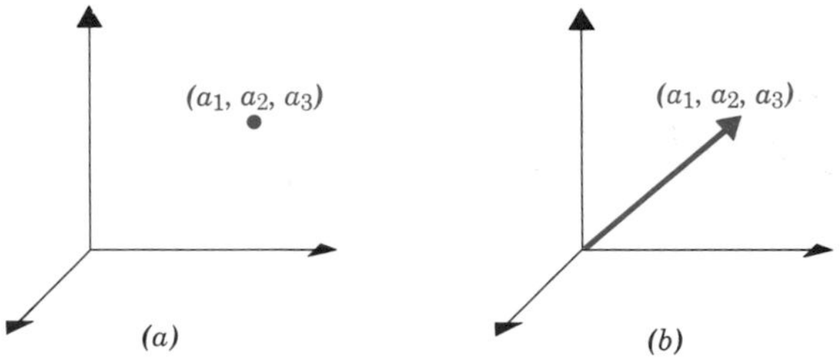

Figure 4.1

Definition. Two vectors $\mathbf{u} = (u_1, u_2, \ldots, u_n)$ and $\mathbf{v} = (v_1, v_2, \ldots, v_n)$ in R^n are called ***equal*** if

$$u_1 = v_1, \qquad u_2 = v_2, \ldots, u_n = v_n$$

The ***sum*** $\mathbf{u} + \mathbf{v}$ is defined by

$$\mathbf{u} + \mathbf{v} = (u_1 + v_1, u_2 + v_2, \ldots, u_n + v_n)$$

and if k is any scalar, the ***scalar multiple*** $k\mathbf{u}$ is defined by

$$k\mathbf{u} = (ku_1, ku_2, \ldots, ku_n)$$

The operations of addition and scalar multiplication in this definition are called the ***standard operations*** on R^n.

We define the ***zero vector*** in R^n to be the vector

$$\mathbf{0} = (0, 0, \ldots, 0)$$

If $\mathbf{u} = (u_1, u_2, \ldots, u_n)$ is any vector in R^n, then the ***negative*** (or ***additive inverse***) of $\mathbf{u}$ is denoted by $-\mathbf{u}$ and is defined by

$$-\mathbf{u} = (-u_1, -u_2, \ldots, -u_n)$$

We define subtraction of vectors in R^n by $\mathbf{v} - \mathbf{u} = \mathbf{v} + (-\mathbf{u})$, or in terms of components

$$\begin{aligned} \mathbf{v} - \mathbf{u} = \mathbf{v} + (-\mathbf{u}) &= (v_1, v_2, \ldots, v_n) + (-u_1, -u_2, \ldots, -u_n) \\ &= (v_1 - u_1, v_2 - u_2, \ldots, v_n - u_n) \end{aligned}$$

The most important arithmetic properties of addition and scalar multiplication of vectors in R^n are listed in the next theorem. The proofs are all easy and are left as exercises.

Theorem 1. *If* $\mathbf{u} = (u_1, u_2, \ldots, u_n)$, $\mathbf{v} = (v_1, v_2, \ldots, v_n)$ *and* $\mathbf{w} = (w_1, w_2, \ldots, w_n)$ *are vectors in* R^n *and* k *and* l *are scalars, then:*

(*a*) $\mathbf{u} + \mathbf{v} = \mathbf{v} + \mathbf{u}$
(*b*) $\mathbf{u} + (\mathbf{v} + \mathbf{w}) = (\mathbf{u} + \mathbf{v}) + \mathbf{w}$
(*c*) $\mathbf{u} + \mathbf{0} = \mathbf{0} + \mathbf{u} = \mathbf{u}$
(*d*) $\mathbf{u} + (-\mathbf{u}) = \mathbf{0}$, *that is,* $\mathbf{u} - \mathbf{u} = \mathbf{0}$

(e) $k(l\mathbf{u}) = (kl)\mathbf{u}$
(f) $k(\mathbf{u} + \mathbf{v}) = k\mathbf{u} + k\mathbf{v}$
(g) $(k + l)\mathbf{u} = k\mathbf{u} + l\mathbf{u}$
(h) $1\mathbf{u} = \mathbf{u}$

This theorem enables us to manipulate vectors in R^n without expressing the vectors in terms of components in much the same way that we manipulate real numbers. For example, to solve the vector equation $\mathbf{x} + \mathbf{u} = \mathbf{v}$ for $\mathbf{x}$, we can add $-\mathbf{u}$ to both sides and proceed as follows.

$$\begin{aligned}(\mathbf{x} + \mathbf{u}) + (-\mathbf{u}) &= \mathbf{v} + (-\mathbf{u})\\ \mathbf{x} + (\mathbf{u} - \mathbf{u}) &= \mathbf{v} - \mathbf{u}\\ \mathbf{x} + \mathbf{0} &= \mathbf{v} - \mathbf{u}\\ \mathbf{x} &= \mathbf{v} - \mathbf{u}\end{aligned}$$

The reader will find it useful to name the parts of Theorem 1 that justify each of the steps in this computation.

To extend the notions of distance, norm, and angle to R^n, we begin with the following generalization of the dot product on R^2 and R^3 (Section 3.3).

Definition. If $\mathbf{u} = (u_1, u_2, \ldots, u_n)$ and $\mathbf{v} = (v_1, v_2, \ldots, v_n)$ are any vectors in R^n, then the ***Euclidean inner product*** $\mathbf{u} \cdot \mathbf{v}$ is defined by

$$\mathbf{u} \cdot \mathbf{v} = u_1v_1 + u_2v_2 + \cdots + u_nv_n$$

Observe that when $n = 2$ or 3, the Euclidean inner product is the ordinary dot product. (Section 3.3.)

Example 1

The Euclidean inner product of the vectors

$$\mathbf{u} = (-1, 3, 5, 7) \qquad \text{and} \qquad \mathbf{v} = (5, -4, 7, 0)$$

in R^4 is

$$\mathbf{u} \cdot \mathbf{v} = (-1)(5) + (3)(-4) + (5)(7) + (7)(0) = 18$$

The main arithmetic properties of the Euclidean inner product are listed in the next theorem.

Theorem 2. *If* $\mathbf{u}$, $\mathbf{v}$, *and* $\mathbf{w}$ *are vectors in* R^n *and* k *is any scalar, then:*

(a) $\mathbf{u} \cdot \mathbf{v} = \mathbf{v} \cdot \mathbf{u}$
(b) $(\mathbf{u} + \mathbf{v}) \cdot \mathbf{w} = \mathbf{u} \cdot \mathbf{w} + \mathbf{v} \cdot \mathbf{w}$
(c) $(k\mathbf{u}) \cdot \mathbf{v} = k(\mathbf{u} \cdot \mathbf{v})$
(d) $\mathbf{v} \cdot \mathbf{v} \geq 0$. *Further,* $\mathbf{v} \cdot \mathbf{v} = 0$ *if and only if* $\mathbf{v} = \mathbf{0}$

We shall prove parts (b) and (d) and leave proof of the rest as exercises.

Proof (b). Let $\mathbf{u} = (u_1, u_2, \ldots, u_n)$, $\mathbf{v} = (v_1, v_2, \ldots, v_n)$ and $\mathbf{w} = (w_1, w_2, \ldots, w_n)$. Then

$$\begin{aligned}(\mathbf{u} + \mathbf{v}) \cdot \mathbf{w} &= (u_1 + v_1, u_2 + v_2, \ldots, u_n + v_n) \cdot (w_1, w_2, \ldots, w_n)\\ &= (u_1 + v_1)w_1 + (u_2 + v_2)w_2 + \cdots + (u_n + v_n)w_n\\ &= (u_1w_1 + u_2w_2 + \cdots + u_nw_n) + (v_1w_1 + v_2w_2 + \cdots + v_nw_n)\\ &= \mathbf{u} \cdot \mathbf{w} + \mathbf{v} \cdot \mathbf{w}\end{aligned}$$

(d). $\mathbf{v} \cdot \mathbf{v} = v_1^2 + v_2^2 + \cdots + v_n^2 \geq 0$. Further, equality holds if and only if $v_1 = v_2 = \cdots = v_n = 0$; that is, if and only if $\mathbf{v} = \mathbf{0}$.

Example 2

Theorem 2 allows us to perform computations with Euclidean inner products in much the same way that we perform them with ordinary arithmetic products. For example,

$$\begin{aligned}(3\mathbf{u} + 2\mathbf{v}) \cdot (4\mathbf{u} + \mathbf{v}) &= (3\mathbf{u}) \cdot (4\mathbf{u} + \mathbf{v}) + (2\mathbf{v}) \cdot (4\mathbf{u} + \mathbf{v})\\ &= (3\mathbf{u}) \cdot (4\mathbf{u}) + (3\mathbf{u}) \cdot \mathbf{v} + (2\mathbf{v}) \cdot (4\mathbf{u}) + (2\mathbf{v}) \cdot \mathbf{v}\\ &= 12(\mathbf{u} \cdot \mathbf{u}) + 3(\mathbf{u} \cdot \mathbf{v}) + 8(\mathbf{v} \cdot \mathbf{u}) + 2(\mathbf{v} \cdot \mathbf{v})\\ &= 12(\mathbf{u} \cdot \mathbf{u}) + 11(\mathbf{u} \cdot \mathbf{v}) + 2(\mathbf{v} \cdot \mathbf{v})\end{aligned}$$

The reader should determine which parts of Theorem 2 were used in each step.

By analogy with the familiar formulas in R^2 and R^3, we define the ***Euclidean norm*** (or ***Euclidean length***) of a vector $\mathbf{u} = (u_1, u_2, \ldots, u_n)$ in R^n by

$$\|\mathbf{u}\| = (\mathbf{u} \cdot \mathbf{u})^{1/2} = \sqrt{u_1^2 + u_2^2 + \cdots + u_n^2}$$

Similarly, the ***Euclidean distance*** between the points $\mathbf{u} = (u_1, u_2, \ldots, u_n)$ and $\mathbf{v} = (v_1, v_2, \ldots, v_n)$ in R^n is defined by

$$d(\mathbf{u}, \mathbf{v}) = \|\mathbf{u} - \mathbf{v}\| = \sqrt{(u_1 - v_1)^2 + (u_2 - v_2)^2 + \cdots + (u_n - v_n)^2}$$

Example 3

If $\mathbf{u} = (1, 3, -2, 7)$ and $\mathbf{v} = (0, 7, 2, 2)$ then

$$\|\mathbf{u}\| = \sqrt{(1)^2 + (3)^2 + (-2)^2 + (7)^2} = \sqrt{63} = 3\sqrt{7}$$

and

$$d(\mathbf{u}, \mathbf{v}) = \sqrt{(1 - 0)^2 + (3 - 7)^2 + (-2 - 2)^2 + (7 - 2)^2} = \sqrt{58}$$

Since so many of the familiar ideas from 2-space and 3-space carry over, it is common to refer to R^n with the operations of addition, scalar multiplication, and inner product that we have defined here as ***Euclidean n-space.***

We conclude this section by noting that many writers prefer to use the matrix notation

$$\mathbf{u} = \begin{bmatrix} u_1 \\ u_2 \\ \vdots \\ u_n \end{bmatrix}$$

rather than the horizontal notation $\mathbf{u} = (u_1, u_2, \ldots, u_n)$ to denote vectors in R^n. This is justified because the matrix operations

$$\mathbf{u} + \mathbf{v} = \begin{bmatrix} u_1 \\ u_2 \\ \vdots \\ u_n \end{bmatrix} + \begin{bmatrix} v_1 \\ v_2 \\ \vdots \\ v_n \end{bmatrix} = \begin{bmatrix} u_1 + v_1 \\ u_2 + v_2 \\ \vdots \\ u_n + v_n \end{bmatrix}$$

$$k\mathbf{u} = k\begin{bmatrix} u_1 \\ u_2 \\ \vdots \\ u_n \end{bmatrix} = \begin{bmatrix} ku_1 \\ ku_2 \\ \vdots \\ ku_n \end{bmatrix}$$

produce the same results as the vector operations

$$\mathbf{u} + \mathbf{v} = (u_1, u_2, \ldots, u_n) + (v_1, v_2, \ldots, v_n) = (u_1 + v_1, u_2 + v_2, \ldots, u_n + v_n)$$
$$k\mathbf{u} = k(u_1, u_2, \ldots, u_n) = (ku_1, ku_2, \ldots, ku_n)$$

The only difference is that results are displayed vertically in one case and horizontally in the other. We will use both notations at various times. However, from here on, we shall denote $n \times 1$ matrices by lowercase boldface letters. Thus a system of linear equations will be written

$$A\mathbf{x} = \mathbf{b}$$

rather than $AX = B$ as before.

EXERCISE SET 4.1

1. Let $\mathbf{u} = (2, 0, -1, 3)$, $\mathbf{v} = (5, 4, 7, -1)$, and $\mathbf{w} = (6, 2, 0, 9)$. Find:
   (a) $\mathbf{u} - \mathbf{v}$ (b) $7\mathbf{v} + 3\mathbf{w}$ (c) $-\mathbf{w} + \mathbf{v}$
   (d) $3(\mathbf{u} - 7\mathbf{v})$ (e) $-3\mathbf{v} - 8\mathbf{w}$ (f) $2\mathbf{v} - (\mathbf{u} + \mathbf{w})$

2. Let $\mathbf{u}$, $\mathbf{v}$, and $\mathbf{w}$ be the vectors in Exercise 1. Find the vector $\mathbf{x}$ that satisfies $2\mathbf{u} - \mathbf{v} + \mathbf{x} = 7\mathbf{x} + \mathbf{w}$.

3. Let $\mathbf{u}_1 = (-1, 3, 2, 0)$, $\mathbf{u}_2 = (2, 0, 4, -1)$, $\mathbf{u}_3 = (7, 1, 1, 4)$, and $\mathbf{u}_4 = (6, 3, 1, 2)$. Find scalars c_1, c_2, c_3, and c_4 such that $c_1\mathbf{u}_1 + c_2\mathbf{u}_2 + c_3\mathbf{u}_3 + c_4\mathbf{u}_4 = (0, 5, 6, -3)$.

4. Show that there do not exist scalars $c_1, c_2,$ and c_3 such that $c_1(1, 0, -2, 1) + c_2(2, 0, 1, 2) + c_3(1, -2, 2, 3) = (1, 0, 1, 0)$.

5. Compute the Euclidean norm of $\mathbf{v}$ when
(a) $\mathbf{v} = (4, -3)$ (b) $\mathbf{v} = (1, -1, 3)$ (c) $\mathbf{v} = (2, 0, 3, -1)$ (d) $\mathbf{v} = (-1, 1, 1, 3, 6)$

6. Let $\mathbf{u} = (3, 0, 1, 2)$, $\mathbf{v} = (-1, 2, 7, -3)$, and $\mathbf{w} = (2, 0, 1, 1)$. Find
(a) $\|\mathbf{u} + \mathbf{v}\|$ (b) $\|\mathbf{u}\| + \|\mathbf{v}\|$ (c) $\|-2\mathbf{u}\| + 2\|\mathbf{u}\|$
(d) $\|3\mathbf{u} - 5\mathbf{v} + \mathbf{w}\|$ (e) $\dfrac{1}{\|\mathbf{w}\|}\mathbf{w}$ (f) $\left\|\dfrac{1}{\|\mathbf{w}\|}\mathbf{w}\right\|$

7. Show that if $\mathbf{v}$ is a nonzero vector in R^n, then $(1/\|\mathbf{v}\|)\mathbf{v}$ has norm 1.

8. Find all scalars k such that $\|k\mathbf{v}\| = 3$, where $\mathbf{v} = (-1, 2, 0, 3)$.

9. Find the Euclidean inner product $\mathbf{u} \cdot \mathbf{v}$ when:
(a) $\mathbf{u} = (-1, 3)$, $\mathbf{v} = (7, 2)$ (b) $\mathbf{u} = (3, 7, 1)$, $\mathbf{v} = (-1, 0, 2)$
(c) $\mathbf{u} = (1, -1, 2, 3)$, $\mathbf{v} = (3, 3, -6, 4)$ (d) $\mathbf{u} = (1, 3, 2, 6, -1)$, $\mathbf{v} = (0, 0, 2, 4, 1)$

10. (a) Find two vectors in R^2 with Euclidean norm 1 whose Euclidean inner products with $(-2, 4)$ are zero.
(b) Show that there are infinitely many vectors in R^3 with Euclidean norm 1 whose Euclidean inner product with $(-1, 7, 2)$ is zero.

11. Find the Euclidean distance between $\mathbf{u}$ and $\mathbf{v}$ when:
(a) $\mathbf{u} = (2, -1)$, $\mathbf{v} = (3, 2)$ (b) $\mathbf{u} = (1, 1, -1)$, $\mathbf{v} = (2, 6, 0)$
(c) $\mathbf{u} = (2, 0, 1, 3)$, $\mathbf{v} = (-1, 4, 6, 6)$ (d) $\mathbf{u} = (6, 0, 1, 3, 0)$, $\mathbf{v} = (-1, 4, 2, 8, 3)$

12. Establish the identity
$$\|\mathbf{u} + \mathbf{v}\|^2 + \|\mathbf{u} - \mathbf{v}\|^2 = 2\|\mathbf{u}\|^2 + 2\|\mathbf{v}\|^2$$
for vectors in R^n. Interpret this result geometrically in R^2.

13. Establish the identity
$$\mathbf{u} \cdot \mathbf{v} = \tfrac{1}{4}\|\mathbf{u} + \mathbf{v}\|^2 - \tfrac{1}{4}\|\mathbf{u} - \mathbf{v}\|^2$$
for vectors in R^n.

14. Verify parts (b), (e), (f), and (g) of Theorem 1 when $\mathbf{u} = (1, 0, -1, 2)$, $\mathbf{v} = (3, -1, 2, 4)$, $\mathbf{w} = (2, 7, 3, 0)$, $k = 6$, and $l = -2$.

15. Verify parts (b) and (c) of Theorem 2 for the values of $\mathbf{u}$, $\mathbf{v}$, $\mathbf{w}$, and k in Exercise 14.

16. Prove (a) through (d) of Theorem 1.

17. Prove (e) through (h) of Theorem 1.

18. Prove (a) and (c) of Theorem 2.

4.2 GENERAL VECTOR SPACES

In this section we generalize the concept of a vector still further. We shall state a set of axioms which, if satisfied by a class of objects, will entitle those objects to be called "vectors." The axioms will be chosen by abstracting the most important properties of vectors in R^n; as consequence, vectors in R^n will automatically satisfy these axioms. Thus our new concept of a vector will include our old vectors and many new kinds of vectors as well.

Definition. Let V be an arbitrary set of objects on which two operations are defined, addition and multiplication by scalars (real numbers). By addition we mean a rule for associating with each pair of objects $\mathbf{u}$ and $\mathbf{v}$ in V an element $\mathbf{u} + \mathbf{v}$, called the ***sum*** of $\mathbf{u}$ and $\mathbf{v}$; and by scalar multiplication we mean a rule for associating with each scalar k and each object $\mathbf{u}$ in V an element $k\mathbf{u}$, called the ***scalar multiple*** of $\mathbf{u}$ by k. If the following axioms are satisfied by all objects $\mathbf{u}, \mathbf{v}, \mathbf{w}$ in V and all scalars k and l, then we call V a ***vector space*** and we call the objects in V ***vectors***:

(1) If $\mathbf{u}$ and $\mathbf{v}$ are objects in V, then $\mathbf{u} + \mathbf{v}$ is in V
(2) $\mathbf{u} + \mathbf{v} = \mathbf{v} + \mathbf{u}$
(3) $\mathbf{u} + (\mathbf{v} + \mathbf{w}) = (\mathbf{u} + \mathbf{v}) + \mathbf{w}$
(4) There is an object $\mathbf{0}$ in V such that $\mathbf{0} + \mathbf{u} = \mathbf{u} + \mathbf{0} = \mathbf{u}$ for all $\mathbf{u}$ in V.
(5) For each $\mathbf{u}$ in V, there is an object $-\mathbf{u}$ in V called the ***negative*** of $\mathbf{u}$ such that $\mathbf{u} + (-\mathbf{u}) = (-\mathbf{u}) + \mathbf{u} = \mathbf{0}$
(6) If k is any real number and $\mathbf{u}$ is any object in V, then $k\mathbf{u}$ is in V
(7) $k(\mathbf{u} + \mathbf{v}) = k\mathbf{u} + k\mathbf{v}$
(8) $(k + l)\mathbf{u} = k\mathbf{u} + l\mathbf{u}$
(9) $k(l\mathbf{u}) = (kl)(\mathbf{u})$
(10) $1\mathbf{u} = \mathbf{u}$

The vector $\mathbf{0}$ in Axiom 4 is called the ***zero vector*** for V.

For some applications it is necessary to consider vector spaces where the scalars are complex numbers rather than real numbers. Such vector spaces are called ***complex vector spaces***. In this text, however, *all our scalars will be real*.

The reader should keep in mind that, in the definition of a vector space, neither the nature of the vectors nor the operations is specified. Any kinds of objects whatsoever can serve as vectors; all that is required is that the vector-space axioms be satisfied. The following examples will give some idea of the diversity of possible vector spaces.

Example 4

The set $V = R^n$ with the standard operations of addition and scalar multiplication defined in the previous section is a vector space. Axioms 1 and 6 follow from the definitions of the standard operations on R^n; the remaining axioms follow from Theorem 1.

Example 5

Let V be any plane through the origin in R^3. We shall show that the points in V form a vector space under the standard addition and scalar multiplication operations for vectors in R^3.

From Example 4, we know that R^3 itself is a vector space under these operations. Thus Axioms 2, 3, 7, 8, 9, and 10 hold for all points in R^3 and consequently for all points in the plane V. We therefore need only show that Axioms 1, 4, 5, and 6 are satisfied.

Since the plane V passes through the origin, it has an equation of the form

$$ax + by + cz = 0 \tag{4.1}$$

(Theorem 6 in Chapter 3.) Thus, if $\mathbf{u} = (u_1, u_2, u_3)$ and $\mathbf{v} = (v_1, v_2, v_3)$ are points in V, then $au_1 + bu_2 + cu_3 = 0$ and $av_1 + bv_2 + cv_3 = 0$. Adding these equations gives

$$a(u_1 + v_1) + b(u_2 + v_2) + c(u_3 + v_3) = 0$$

This equality tells us that the coordinates of the point $\mathbf{u} + \mathbf{v} = (u_1 + v_1, u_2 + v_2, u_3 + v_3)$ satisfy (4.1); thus $\mathbf{u} + \mathbf{v}$ lies in the plane V. This proves that Axiom 1 is satisfied. Multiplying $au_1 + bu_2 + cu_3 = 0$ through by -1 gives

$$a(-u_1) + b(-u_2) + c(-u_3) = 0$$

Thus $-\mathbf{u} = (-u_1, -u_2, -u_3)$ lies in V. This establishes Axiom 5. The verifications of Axioms 4 and 6 are left as exercises.

Example 6

The points on a line V passing through the origin in R^3 form a vector space under the standard addition and scalar multiplication operations for vectors in R^3.

The argument is similar to that used in Example 5 and is based on the fact that the points of V satisfy parametric equations of the form

$$\begin{aligned} x &= at \\ y &= bt \qquad -\infty < t < +\infty \\ z &= ct \end{aligned}$$

(Section 3.5). The details are left as an exercise.

Example 7

The set V of all $m \times n$ matrices with real entries, together with the operations of matrix addition and scalar multiplication, is a vector space. The $m \times n$ zero matrix is the zero vector $\mathbf{0}$, and if $\mathbf{u}$ is the $m \times n$ matrix A, then the matrix $-A$ is the vector $-\mathbf{u}$ in Axiom 5. Most of the remaining axioms are satisfied by virtue of Theorem 2 in Section 1.5. We shall denote this vector space by the symbol M_{mn}.

Example 8

Let V be the set of real-valued functions defined on the entire real line. If $\mathbf{f} = f(x)$ and $\mathbf{g} = g(x)$ are two such functions and k is any real number, define the sum function $\mathbf{f} + \mathbf{g}$ and the scalar multiple $k\mathbf{f}$ by

$$(\mathbf{f} + \mathbf{g})(x) = f(x) + g(x)$$
$$(k\mathbf{f})(x) = kf(x)$$

In other words, the value of the function $\mathbf{f} + \mathbf{g}$ at x is obtained by adding together the values of $\mathbf{f}$ and $\mathbf{g}$ at x (Figure 4.2*a*). Similarly, the value of $k\mathbf{f}$ at x is k times the value of $\mathbf{f}$ at x (Figure 4.2*b*). The set V is a vector space under these operations.

The zero vector in this space is the zero constant function, that is, the function whose graph is a horizontal line through the origin. The verification of the remaining axioms is an exercise.

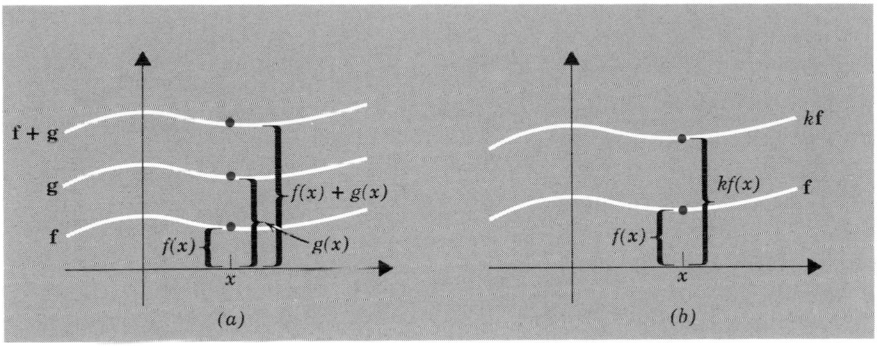

Figure 4.2

Example 9

Let V be the set of all points (x, y) in R^2 that lie in the first quadrant; that is, such that $x \geq 0$ and $y \geq 0$. The set V *fails* to be a vector space under the standard operations on R^2, since Axioms 5 and 6 are not satisfied. To see this, observe that $\mathbf{v} = (1, 1)$ lies in V, but $(-1)\mathbf{v} = -\mathbf{v} = (-1, -1)$ does not (Figure 4.3).

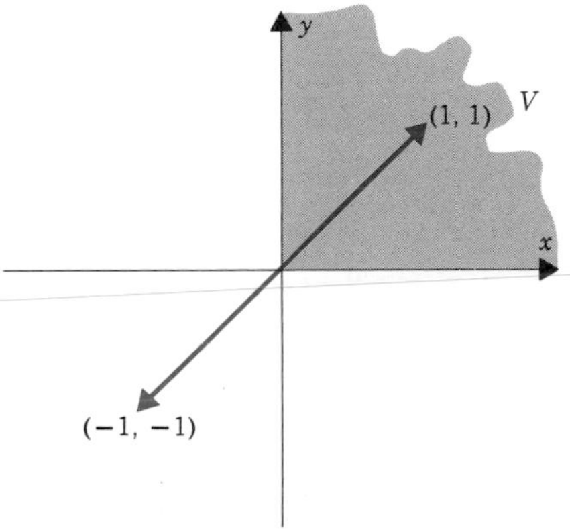

Figure 4.3

Example 10

Let V consist of a single object, which we denote by $\mathbf{0}$, and define

$$\mathbf{0} + \mathbf{0} = \mathbf{0}$$
$$k\mathbf{0} = \mathbf{0}$$

for all scalars k. It is easy to check that all the vector space axioms are satisfied. We call this the ***zero vector space***.

As we progress, we shall add more examples of vector spaces to our list. We conclude this section with a theorem that gives a useful list of vector properties.

Theorem 3. *Let V be a vector space, $\mathbf{u}$ a vector in V, and k a scalar; then:*

(*a*) $0\mathbf{u} = \mathbf{0}$
(*b*) $k\mathbf{0} = \mathbf{0}$
(*c*) $(-1)\mathbf{u} = -\mathbf{u}$
(*d*) *If* $k\mathbf{u} = \mathbf{0}$, *then* $k = 0$ *or* $\mathbf{u} = \mathbf{0}$

Proof. We shall prove parts (*a*) and (*c*) and leave proofs of the remaining parts as exercises.

(*a*) We can write

$$\begin{aligned} 0\mathbf{u} + 0\mathbf{u} &= (0 + 0)\mathbf{u} && \text{(Axiom 8)} \\ &= 0\mathbf{u} && \text{(Property of the number 0)} \end{aligned}$$

By Axiom 5 the vector $0\mathbf{u}$ has a negative, $-0\mathbf{u}$. Adding this negative to both sides above yields

$$[0\mathbf{u} + 0\mathbf{u}] + (-0\mathbf{u}) = 0\mathbf{u} + (-0\mathbf{u})$$

or

$$0\mathbf{u} + [0\mathbf{u} + (-0\mathbf{u})] = 0\mathbf{u} + (-0\mathbf{u}) \qquad \text{(Axiom 3)}$$

or

$$0\mathbf{u} + \mathbf{0} = \mathbf{0} \qquad \text{(Axiom 5)}$$

or

$$0\mathbf{u} = \mathbf{0} \qquad \text{(Axiom 4)}$$

(*c*) To show $(-1)\mathbf{u} = -\mathbf{u}$, we must demonstrate that $\mathbf{u} + (-1)\mathbf{u} = \mathbf{0}$. To see this, observe that

$$\begin{aligned} \mathbf{u} + (-1)\mathbf{u} &= 1\mathbf{u} + (-1)\mathbf{u} && \text{(Axiom 10)} \\ &= (1 + (-1))\mathbf{u} && \text{(Axiom 8)} \\ &= 0\mathbf{u} && \text{(Property of numbers)} \\ &= \mathbf{0} \quad \blacksquare && \text{(Part } a \text{ above)} \end{aligned}$$

EXERCISE SET 4.2

In Exercises 1–14 a set of objects is given together with operations of addition and scalar multiplication. Determine which sets are vector spaces under the given operations. For those that are not, list all axioms that fail to hold.

1. The set of all triples of real numbers (x, y, z) with the operations $(x, y, z) + (x', y', z') = (x + x', y + y', z + z')$ and $k(x, y, z) = (kx, y, z)$.

2. The set of all triples of real numbers (x, y, z) with the operations $(x, y, z) + (x', y', z') = (x + x', y + y', z + z')$ and $k(x, y, z) = (0, 0, 0)$.

3. The set of all pairs of real numbers (x, y) with the operations $(x, y) + (x', y') = (x + x', y + y')$ and $k(x, y) = (2kx, 2ky)$.

4. The set of all real numbers x with the standard operations of addition and multiplication.

5. The set of all pairs of real numbers of the form $(x, 0)$ with the standard operations on R^2.

6. The set of all pairs of real numbers of the form (x, y) where $x \geq 0$, with the standard operations on R^2.

7. The set of all n-tuples of real numbers of the form $(x, x, \ldots, x)$ with the standard operations on R^n.

8. The set of all pairs of real numbers (x, y) with the operations $(x, y) + (x', y') = (x + x' + 1, y + y' + 1)$ and $k(x, y) = (kx, ky)$.

9. The set of all positive real numbers x with the operations $x + x' = xx'$ and $kx = x^k$.

10. The set of all 2×2 matrices of the form

$$\begin{bmatrix} a & 1 \\ 1 & b \end{bmatrix}$$

with matrix addition and scalar multiplication.

11. The set of all 2×2 matrices of the form

$$\begin{bmatrix} a & 0 \\ 0 & b \end{bmatrix}$$

with matrix addition and scalar multiplication.

12. The set of all real-valued functions f defined everywhere on the real line and such that $f(1) = 0$, with the operations defined in Example 8.

13. The set of all 2×2 matrices of the form

$$\begin{bmatrix} a & a + b \\ a + b & b \end{bmatrix}$$

with matrix addition and scalar multiplication.

14. The set whose only element is the moon. The operations are moon + moon = moon and k(moon) = moon, where k is a real number.

15. Prove that a line passing through the origin in R^3 is a vector space under the standard operations on R^3.

16. Complete the unfinished details in Example 5.

17. Complete the unfinished details in Example 8.

18. Prove part (*c*) of Theorem 3.

19. Prove part (*d*) of Theorem 3.

20. Prove that a vector space cannot have more than one zero vector.

21. Prove that a vector has exactly one negative.

4.3 SUBSPACES

If V is a vector space, then certain subsets of V themselves form vector spaces under the vector addition and scalar multiplication defined on V. In this section we shall study such subsets in detail.

Definition. A subset W of a vector space V is called a ***subspace*** of V if W is itself a vector space under the addition and scalar multiplication defined on V.

For example, lines and planes passing through the origin are subspaces of R^3 (Examples 5 and 6).

In general, one must verify the 10 vector space axioms to show that a set W with addition and scalar multiplication forms a vector space. However, if W is part of a larger set V which is already known to be a vector space, then certain axioms need not be verified for W because they are "inherited" from V. For example, there is no need to check that $\mathbf{u} + \mathbf{v} = \mathbf{v} + \mathbf{u}$ (Axiom 2) for W because this holds for all vectors in V and consequently for all vectors in W. Other axioms inherited by W from V are: 3, 7, 8, 9, and 10. Thus, to show that a set W is a subspace of a vector space V, we need only verify Axioms 1, 4, 5, and 6. The following theorem shows that even Axioms 4 and 5 can be dispensed with.

Theorem 4. *If W is a set of one or more vectors from a vector space V, then W is a subspace of V if and only if the following conditions hold.*

(*a*) *If $\mathbf{u}$ and $\mathbf{v}$ are vectors in W, then $\mathbf{u} + \mathbf{v}$ is in W.*
(*b*) *If k is any scalar and $\mathbf{u}$ is any vector in W, then $k\mathbf{u}$ is in W.*

[Conditions (*a*) and (*b*) are often described by saying that W is ***closed under addition*** and ***closed under scalar multiplication***.]

Proof. If W is a subspace of V, then all the vector-space axioms are satisfied; in particular Axioms 1 and 6 hold. But these are precisely conditions (*a*) and (*b*).

Conversely, assume conditions (*a*) and (*b*) hold. Since these conditions are vector-space Axioms 1 and 6, we need only show that W satisfies the remaining eight axioms. Axioms 2, 3, 7, 8, 9, and 10 are automatically satisfied by the vectors in W since they are satisfied by all vectors in V. Therefore, to complete the proof, we need only verify that Axioms 4 and 5 are satisfied by W.

Let $\mathbf{u}$ be any vector in W. By condition (b), $k\mathbf{u}$ is in W for every scalar k. Setting $k = 0$, it follows that $0\mathbf{u} = \mathbf{0}$ is in W, and setting $k = -1$, it follows that $(-1)\mathbf{u} = -\mathbf{u}$ is in W. ∎

Every vector space V has at least two subspaces. V itself is a subspace and the set $\{\mathbf{0}\}$ consisting of just the zero vector in V is a subspace called the ***zero subspace***. The following examples provide less trivial illustrations of subspaces.

Example 11

In Example 5 of Section 4.2 we showed that all vectors in any plane through the origin of R^3 form a vector space; that is, planes through the origin are subspaces of R^3. We can also prove this result geometrically using Theorem 4.

Let W be any plane through the origin and let $\mathbf{u}$ and $\mathbf{v}$ be any vectors in W. Then $\mathbf{u} + \mathbf{v}$ must lie in W because it is the diagonal of the parallelogram determined by $\mathbf{u}$ and $\mathbf{v}$ (Figure 4.4) and $k\mathbf{u}$ must lie in W for any scalar k because $k\mathbf{u}$ lies on a line through $\mathbf{u}$. Thus W is a subspace of R^3.

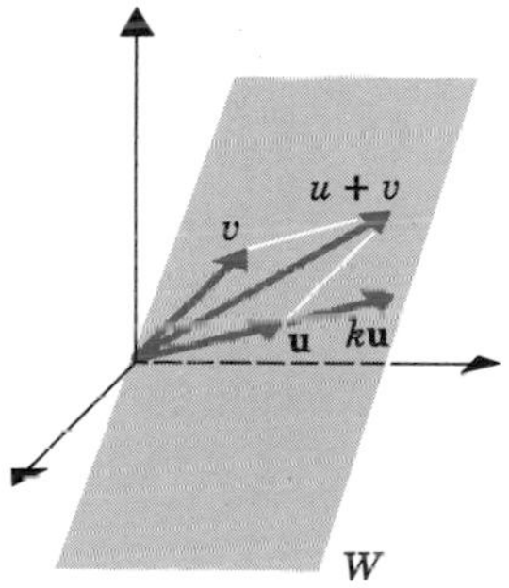

Figure 4.4

REMARK. A geometric argument like the one in this example can be used to show that lines through the origin are subspaces of R^3. It can be shown (Exercise 20, Section 4.5) that the only subspaces of R^3 are: $\{\mathbf{0}\}$, R^3, lines through the origin, and planes through the origin. Also, the only subspaces of R^2 are: $\{\mathbf{0}\}$, R^2, and lines through the origin.

Example 12

Show that the set W of all 2×2 matrices having zeros on the main diagonal is a subspace of the vector space M_{22} of all 2×2 matrices.

Solution. Let

$$A = \begin{bmatrix} 0 & a_{12} \\ a_{21} & 0 \end{bmatrix} \qquad B = \begin{bmatrix} 0 & b_{12} \\ b_{21} & 0 \end{bmatrix}$$

be any two matrices in W and k any scalar. Then

$$kA = \begin{bmatrix} 0 & ka_{12} \\ ka_{21} & 0 \end{bmatrix} \quad \text{and} \quad A + B = \begin{bmatrix} 0 & a_{12} + b_{12} \\ a_{21} + b_{21} & 0 \end{bmatrix}$$

Since kA and $A + B$ have zeros on the main diagonal, they lie in W. Thus W is a subspace of M_{22}.

Example 13

Let n be a positive integer, and let W consist of the zero function and all real polynomial functions having degree $\leq n$; that is, all functions expressible in the form

$$p(x) = a_0 + a_1 x + \cdots + a_n x^n \tag{4.2}$$

where $a_0, \ldots, a_n$ are real numbers. The set W is a subspace of the vector space of all real-valued functions discussed in Example 8. To see this, let **p** and **q** be the polynomials

$$p(x) = a_0 + a_1 x + \cdots + a_n x^n$$

and

$$q(x) = b_0 + b_1 x + \cdots + b_n x^n$$

Then

$$(\mathbf{p} + \mathbf{q})(x) = p(x) + q(x) = (a_0 + b_0) + (a_1 + b_1)x + \cdots + (a_n + b_n)x^n$$

and

$$(k\mathbf{p})(x) = kp(x) = (ka_0) + (ka_1)x + \cdots + (ka_n)x^n$$

have the form given in (4.2). Therefore, $\mathbf{p} + \mathbf{q}$ and $k\mathbf{p}$ lie in W. We shall denote the vector space W in this example by the symbol P_n.

Example 14

(For readers who have studied calculus.)

Recall from calculus that if **f** and **g** are continuous functions and k is a constant, then $\mathbf{f} + \mathbf{g}$ and $k\mathbf{f}$ are also continuous functions. It follows that the set of all continuous functions is a subspace of the vector space of all real-valued functions. This space is denoted by $C(-\infty, +\infty)$. A closely related example is the vector space of all functions that are continuous on a closed interval $a \leq x \leq b$. This space is denoted by $C[a, b]$.

Example 15

Consider a system of m linear equations in n unknowns

$$\begin{array}{ccccccccc}
a_{11}x_1 & + & a_{12}x_2 & + & \cdots & + & a_{1n}x_n & = & b_1 \\
a_{21}x_1 & + & a_{22}x_2 & + & \cdots & + & a_{2n}x_n & = & b_2 \\
\vdots & & \vdots & & & & \vdots & & \vdots \\
a_{m1}x_1 & + & a_{m2}x_2 & + & \cdots & + & a_{mn}x_n & = & b_m
\end{array}$$

or, in matrix notation, $A\mathbf{x} = \mathbf{b}$. A vector*

$$\mathbf{s} = \begin{bmatrix} s_1 \\ s_2 \\ \vdots \\ s_n \end{bmatrix}$$

in R^n is called a ***solution vector*** of the system if $x_1 = s_1, x_2 = s_2, \ldots, x_n = s_n$ is a solution of the system. In this example we shall show that the set of solution vectors of a *homogeneous* system is a subspace of R^n.

Let $A\mathbf{x} = \mathbf{0}$ be a homogeneous system of m linear equations in n unknowns, let W be the set of solution vectors, and let $\mathbf{s}$ and $\mathbf{s}'$ be vectors in W. To show that W is closed under addition and scalar multiplication, we must demonstrate that $\mathbf{s} + \mathbf{s}'$ and $k\mathbf{s}$ are also solution vectors, where k is any scalar. Since $\mathbf{s}$ and $\mathbf{s}'$ are solution vectors we have

$$A\mathbf{s} = \mathbf{0} \qquad \text{and} \qquad A\mathbf{s}' = \mathbf{0}$$

Therefore

$$A(\mathbf{s} + \mathbf{s}') = A\mathbf{s} + A\mathbf{s}' = \mathbf{0} + \mathbf{0} = \mathbf{0}$$

and

$$A(k\mathbf{s}) = k(A\mathbf{s}) = k\mathbf{0} = \mathbf{0}$$

These equations show that $\mathbf{s} + \mathbf{s}'$ and $k\mathbf{s}$ satisfy the equation $A\mathbf{x} = \mathbf{0}$. Thus $\mathbf{s} + \mathbf{s}'$ and $k\mathbf{s}$ are solution vectors.

The subspace W in this example is called the ***solution space*** of the system $A\mathbf{x} = \mathbf{0}$.

In many problems a vector space V is given, and it is of importance to find the "smallest" subspace of V that contains a specified set of vectors $\{\mathbf{v}_1, \mathbf{v}_2, \ldots, \mathbf{v}_r\}$. The following definition provides the key to the construction of such subspaces.

Definition. A vector $\mathbf{w}$ is called a ***linear combination*** of the vectors $\mathbf{v}_1, \mathbf{v}_2, \ldots, \mathbf{v}_r$ if it can be expressed in the form

$$\mathbf{w} = k_1\mathbf{v}_1 + k_2\mathbf{v}_2 + \cdots + k_r\mathbf{v}_r$$

where $k_1, k_2, \ldots, k_r$ are scalars.

Example 16

Consider the vectors $\mathbf{u} = (1, 2, -1)$ and $\mathbf{v} = (6, 4, 2)$ in R^3. Show that $\mathbf{w} = (9, 2, 7)$ is a linear combination of $\mathbf{u}$ and $\mathbf{v}$ and that $\mathbf{w}' = (4, -1, 8)$ is *not* a linear combination of $\mathbf{u}$ and $\mathbf{v}$.

Solution. In order for $\mathbf{w}$ to be a linear combination of $\mathbf{u}$ and $\mathbf{v}$, there must be scalars k_1 and k_2 such that $\mathbf{w} = k_1\mathbf{u} + k_2\mathbf{v}$; that is

$$(9, 2, 7) = k_1(1, 2, -1) + k_2(6, 4, 2)$$

* In this example we are using the matrix notation for vectors in R^n.

or

$$(9, 2, 7) = (k_1 + 6k_2, 2k_1 + 4k_2, -k_1 + 2k_2)$$

Equating corresponding components gives

$$\begin{aligned} k_1 + 6k_2 &= 9 \\ 2k_1 + 4k_2 &= 2 \\ -k_1 + 2k_2 &= 7 \end{aligned}$$

Solving this system yields $k_1 = -3$, $k_2 = 2$ so that

$$\mathbf{w} = -3\mathbf{u} + 2\mathbf{v}$$

Similarly, for $\mathbf{w}'$ to be a linear combination of $\mathbf{u}$ and $\mathbf{v}$, there must be scalars k_1 and k_2 such that $\mathbf{w}' = k_1\mathbf{u} + k_2\mathbf{v}$; that is

$$(4, -1, 8) = k_1(1, 2, -1) + k_2(6, 4, 2)$$

or

$$(4, -1, 8) = (k_1 + 6k_2, 2k_1 + 4k_2, -k_1 + 2k_2)$$

Equating corresponding components gives

$$\begin{aligned} k_1 + 6k_2 &= 4 \\ 2k_1 + 4k_2 &= -1 \\ -k_1 + 2k_2 &= 8 \end{aligned}$$

This system of equations is inconsistent (verify), so that no such scalars exist. Consequently, $\mathbf{w}$ is not a linear combination of $\mathbf{u}$ and $\mathbf{v}$.

Definition. If $\mathbf{v}_1, \mathbf{v}_2, \ldots, \mathbf{v}_r$ are vectors in a vector space V and if every vector in V is expressible as a linear combination of $\mathbf{v}_1, \mathbf{v}_2, \ldots, \mathbf{v}_r$ then we say that these vectors ***span*** V.

Example 17

The vectors $\mathbf{i} = (1, 0, 0)$, $\mathbf{j} = (0, 1, 0)$, and $\mathbf{k} = (0, 0, 1)$ span R^3 because every vector (a, b, c) in R^3 can be written as

$$(a, b, c) = a\mathbf{i} + b\mathbf{j} + c\mathbf{k}$$

which is a linear combination of $\mathbf{i}$, $\mathbf{j}$, and $\mathbf{k}$.

Example 18

The polynomials $1, x, x^2, \ldots, x^n$ span the vector space P_n (see Example 13) since each polynomial $\mathbf{p}$ in P_n can be written as

$$\mathbf{p} = a_0 + a_1x + \cdots + a_nx^n$$

which is a linear combination of $1, x, x^2, \ldots, x^n$.

Example 19

Determine if $\mathbf{v}_1 = (1, 1, 2)$, $\mathbf{v}_2 = (1, 0, 1)$, and $\mathbf{v}_3 = (2, 1, 3)$ span R^3.

Solution. We must determine whether an arbitrary vector $\mathbf{b} = (b_1, b_2, b_3)$ in R^3 can be expressed as a linear combination

$$\mathbf{b} = k_1\mathbf{v}_1 + k_2\mathbf{v}_2 + k_3\mathbf{v}_3$$

of the vectors $\mathbf{v}_1$, $\mathbf{v}_2$, and $\mathbf{v}_3$. Expressing this equation in terms of components gives

$$(b_1, b_2, b_3) = k_1(1, 1, 2) + k_2(1, 0, 1) + k_3(2, 1, 3)$$

or

$$(b_1, b_2, b_3) = (k_1 + k_2 + 2k_3,\quad k_1 + k_3,\quad 2k_1 + k_2 + 3k_3)$$

or

$$\begin{aligned} k_1 + k_2 + 2k_3 &= b_1 \\ k_1 \phantom{{}+k_2} + k_3 &= b_2 \\ 2k_1 + k_2 + 3k_3 &= b_3 \end{aligned}$$

The problem thus reduces to determining whether or not this system is consistent for all values of b_1, b_2, and b_3. By parts (*a*) and (*d*) of Theorem 13 in Section 1.7, this system will be consistent for all b_1, b_2, and b_3 if and only if the matrix of coefficients

$$A = \begin{bmatrix} 1 & 1 & 2 \\ 1 & 0 & 1 \\ 2 & 1 & 3 \end{bmatrix}$$

is invertible. But $\det(A) = 0$ (verify), so that A is not invertible, and consequently $\mathbf{v}_1$, $\mathbf{v}_2$, and $\mathbf{v}_3$ do not span R^3.

In general, a given set of vectors $\{\mathbf{v}_1, \mathbf{v}_2, \ldots, \mathbf{v}_r\}$ in a vector space V may or may not span V. If they span then every vector in V is expressible as linear combination of $\mathbf{v}_1, \mathbf{v}_2, \ldots, \mathbf{v}_r$ and if they do not span then some vectors are so expressible, while others are not. The following theorem shows that if we group together all vectors in V that are expressible as linear combinations of $\mathbf{v}_1, \mathbf{v}_2, \ldots, \mathbf{v}_r$ then we obtain a subspace of V. It is called the ***linear space spanned*** by $\{\mathbf{v}_1, \mathbf{v}_2, \ldots, \mathbf{v}_r\}$, or more simply the ***space spanned*** by $\{\mathbf{v}_1, \mathbf{v}_2, \ldots, \mathbf{v}_r\}$.

Theorem 5. *If $\mathbf{v}_1, \mathbf{v}_2, \ldots, \mathbf{v}_r$ are vectors in a vector space V, then:*

(*a*) *The set W of all linear combinations of $\mathbf{v}_1, \mathbf{v}_2, \ldots, \mathbf{v}_r$ is a subspace of V.*
(*b*) *W is the smallest subspace of V that contains $\mathbf{v}_1, \mathbf{v}_2, \ldots, \mathbf{v}_r$ in the sense that every other subspace of V that contains $\mathbf{v}_1, \mathbf{v}_2, \ldots, \mathbf{v}_r$ must contain W.*

Proof.

(*a*) To show that W is a subspace of V, we must prove it is closed under addition and scalar multiplication. If $\mathbf{u}$ and $\mathbf{v}$ are vectors in W, then

$$\mathbf{u} = c_1\mathbf{v}_1 + c_2\mathbf{v}_2 + \cdots + c_r\mathbf{v}_r$$

and

$$\mathbf{v} = k_1\mathbf{v}_1 + k_2\mathbf{v}_2 + \cdots + k_r\mathbf{v}_r$$

where $c_1, c_2, \ldots, c_r, k_1, k_2, \ldots, k_r$ are scalars. Therefore

$$\mathbf{u} + \mathbf{v} = (c_1 + k_1)\mathbf{v}_1 + (c_2 + k_2)\mathbf{v}_2 + \cdots + (c_r + k_r)\mathbf{v}_r$$

and, for any scalar k

$$k\mathbf{u} = (kc_1)\mathbf{v}_1 + (kc_2)\mathbf{v}_2 + \cdots + (kc_r)\mathbf{v}_r$$

Thus $\mathbf{u} + \mathbf{v}$ and $k\mathbf{u}$ are linear combinations of $\mathbf{v}_1, \mathbf{v}_2, \ldots, \mathbf{v}_r$, and consequently lie in W. Therefore W is closed under addition and scalar multiplication.

(*b*) Each vector $\mathbf{v}_i$ is a linear combination of $\mathbf{v}_1, \mathbf{v}_2, \ldots, \mathbf{v}_r$ since we can write

$$\mathbf{v}_i = 0\mathbf{v}_1 + 0\mathbf{v}_2 + \cdots + 1\mathbf{v}_i + \cdots + 0\mathbf{v}_r$$

The subspace W therefore contains each of the vectors $\mathbf{v}_1, \mathbf{v}_2, \ldots, \mathbf{v}_r$. Let W' be any other subspace that contains $\mathbf{v}_1\ \mathbf{v}_2, \ldots, \mathbf{v}_r$. Since W' is closed under addition and scalar multiplication, it must contain all linear combinations

$$c_1\mathbf{v}_1 + c_2\mathbf{v}_2 + \cdots + c_r\mathbf{v}_r \qquad \text{of} \qquad \mathbf{v}_1, \mathbf{v}_2, \ldots, \mathbf{v}_r;$$

thus W' contains each vector of W. ▪

The linear space W spanned by a set of vectors $S = \{\mathbf{v}_1, \mathbf{v}_2, \ldots, \mathbf{v}_r\}$ will be denoted by

$$\text{lin}(S) \qquad \text{or} \qquad \text{lin}\{\mathbf{v}_1, \mathbf{v}_2, \ldots, \mathbf{v}_r\}$$

Example 20

If $\mathbf{v}_1$ and $\mathbf{v}_2$ are noncollinear vectors in R^3 with initial points at the origin, then $\text{lin}\{\mathbf{v}_1, \mathbf{v}_2\}$, which consists of all linear combinations $k_1\mathbf{v}_1 + k_2\mathbf{v}_2$, is the plane determined by $\mathbf{v}_1$ and $\mathbf{v}_2$ (Figure 4.5*a*).

Similarly, if $\mathbf{v}$ is a nonzero vector in R^2 or R^3, then $\text{lin}\{\mathbf{v}\}$, which is the set of all scalar multiples $k\mathbf{v}$, is the line determined by $\mathbf{v}$ (Figure 4.5*b*).

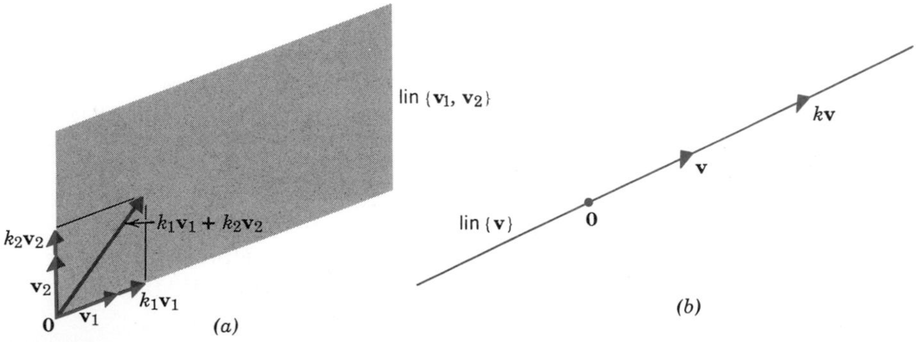

Figure 4.5

EXERCISE SET 4.3

1. Use Theorem 4 to determine which of the following are subspaces of R^3.
(a) all vectors of the form $(a, 0, 0)$
(b) all vectors of the form $(a, 1, 1)$
(c) all vectors of the form (a, b, c), where $b = a + c$
(d) all vectors of the form (a, b, c), where $b = a + c + 1$

2. Use Theorem 4 to determine which of the following are subspaces of M_{22}.
(a) all matrices of the form

$$\begin{bmatrix} a & b \\ c & d \end{bmatrix}$$

where a, b, c, and d are integers
(b) all matrices of the form

$$\begin{bmatrix} a & b \\ c & d \end{bmatrix}$$

where $a + d = 0$
(c) all 2×2 matrices A such that $A = A^t$
(d) all 2×2 matrices A such that $\det(A) = 0$

3. Use Theorem 4 to determine which of the following are subspaces of P_3.
(a) all polynomials $a_0 + a_1x + a_2x^2 + a_3x^3$ for which $a_0 = 0$
(b) all polynomials $a_0 + a_1x + a_2x^2 + a_3x^3$ for which $a_0 + a_1 + a_2 + a_3 = 0$
(c) all polynomials $a_0 + a_1x + a_2x^2 + a_3x^3$ for which a_0, a_1, a_2, and a_3 are integers
(d) all polynomials of the form $a_0 + a_1x$, where a_0 and a_1 are real numbers.

4. Use Theorem 4 to determine which of the following are subspaces of the space of all real-valued functions f defined on the entire real line.
(a) all f such that $f(x) \leq 0$ for all x
(b) all f such that $f(0) = 0$
(c) all f such that $f(0) = 2$
(d) all constant functions
(e) all f of the form $k_1 + k_2 \sin x$, where k_1 and k_2 are real numbers

5. Which of the following are linear combinations of $\mathbf{u} = (1, -1, 3)$ and $\mathbf{v} = (2, 4, 0)$?
(a) $(3, 3, 3)$ (b) $(4, 2, 6)$ (c) $(1, 5, 6)$ (d) $(0, 0, 0)$

6. Express the following as linear combinations of $\mathbf{u} = (2, 1, 4)$, $\mathbf{v} = (1, -1, 3)$, and $\mathbf{w} = (3, 2, 5)$.
(a) $(5, 9, 5)$ (b) $(2, 0, 6)$ (c) $(0, 0, 0)$ (d) $(2, 2, 3)$

7. Express the following as linear combinations of $\mathbf{p}_1 = 2 + x + 4x^2$, $\mathbf{p}_2 = 1 - x + 3x^2$, and $\mathbf{p}_3 = 3 + 2x + 5x^2$?
(a) $5 + 9x + 5x^2$ (b) $2 + 6x^2$ (c) 0 (d) $2 + 2x + 3x^2$

8. Which of the following are linear combinations of

$$A = \begin{bmatrix} 1 & 2 \\ -1 & 3 \end{bmatrix}, B = \begin{bmatrix} 0 & 1 \\ 2 & 4 \end{bmatrix}, \text{ and } C = \begin{bmatrix} 4 & -2 \\ 0 & -2 \end{bmatrix}?$$

(a) $\begin{bmatrix} 6 & 3 \\ 0 & 8 \end{bmatrix}$ (b) $\begin{bmatrix} -1 & 7 \\ 5 & 1 \end{bmatrix}$ (c) $\begin{bmatrix} 0 & 0 \\ 0 & 0 \end{bmatrix}$ (d) $\begin{bmatrix} 6 & -1 \\ -8 & -8 \end{bmatrix}$

9. In each part determine whether the given vectors span R^3.
(a) $\mathbf{v}_1 = (1, 1, 1)$, $\mathbf{v}_2 = (2, 2, 0)$, $\mathbf{v}_3 = (3, 0, 0)$
(b) $\mathbf{v}_1 = (2, -1, 3)$, $\mathbf{v}_2 = (4, 1, 2)$, $\mathbf{v}_3 = (8, -1, 8)$
(c) $\mathbf{v}_1 = (3, 1, 4)$, $\mathbf{v}_2 = (2, -3, 5)$, $\mathbf{v}_3 = (5, -2, 9)$, $\mathbf{v}_4 = (1, 4, -1)$
(d) $\mathbf{v}_1 = (1, 3, 3)$, $\mathbf{v}_2 = (1, 3, 4)$, $\mathbf{v}_3 = (1, 4, 3)$, $\mathbf{v}_4 = (6, 2, 1)$

10. Determine which of the following lie in the space spanned by

$$\mathbf{f} = \cos^2 x \qquad \text{and} \qquad \mathbf{g} = \sin^2 x$$

(a) $\cos 2x$ (b) $3 + x^2$ (c) 1 (d) $\sin x$

11. Determine if the following polynomials span P_3

$$\mathbf{p}_1 = 1 + 2x - x^2 \qquad \mathbf{p}_2 = 3 + x^2$$
$$\mathbf{p}_3 = 5 + 4x - x^2 \qquad \mathbf{p}_4 = -2 + 2x - 2x^2$$

12. Let $\mathbf{v}_1 = (2, 1, 0, 3)$, $\mathbf{v}_2 = (3, -1, 5, 2)$, and $\mathbf{v}_3 = (-1, 0, 2, 1)$. Which of the following vectors are in $\text{lin}\{\mathbf{v}_1, \mathbf{v}_2, \mathbf{v}_3\}$?
(a) $(2, 3, -7, 3)$ (b) $(0, 0, 0, 0)$ (c) $(1, 1, 1, 1)$ (d) $(-4, 6, -13, 4)$

13. Find an equation for the plane spanned by the vectors $\mathbf{u} = (1, 1, -1)$ and $\mathbf{v} = (2, 3, 5)$.

14. Find parametric equations for the line spanned by the vector $\mathbf{u} = (2, 7, -1)$.

15. Show that the solution vectors of a consistent nonhomogeneous system of m linear equations in n unknowns do not form a subspace of R^n.

16. **(For readers who have studied calculus.)** Show that the following sets of functions are subspaces of the vector space in Example 8.
(a) all everywhere continuous functions
(b) all everywhere differentiable functions
(c) all everywhere differentiable functions that satisfy $\mathbf{f}' + 2\mathbf{f} = \mathbf{0}$

4.4 LINEAR INDEPENDENCE

From Section 4.3, a vector space V is spanned by a set of vectors $S = \{\mathbf{v}_1, \mathbf{v}_2, \ldots, \mathbf{v}_r\}$ if each vector in V is a linear combination of $\mathbf{v}_1, \mathbf{v}_2, \ldots, \mathbf{v}_r$. Spanning sets are useful in a variety of problems since it is often possible to study a vector space V by first studying the vectors in a spanning set S, and then extending the results to the rest of V. Therefore, it is desirable to keep the spanning set S as small as possible. The problem of finding smallest spanning sets for a vector space depends on the notion of linear independence, which we shall study in this section.

If $S = \{\mathbf{v}_1, \mathbf{v}_2, \ldots, \mathbf{v}_r\}$ is a set of vectors, then the vector equation

$$k_1\mathbf{v}_1 + k_2\mathbf{v}_2 + \cdots + k_r\mathbf{v}_r = \mathbf{0}$$

has at least one solution, namely

$$k_1 = 0, \qquad k_2 = 0, \ldots, k_r = 0$$

If this is the only solution, then S is called a ***linearly independent*** set. If there are other solutions, then S is called a ***linearly dependent*** set.

Example 21

The set of vectors $S = \{\mathbf{v}_1, \mathbf{v}_2, \mathbf{v}_3\}$, where $\mathbf{v}_1 = (2, -1, 0, 3)$, $\mathbf{v}_2 = (1, 2, 5, -1)$, $\mathbf{v}_3 = (7, -1, 5, 8)$ is linearly dependent, since $3\mathbf{v}_1 + \mathbf{v}_2 - \mathbf{v}_3 = \mathbf{0}$.

Example 22

The polynomials $\mathbf{p}_1 = 1 - x$, $\mathbf{p}_2 = 5 + 3x - 2x^2$, and $\mathbf{p}_3 = 1 + 3x - x^2$ form a linearly dependent set in P_2 since $3\mathbf{p}_1 - \mathbf{p}_2 + 2\mathbf{p}_3 = \mathbf{0}$.

Example 23

Consider the vectors $\mathbf{i} = (1, 0, 0)$, $\mathbf{j} = (0, 1, 0)$, and $\mathbf{k} = (0, 0, 1)$ in R^3. In terms of components the vector equation

$$k_1\mathbf{i} + k_2\mathbf{j} + k_3\mathbf{k} = \mathbf{0}$$

becomes

$$k_1(1, 0, 0) + k_2(0, 1, 0) + k_3(0, 0, 1) = (0, 0, 0)$$

or equivalently

$$(k_1, k_2, k_3) = (0, 0, 0)$$

Thus $k_1 = 0$, $k_2 = 0$, $k_3 = 0$; the set $S = \{\mathbf{i}, \mathbf{j}, \mathbf{k}\}$ is therefore linearly independent. A similar argument can be used to show that the vectors $\mathbf{e}_1 = (1, 0, 0, \ldots, 0)$, $\mathbf{e}_2 = (0, 1, 0, \ldots, 0), \ldots, \mathbf{e}_n = (0, 0, 0, \ldots, 1)$ form a linearly independent set in R^n.

Example 24

Determine whether the vectors

$$\mathbf{v}_1 = (1, -2, 3) \qquad \mathbf{v}_2 = (5, 6, -1) \qquad \mathbf{v}_3 = (3, 2, 1)$$

form a linearly dependent or linearly independent set.

Solution. In terms of components the vector equation

$$k_1\mathbf{v}_1 + k_2\mathbf{v}_2 + k_3\mathbf{v}_3 = \mathbf{0}$$

becomes

$$k_1(1, -2, 3) + k_2(5, 6, -1) + k_3(3, 2, 1) = (0, 0, 0)$$

or equivalently

$$(k_1 + 5k_2 + 3k_3, -2k_1 + 6k_2 + 2k_3, 3k_1 - k_2 + k_3) = (0, 0, 0)$$

Equating corresponding components gives

$$\begin{aligned} k_1 + 5k_2 + 3k_3 &= 0 \\ -2k_1 + 6k_2 + 2k_3 &= 0 \\ 3k_1 - k_2 + k_3 &= 0 \end{aligned}$$

Thus $\mathbf{v}_1$, $\mathbf{v}_2$, and $\mathbf{v}_3$ form a linearly dependent set if this system has a nontrivial solution, or a linearly independent set if it has only the trivial solution. Solving this system yields

$$k_1 = -\tfrac{1}{2}t \qquad k_2 = -\tfrac{1}{2}t \qquad k_3 = t$$

Thus the system has nontrivial solutions and $\mathbf{v}_1$, $\mathbf{v}_2$, and $\mathbf{v}_3$ form a linearly dependent set. Alternatively, we could have shown the existence of nontrivial solutions without solving the system by showing that the coefficient matrix has determinant zero and consequently is not invertible (verify).

The term "linearly dependent" suggests that the vectors "depend" on each other in some way. To see that this is indeed the case, assume $S = \{\mathbf{v}_1, \mathbf{v}_2, \ldots, \mathbf{v}_r\}$ is a linearly dependent set, so that the vector equation

$$k_1\mathbf{v}_1 + k_2\mathbf{v}_2 + \cdots + k_r\mathbf{v}_r = \mathbf{0}$$

has a solution other than $k_1 = k_2 = \cdots = k_r = 0$. To be specific, assume $k_1 \neq 0$. Multiplying both sides by $1/k_1$ and solving for $\mathbf{v}_1$ yields

$$\mathbf{v}_1 = \left(-\frac{k_2}{k_1}\right)\mathbf{v}_2 + \cdots + \left(-\frac{k_r}{k_1}\right)\mathbf{v}_r$$

Thus $\mathbf{v}_1$ has been expressed as a linear combination of the remaining vectors $\mathbf{v}_2, \ldots, \mathbf{v}_r$. We leave it as an exercise to show that *a set with two or more vectors is linearly dependent if and only if at least one of the vectors is a linear combination of the remaining vectors.*

The next two examples give a geometric interpretation of linear dependence in R^2 and R^3.

Example 25

Two vectors, $\mathbf{v}_1$ and $\mathbf{v}_2$, form a linearly dependent set if and only if one of the vectors is a scalar multiple of the other. To see why, assume $S = \{\mathbf{v}_1, \mathbf{v}_2\}$ is linearly dependent. Since the vector equation $k_1\mathbf{v}_1 + k_2\mathbf{v}_2 = \mathbf{0}$ has a solution other than $k_1 = k_2 = 0$, this equation can be rewritten as

$$\mathbf{v}_1 = \left(-\frac{k_2}{k_1}\right)\mathbf{v}_2 \qquad \text{or} \qquad \mathbf{v}_2 = \left(-\frac{k_1}{k_2}\right)\mathbf{v}_1$$

But this tells us that $\mathbf{v}_1$ is a scalar multiple of $\mathbf{v}_2$ or $\mathbf{v}_2$ is a scalar multiple of $\mathbf{v}_1$. The converse is left as an exercise.

It follows that two vectors in R^2 or R^3 are linearly dependent if and only if they lie on the same line through the origin (Figure 4.6).

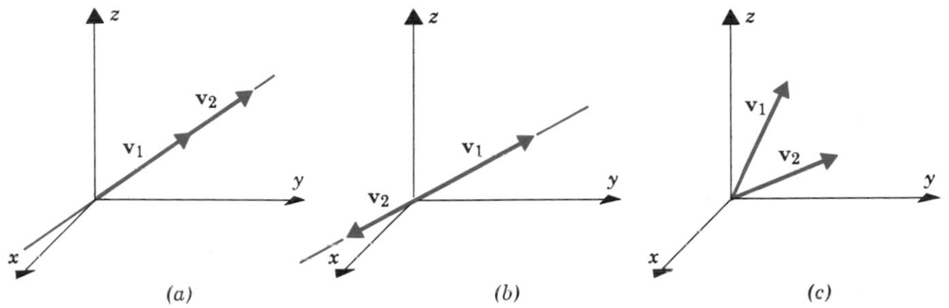

Figure 4.6 (*a*) Linearly dependent. (*b*) Linearly dependent. (*c*) Linearly independent.

Example 26

If $\mathbf{v}_1$, $\mathbf{v}_2$, and $\mathbf{v}_3$ are three vectors in R^3, then the set $S = \{\mathbf{v}_1, \mathbf{v}_2, \mathbf{v}_3\}$ is linearly dependent if and only if the three vectors lie in the same plane through the origin when the vectors are placed with initial points at the origin (Figure 4.7). To see this, recall that $\mathbf{v}_1$, $\mathbf{v}_2$, and $\mathbf{v}_3$ are linearly dependent if and only if at least one of the vectors is a linear combination of the remaining two, or equivalently if and only if at least one of the vectors is in the space spanned by the remaining two. But the space spanned by any two vectors in R^3 is either a line through the origin, a plane through the origin, or the origin itself (Exercise 17.). In any case the space spanned by two vectors in R^3 always lies in a plane through the origin.

We conclude this section with a theorem that shows that a linearly independent set in R^n can contain at most n vectors.

Figure 4.7 (*a*) Linearly dependent. (*b*) Linearly dependent. (*c*) Linearly independent.

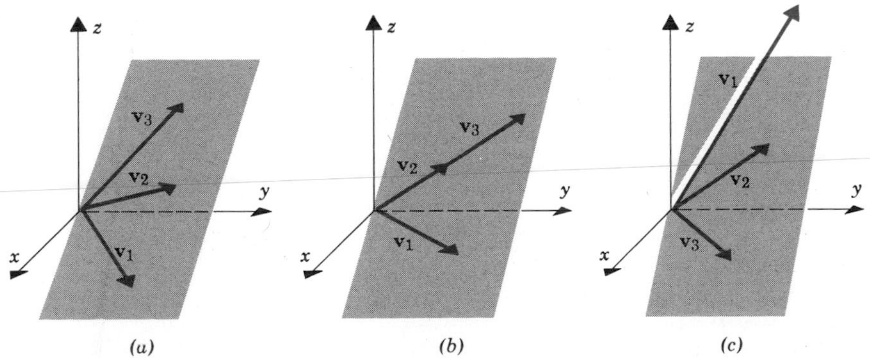

Theorem 6. *Let $S = \{\mathbf{v}_1, \mathbf{v}_2, \ldots, \mathbf{v}_r\}$ be a set of vectors in R^n. If $r > n$, then S is linearly dependent.*

Proof. Suppose

$$\begin{aligned} \mathbf{v}_1 &= (v_{11}, v_{12}, \ldots, v_{1n}) \\ \mathbf{v}_2 &= (v_{21}, v_{22}, \ldots, v_{2n}) \\ &\vdots \\ \mathbf{v}_r &= (v_{r1}, v_{r2}, \ldots, v_{rn}) \end{aligned}$$

Consider the equation

$$k_1\mathbf{v}_1 + k_2\mathbf{v}_2 + \cdots + k_r\mathbf{v}_r = \mathbf{0}$$

If, as illustrated in Example 24, we express both sides of this equation in terms of components and then equate corresponding components, we obtain the system

$$\begin{aligned} v_{11}k_1 + v_{21}k_2 + \cdots + v_{r1}k_r &= 0 \\ v_{12}k_1 + v_{22}k_2 + \cdots + v_{r2}k_r &= 0 \\ \vdots \qquad \vdots \qquad\qquad \vdots \quad &\quad \vdots \\ v_{1n}k_1 + v_{2n}k_2 + \cdots + v_{rn}k_r &= 0 \end{aligned}$$

This is a homogeneous system of n equations in the r unknowns $k_1, \ldots, k_r$. Since $r > n$, it follows from Theorem 1 in Section 1.3 that the system has nontrivial solutions. Therefore, $S = \{\mathbf{v}_1, \mathbf{v}_2, \ldots, \mathbf{v}_r\}$ is a linearly dependent set. ∎

In particular, this theorem tells us that a set in R^2 with more than two vectors is linearly dependent, and a set in R^3 with more than three vectors is linearly dependent.

EXERCISE SET 4.4

1. Explain why the following are linearly dependent sets of vectors. (Solve this problem by inspection.)
   (a) $\mathbf{u}_1 = (1, 2)$ and $\mathbf{u}_2 = (-3, -6)$ in R^2
   (b) $\mathbf{u}_1 = (2, 3)$, $\mathbf{u}_2 = (-5, 8)$, $\mathbf{u}_3 = (6, 1)$ in R^2
   (c) $\mathbf{p}_1 = 2 + 3x - x^2$ and $\mathbf{p}_2 = 6 + 9x - 3x^2$ in P_2
   (d) $A = \begin{bmatrix} 1 & 3 \\ 2 & 0 \end{bmatrix}$ and $B = \begin{bmatrix} -1 & -3 \\ -2 & 0 \end{bmatrix}$ in M_{22}

2. Which of the following sets of vectors in R^3 are linearly dependent?
   (a) $(2, -1, 4)$, $(3, 6, 2)$, $(2, 10, -4)$
   (b) $(3, 1, 1)$, $(2, -1, 5)$, $(4, 0, -3)$
   (c) $(6, 0, -1)$, $(1, 1, 4)$
   (d) $(1, 3, 3)$, $(0, 1, 4)$, $(5, 6, 3)$, $(7, 2, -1)$

3. Which of the following sets of vectors in R^4 are linearly dependent?
(a) $(1, 2, 1, -2)$, $(0, -2, -2, 0)$, $(0, 2, 3, 1)$, $(3, 0, -3, 6)$
(b) $(4, -4, 8, 0)$, $(2, 2, 4, 0)$, $(6, 0, 0, 2)$, $(6, 3, -3, 0)$
(c) $(4, 4, 0, 0)$, $(0, 0, 6, 6)$, $(-5, 0, 5, 5)$
(d) $(3, 0, 4, 1)$, $(6, 2, -1, 2)$, $(-1, 3, 5, 1)$, $(-3, 7, 8, 3)$

4. Which of the following sets of vectors in P_2 are linearly dependent?
(a) $2 - x + 4x^2$, $3 + 6x + 2x^2$, $2 + 10x - 4x^2$
(b) $3 + x + x^2$, $2 - x + 5x^2$, $4 - 3x^2$
(c) $6 - x^2$, $1 + x + 4x^2$
(d) $1 + 3x + 3x^2$, $x + 4x^2$, $5 + 6x + 3x^2$, $7 + 2x - x^2$

5. Let V be the vector space of all real-valued functions defined on the entire real line. Which of the following sets of vectors in V are linearly dependent?
(a) $2, 4 \sin^2 x, \cos^2 x$
(b) $x, \cos x$
(c) $1, \sin x, \sin 2x$
(d) $\cos 2x, \sin^2 x, \cos^2 x$
(e) $(1 + x)^2, x^2 + 2x, 3$
(f) $0, x, x^2$

6. Assume that $\mathbf{v}_1$, $\mathbf{v}_2$, and $\mathbf{v}_3$ are vectors in R^3 that have their initial points at the origin. In each part determine whether the three vectors lie in a plane.
(a) $\mathbf{v}_1 = (1, 0, -2)$, $\mathbf{v}_2 = (3, 1, 2)$, $\mathbf{v}_3 = (1, -1, 0)$
(b) $\mathbf{v}_1 = (2, -1, 4)$, $\mathbf{v}_2 = (4, 2, 3)$, $\mathbf{v}_3 = (2, 7, -6)$

7. Assume that $\mathbf{v}_1$, $\mathbf{v}_2$, and $\mathbf{v}_3$ are vectors in R^3 that have their initial points at the origin. In each part determine whether the three vectors lie on the same line.
(a) $\mathbf{v}_1 = (3, -6, 9)$, $\mathbf{v}_2 = (2, -4, 6)$, $\mathbf{v}_3 = (1, 1, 1)$
(b) $\mathbf{v}_1 = (2, -1, 4)$, $\mathbf{v}_2 = (4, 2, 3)$, $\mathbf{v}_3 = (2, 7, -6)$
(c) $\mathbf{v}_1 = (4, 6, 8)$, $\mathbf{v}_2 = (2, 3, 4)$, $\mathbf{v}_3 = (-2, -3, -4)$

8. For which real values of λ do the following vectors form a linearly dependent set in R^3?

$$\mathbf{v}_1 = (\lambda, -\tfrac{1}{2}, -\tfrac{1}{2}), \mathbf{v}_2 = (-\tfrac{1}{2}, \lambda, -\tfrac{1}{2}), \mathbf{v}_3 = (-\tfrac{1}{2}, -\tfrac{1}{2}, \lambda)$$

9. Let $S = \{\mathbf{v}_1, \mathbf{v}_2, \ldots, \mathbf{v}_n\}$ be a set of vectors in a vector space V. Show that if one of the vectors is zero, then S is linearly dependent.

10. If $\{\mathbf{v}_1, \mathbf{v}_2, \mathbf{v}_3\}$ is a linearly independent set of vectors, show that $\{\mathbf{v}_1, \mathbf{v}_2\}$, $\{\mathbf{v}_1, \mathbf{v}_3\}$, $\{\mathbf{v}_2, \mathbf{v}_3\}$, $\{\mathbf{v}_1\}$, $\{\mathbf{v}_2\}$, and $\{\mathbf{v}_3\}$ are also linearly independent sets.

11. If $S = \{\mathbf{v}_1, \mathbf{v}_2, \ldots, \mathbf{v}_n\}$ is a linearly independent set of vectors, show that every subset of S with one or more vectors is also linearly independent.

12. If $\{\mathbf{v}_1, \mathbf{v}_2, \mathbf{v}_3\}$ is a linearly dependent set of vectors in a vector space V, show that $\{\mathbf{v}_1, \mathbf{v}_2, \mathbf{v}_3, \mathbf{v}_4\}$ is also linearly dependent, where $\mathbf{v}_4$ is any other vector in V.

13. If $\{\mathbf{v}_1, \mathbf{v}_2, \ldots, \mathbf{v}_r\}$ is a linearly dependent set of vectors in a vector space V, show that $\{\mathbf{v}_1, \mathbf{v}_2, \ldots, \mathbf{v}_{r+1}, \ldots, \mathbf{v}_n\}$ is also linearly dependent, where $\mathbf{v}_{r+1}, \ldots, \mathbf{v}_n$ are any other vectors in V.

14. Show that every set with more than three vectors from P_2 is linearly dependent.

15. Show that if $\{\mathbf{v}_1, \mathbf{v}_2\}$ is linearly independent and $\mathbf{v}_3$ does not lie in $\text{lin}\{\mathbf{v}_1, \mathbf{v}_2\}$, then $\{\mathbf{v}_1, \mathbf{v}_2, \mathbf{v}_3\}$ is linearly independent.

16. Prove that a set with two or more vectors is linearly dependent if and only if at least one of the vectors is expressible as a linear combination of the remaining vectors.

17. Prove: The space spanned by two vectors in R^3 is either a line through the origin, a plane through the origin, or the origin itself.

18. **(For readers who have studied calculus.)** Let V be the vector space of real-valued functions defined on the entire real line. If **f**, **g**, and **h** are vectors in V that are twice differentiable, then the function $\mathbf{w} = w(x)$ defined by

$$w(x) = \begin{vmatrix} f(x) & g(x) & h(x) \\ f'(x) & g'(x) & h'(x) \\ f''(x) & g''(x) & h''(x) \end{vmatrix}$$

is called the ***Wronskian*** of **f**, **g**, and **h**. Prove that **f**, **g**, and **h** form a linearly independent set if the Wronskian is not the zero vector in V [i.e., $w(x)$ is not identically zero].

19. **(For readers who have studied calculus.)** Use the Wronskian (Exercise 18) to show that the following sets of vectors are linearly independent.
(a) $1, x, e^x$ (b) $\sin x, \cos x, x \sin x$ (c) e^x, xe^x, x^2e^x (d) $1, x, x^2$

4.5 BASIS AND DIMENSION

We usually think of a line as being one dimensional, a plane as two dimensional, and space around us as three dimensional. It is the primary purpose of this section to make this intuitive notion of dimension precise.

Definition. If V is any vector space and $S = \{\mathbf{v}_1, \mathbf{v}_2, \ldots, \mathbf{v}_r\}$ is a finite set of vectors in V, then S is called a ***basis*** for V if

(i) S is linearly independent;
(ii) S spans V

Example 27

Let $\mathbf{e}_1 = (1, 0, 0, \ldots, 0), \mathbf{e}_2 = (0, 1, 0, \ldots, 0), \ldots, \mathbf{e}_n = (0, 0, 0, \ldots, 1)$. In Example 23 we pointed out that $S = \{\mathbf{e}_1, \mathbf{e}_2, \ldots, \mathbf{e}_n\}$ is a linearly independent set in R^n. Since any vector $\mathbf{v} = (v_1, v_2, \ldots, v_n)$ in R^n can be written as $\mathbf{v} = v_1\mathbf{e}_1 + v_2\mathbf{e}_2 + \cdots + v_n\mathbf{e}_n$, S spans R^n and is therefore a basis. It is called the ***standard basis for R^n***.

Example 28

Let $\mathbf{v}_1 = (1, 2, 1), \mathbf{v}_2 = (2, 9, 0)$, and $\mathbf{v}_3 = (3, 3, 4)$. Show that the set $S = \{\mathbf{v}_1, \mathbf{v}_2, \mathbf{v}_3\}$ is a basis for R^3.

Solution. To show that S spans R^3, we must show that an arbitrary vector $\mathbf{b} = (b_1, b_2, b_3)$ can be expressed as a linear combination

$$\mathbf{b} = k_1\mathbf{v}_1 + k_2\mathbf{v}_2 + k_3\mathbf{v}_3 \tag{4.3}$$

of the vectors in S. Expressing (4.3) in terms of components gives

$$(b_1, b_2, b_3) = k_1(1, 2, 1) + k_2(2, 9, 0) + k_3(3, 3, 4)$$

or

$$(b_1, b_2, b_3) = (k_1 + 2k_2 + 3k_3, 2k_1 + 9k_2 + 3k_3, k_1 + 4k_3)$$

or

$$\begin{aligned} k_1 + 2k_2 + 3k_3 &= b_1 \\ 2k_1 + 9k_2 + 3k_3 &= b_2 \\ k_1 \qquad\quad + 4k_3 &= b_3 \end{aligned} \tag{4.4}$$

Thus, to show S spans V, we must demonstrate that system (4.4) has a solution for all choices of $\mathbf{b} = (b_1, b_2, b_3)$. To prove that S is linearly independent, we must show that the only solution of

$$k_1\mathbf{v}_1 + k_2\mathbf{v}_2 + k_3\mathbf{v}_3 = \mathbf{0} \tag{4.5}$$

is $k_1 = k_2 = k_3 = 0$.

As above, if (4.5) is expressed in terms of components, the verification of independence reduces to showing the homogeneous system

$$\begin{aligned} k_1 + 2k_2 + 3k_3 &= 0 \\ 2k_1 + 9k_2 + 3k_3 &= 0 \\ k_1 \qquad\quad + 4k_3 &= 0 \end{aligned} \tag{4.6}$$

has only the trivial solution. Observe that systems (4.4) and (4.6) have the same coefficient matrix. Thus by parts (*a*), (*b*), and (*d*) of Theorem 13 in Section 1.7, we can simultaneously prove S is linearly independent and spans R^3 by demonstrating that the matrix of coefficients

$$A = \begin{bmatrix} 1 & 2 & 3 \\ 2 & 9 & 3 \\ 1 & 0 & 4 \end{bmatrix}$$

in systems (4.4) and (4.6) is invertible. Since

$$\det(A) = \begin{vmatrix} 1 & 2 & 3 \\ 2 & 9 & 3 \\ 1 & 0 & 4 \end{vmatrix} = -1$$

it follows from Theorem 6 in Section 2.3 that A is invertible. Thus S is a basis for R^3.

Example 29

The set $S = \{1, x, x^2, \ldots, x^n\}$ is a basis for the vector space P_n introduced in Example 13. From Example 18, the vectors in S span P_n. To see that S is linearly independent, assume that some linear combination of vectors in S is the zero vector, that is

$$c_0 + c_1 x + \cdots + c_n x^n \equiv 0 \tag{4.7}$$

We must show that $c_0 = c_1 = \cdots = c_n = 0$. From algebra a nonzero polynomial of degree n has at most n distinct roots. Since (4.7) is an identity, every value of x is a root of the left-hand side. This implies that $c_1 = c_2 = \cdots = c_n = 0$; for otherwise, $c_0 + c_1 x + \cdots + c_n x^n$ could have at most n roots. The set S is therefore linearly independent.

The basis S in this example is called the ***standard basis for P_n***.

Example 30

Let

$$M_1 = \begin{bmatrix} 1 & 0 \\ 0 & 0 \end{bmatrix} \quad M_2 = \begin{bmatrix} 0 & 1 \\ 0 & 0 \end{bmatrix} \quad M_3 = \begin{bmatrix} 0 & 0 \\ 1 & 0 \end{bmatrix} \quad M_4 = \begin{bmatrix} 0 & 0 \\ 0 & 1 \end{bmatrix}$$

The set $S = \{M_1, M_2, M_3, M_4\}$ is a basis for the vector space M_{22} of 2×2 matrices. To see that S spans M_{22}, note that a typical vector (matrix)

$$\begin{bmatrix} a & b \\ c & d \end{bmatrix}$$

can be written as

$$\begin{aligned} \begin{bmatrix} a & b \\ c & d \end{bmatrix} &= a\begin{bmatrix} 1 & 0 \\ 0 & 0 \end{bmatrix} + b\begin{bmatrix} 0 & 1 \\ 0 & 0 \end{bmatrix} + c\begin{bmatrix} 0 & 0 \\ 1 & 0 \end{bmatrix} + d\begin{bmatrix} 0 & 0 \\ 0 & 1 \end{bmatrix} \\ &= aM_1 + bM_2 + cM_3 + dM_4 \end{aligned}$$

To see that S is linearly independent, assume that

$$aM_1 + bM_2 + cM_3 + dM_4 = 0$$

that is

$$a\begin{bmatrix} 1 & 0 \\ 0 & 0 \end{bmatrix} + b\begin{bmatrix} 0 & 1 \\ 0 & 0 \end{bmatrix} + c\begin{bmatrix} 0 & 0 \\ 1 & 0 \end{bmatrix} + d\begin{bmatrix} 0 & 0 \\ 0 & 1 \end{bmatrix} = \begin{bmatrix} 0 & 0 \\ 0 & 0 \end{bmatrix}$$

Then

$$\begin{bmatrix} a & b \\ c & d \end{bmatrix} = \begin{bmatrix} 0 & 0 \\ 0 & 0 \end{bmatrix}$$

Thus $a = b = c = d = 0$ so that S is linearly independent.

Example 31

If $S = \{\mathbf{v}_1, \mathbf{v}_2, \ldots, \mathbf{v}_r\}$ is a *linearly independent* set in a vector space V, then S is a basis for the subspace $\text{lin}(S)$ since S is independent and by definition of $\text{lin}(S)$, S spans $\text{lin}(S)$.

A non-zero vector space V is called ***finite dimensional*** if it contains a finite set of vectors $\{\mathbf{v}_1, \mathbf{v}_2, \ldots, \mathbf{v}_n\}$ which forms a basis. If no such set exists, V is called

infinite dimensional. In addition, we shall regard the zero vector space as finite dimensional even though it has no linearly independent sets, and consequently no basis.)

Example 32

By examples 27, 29, and 30, R^n, P_n and M_{22} are finite dimensional vector spaces.

The next theorem provides the key to the concept of dimension for vector spaces. From it we shall obtain one of the most important results in linear algebra.

Theorem 7. *If $S = \{\mathbf{v}_1, \mathbf{v}_2, \ldots, \mathbf{v}_n\}$ is a basis for a vector space V, then every set with more than n vectors is linearly dependent.*

Proof. Let $S' = \{\mathbf{w}_1, \mathbf{w}_2, \ldots, \mathbf{w}_m\}$ be any set of m vectors in V, where $m > n$. We wish to show that S' is linearly dependent. Since $S = \{\mathbf{v}_1, \mathbf{v}_2, \ldots, \mathbf{v}_n\}$ is a basis, each $\mathbf{w}_i$ can be expressed as a linear combination of the vectors in S, say,

$$\begin{aligned} \mathbf{w}_1 &= a_{11}\mathbf{v}_1 + a_{21}\mathbf{v}_2 + \cdots + a_{n1}\mathbf{v}_n \\ \mathbf{w}_2 &= a_{12}\mathbf{v}_1 + a_{22}\mathbf{v}_2 + \cdots + a_{n2}\mathbf{v}_n \\ &\vdots \\ \mathbf{w}_m &= a_{1m}\mathbf{v}_1 + a_{2m}\mathbf{v}_2 + \cdots + a_{nm}\mathbf{v}_n \end{aligned} \tag{4.8}$$

To show that S' is linearly dependent, we must find scalars $k_1, k_2, \ldots, k_m$, not all zero, such that

$$k_1\mathbf{w}_1 + k_2\mathbf{w}_2 + \cdots + k_m\mathbf{w}_m = \mathbf{0} \tag{4.9}$$

Using the equations in (4.8), we can rewrite (4.9) as

$$\begin{aligned} &(k_1a_{11} + k_2a_{12} + \cdots + k_ma_{1m})\mathbf{v}_1 \\ &\quad + (k_1a_{21} + k_2a_{22} + \cdots + k_ma_{2m})\mathbf{v}_2 \\ &\quad \vdots \\ &\quad + (k_1a_{n1} + k_2a_{n2} + \cdots + k_ma_{nm})\mathbf{v}_n = \mathbf{0} \end{aligned}$$

The problem of proving that S' is a linearly dependent set thus reduces to showing there are $k_1, k_2, \ldots, k_m$, not all zero, that satisfy

$$\begin{aligned} a_{11}k_1 + a_{12}k_2 + \cdots + a_{1m}k_m &= 0 \\ a_{21}k_1 + a_{22}k_2 + \cdots + a_{2m}k_m &= 0 \\ &\vdots \\ a_{n1}k_1 + a_{n2}k_2 + \cdots + a_{nm}k_m &= 0 \end{aligned} \tag{4.10}$$

Since (4.10) has more unknowns than equations, the proof is complete, since Theorem 1 in Section 1.3 guarantees the existence of nontrivial solutions. ∎

As a consequence, we obtain the following result.

Theorem 8. *Any two bases for a finite dimensional vector space have the same number of vectors.*

Proof. Let $S = \{\mathbf{v}_1, \mathbf{v}_2, \ldots, \mathbf{v}_n\}$ and $S' = \{\mathbf{w}_1, \mathbf{w}_2, \ldots, \mathbf{w}_m\}$ be two bases for a finite dimensional vector space V. Since S is a basis and S' is a linearly independent set, Theorem 7 implies that $m \leq n$. Similarly, since S' is a basis and S is linearly independent, we also have $n \leq m$. Therefore $m = n$. ∎

Example 33

The standard basis for R^n contains n vectors (Example 27). Therefore every basis for R^n contains n vectors.

Example 34

The standard basis for P_n (Example 29) contains $n + 1$ vectors; thus every basis for P_n contains $n + 1$ vectors.

The number of vectors in a basis for a finite dimensional vector space is a particularly important quantity. By Example 33, every basis for R^2 has two vectors and every basis for R^3 has three vectors. Since R^2 (the plane) is intuitively two dimensional and R^3 is intuitively three dimensional, the dimension of these spaces is the same as the number of vectors that occur in their bases. This suggests the following definition.

Definition. The ***dimension*** of a finite dimensional vector space V is defined to be the number of vectors in a basis for V. In addition, we define the zero vector space to have dimension zero.

From Examples 33 and 34, R^n is an n-dimensional vector space and P_n is an $n + 1$-dimensional vector space.

Example 35

Determine a basis and the dimension for the solution space of the homogeneous system

$$\begin{aligned}
2x_1 + 2x_2 - x_3 \phantom{{}- 3x_4} + x_5 &= 0 \\
-x_1 - x_2 + 2x_3 - 3x_4 + x_5 &= 0 \\
x_1 + x_2 - 2x_3 \phantom{{}- 3x_4} - x_5 &= 0 \\
x_3 + x_4 + x_5 &= 0
\end{aligned}$$

Solution. In Example 8 of Section 1.3 it was shown that the solutions are given by

$$x_1 = -s - t \qquad x_2 = s \qquad x_3 = -t \qquad x_4 = 0 \qquad x_5 = t$$

Therefore the solution vectors can be written

$$\begin{bmatrix} x_1 \\ x_2 \\ x_3 \\ x_4 \\ x_5 \end{bmatrix} = \begin{bmatrix} -s-t \\ s \\ -t \\ 0 \\ t \end{bmatrix} = \begin{bmatrix} -s \\ s \\ 0 \\ 0 \\ 0 \end{bmatrix} + \begin{bmatrix} -t \\ 0 \\ -t \\ 0 \\ t \end{bmatrix} = s\begin{bmatrix} -1 \\ -1 \\ 0 \\ 0 \\ 0 \end{bmatrix} + t\begin{bmatrix} -1 \\ 0 \\ -1 \\ 0 \\ 1 \end{bmatrix}$$

which shows that the vectors

$$\mathbf{v}_1 = \begin{bmatrix} -1 \\ 1 \\ 0 \\ 0 \\ 0 \end{bmatrix} \quad \text{and} \quad \mathbf{v}_2 = \begin{bmatrix} -1 \\ 0 \\ -1 \\ 0 \\ 1 \end{bmatrix}$$

span the solution space. Since they are also linearly independent (verify), $\{\mathbf{v}_1, \mathbf{v}_2\}$ is a basis, and the solution space is two dimensional.

In general, to show that a set of vectors $\{\mathbf{v}_1, \mathbf{v}_2, \ldots, \mathbf{v}_n\}$ is a basis for a vector space V we must show that the vectors are linearly independent and span V. However, if we happen to know that V has dimension n (so that $\{\mathbf{v}_1, \mathbf{v}_2, \ldots, \mathbf{v}_n\}$ contains the right number of vectors for a basis) then it suffices to check *either* linear independence *or* spanning—the remaining condition will hold automatically. This is the content of parts (*a*) and (*b*) of the following theorem. Part (*c*) of this theorem states that every linearly independent set forms part of some basis for V.

Theorem 9

(*a*) *If $S = \{\mathbf{v}_1, \mathbf{v}_2, \ldots, \mathbf{v}_n\}$ is a set of n linearly independent rectors in an n-dimensional space V, then S is a basis for V*

(*b*) *If $S = \{\mathbf{v}_1, \mathbf{v}_2, \ldots, \mathbf{v}_n\}$ is a set of n vectors that spans an n-dimensional space V, then S is a basis for V.*

(*c*) *If $S = \{\mathbf{v}_1, \mathbf{v}_2, \ldots, \mathbf{v}_r\}$ is a linearly independent set in an n-dimensional space V and $r < n$, then S can be enlarged to a basis for V; that is, there are vectors $\mathbf{v}_{r+1}, \ldots, \mathbf{v}_n$ such that $\{\mathbf{v}_1, \mathbf{v}_2, \ldots, \mathbf{v}_r, \mathbf{v}_{r+1}, \ldots, \mathbf{v}_n\}$ is a basis for V.*

The proofs are left as exercises.

Example 36

Show that $\mathbf{v}_1 = (-3, 7)$ and $\mathbf{v}_2 = (5, 5)$ form a basis for R^2.

Solution. Since neither vector is a scalar multiple of the other, $S = \{\mathbf{v}_1, \mathbf{v}_2\}$ is linearly independent. Since R^2 is two-dimensional, S is a basis for R^2 by part (*a*) of Theorem 9.

EXERCISE SET 4.5

1. Explain why the following sets of vectors are *not* bases for the indicated vector spaces. (Solve this problem by inspection.)
   (a) $\mathbf{u}_1 = (1, 2)$, $\mathbf{u}_2 = (0, 3)$, $\mathbf{u}_3 = (2, 7)$ for R^2
   (b) $\mathbf{u}_1 = (-1, 3, 2)$, $\mathbf{u}_2 = (6, 1, 1)$ for R^3
   (c) $\mathbf{p}_1 = 1 + x + x^2$, $\mathbf{p}_2 = x - 1$ for P_2
   (d) $A = \begin{bmatrix} 1 & 1 \\ 2 & 3 \end{bmatrix}$ $B = \begin{bmatrix} 6 & 0 \\ -1 & 4 \end{bmatrix}$ $C = \begin{bmatrix} 3 & 0 \\ 1 & 7 \end{bmatrix}$
   $D = \begin{bmatrix} 5 & 1 \\ 4 & 2 \end{bmatrix}$ $E = \begin{bmatrix} 7 & 1 \\ 2 & 9 \end{bmatrix}$ for M_{22}

2. Which of the following sets of vectors are bases for R^2?
   (a) $(2, 1), (3, 0)$ (b) $(4, 1), (-7, -8)$ (c) $(0, 0), (1, 3)$ (d) $(3, 9), (-4, -12)$

3. Which of the following sets of vectors are bases for R^3?
   (a) $(1, 0, 0), (2, 2, 0), (3, 3, 3)$ (b) $(3, 1, -4), (2, 5, 6), (1, 4, 8)$
   (c) $(2, -3, 1), (4, 1, 1), (0, -7, 1)$ (d) $(1, 6, 4), (2, 4, -1), (-1, 2, 5)$

4. Which of the following sets of vectors are bases for P_2?
   (a) $1 - 3x + 2x^2$, $1 + x + 4x^2$, $1 - 7x$
   (b) $4 + 6x + x^2$, $-1 + 4x + 2x^2$, $5 + 2x - x^2$
   (c) $1 + x + x^2$, $x + x^2$, x^2
   (d) $-4 + x + 3x^2$, $6 + 5x + 2x^2$, $8 + 4x + x^2$

5. Show that the following set of vectors is a basis for M_{22}.

$$\begin{bmatrix} 3 & 6 \\ 3 & -6 \end{bmatrix}, \quad \begin{bmatrix} 0 & -1 \\ -1 & 0 \end{bmatrix}, \quad \begin{bmatrix} 0 & -8 \\ -12 & -4 \end{bmatrix}, \quad \begin{bmatrix} 1 & 0 \\ -1 & 2 \end{bmatrix}$$

6. Let V be the space spanned by $\mathbf{v}_1 = \cos^2 x$, $\mathbf{v}_2 = \sin^2 x$, $\mathbf{v}_3 = \cos 2x$.
   (a) Show that $S = \{\mathbf{v}_1, \mathbf{v}_2, \mathbf{v}_3\}$ is not a basis for V.
   (b) Find a basis for V.

In Exercises 7–12 determine the dimension of and a basis for the solution space of the system.

7. $$\begin{aligned} 2x_1 + x_2 + 3x_3 &= 0 \\ x_1 + 2x_2 &= 0 \\ x_2 + x_3 &= 0 \end{aligned}$$

8. $$\begin{aligned} 3x_1 + x_2 + x_3 + x_4 &= 0 \\ 5x_1 - x_2 + x_3 - x_4 &= 0 \end{aligned}$$

9. $$\begin{aligned} 3x_1 + x_2 + 2x_3 &= 0 \\ 4x_1 + 5x_3 &= 0 \\ x_1 - 3x_2 + 4x_3 &= 0 \end{aligned}$$

10. $$\begin{aligned} x_1 - 3x_2 + x_3 &= 0 \\ 2x_1 - 6x_2 + 2x_3 &= 0 \\ 3x_1 - 9x_2 + 3x_3 &= 0 \end{aligned}$$

11. $$\begin{aligned} 2x_1 - 4x_2 + x_3 + x_4 &= 0 \\ x_1 - 5x_2 + 2x_3 &= 0 \\ -2x_2 - 2x_3 - x_4 &= 0 \\ x_1 + 3x_2 + x_4 &= 0 \\ x_1 - 2x_2 - x_3 + x_4 &= 0 \end{aligned}$$

12. $$\begin{aligned} x + y + z &= 0 \\ 3x + 2y - z &= 0 \\ 2x - 4y + z &= 0 \\ 4x + 8y - 3z &= 0 \\ 2x + y - 2z &= 0 \end{aligned}$$

13. Determine bases for the following subspaces of R^3.
    (a) The plane $3x - 2y + 5z = 0$

(b) The plane $x - y = 0$

$$\text{(c) The line } \begin{aligned} x &= 2t \\ y &= -t \qquad -\infty < t < +\infty \\ z &= 4t \end{aligned}$$

(d) All vectors of the form (a, b, c), where $b = a + c$

14. Determine the dimensions of the following subspaces of R^4.
(a) All vectors of the form $(a, b, c, 0)$
(b) All vectors of the form (a, b, c, d), where $d = a + b$ and $c = a - b$.
(c) All vectors of the form (a, b, c, d), where $a = b = c = d$

15. Determine the dimension of the subspace of P_3 consisting of all polynomials $a_0 + a_1x + a_2x^2 + a_3x^3$ for which $a_0 = 0$.

16. Let $\{\mathbf{v}_1, \mathbf{v}_2, \mathbf{v}_3\}$ be a basis for a vector space V. Show that $\{\mathbf{u}_1, \mathbf{u}_2, \mathbf{u}_3\}$ is also a basis, where $\mathbf{u}_1 = \mathbf{v}_1$, $\mathbf{u}_2 = \mathbf{v}_1 + \mathbf{v}_2$, and $\mathbf{u}_3 = \mathbf{v}_1 + \mathbf{v}_2 + \mathbf{v}_3$.

17. Show that the vector space of all real-valued functions defined on the entire real line is infinite dimensional. (***Hint.*** Assume it is finite dimensional with dimension n, and obtain a contradiction by producing $n + 1$ linearly independent vectors.)

18. Show that a subspace of a finite dimensional vector space is finite dimensional.

19. Let V be a subspace of a finite dimensional vector space W. Show that $\dim(V) \leq \dim(W)$.

20. Show that the only subspaces of R^3 are lines through the origin, planes through the origin, the zero subspace, and R^3 itself. (***Hint.*** By Exercise 19 the subspaces of R^3 must be 0-dimensional, 1-dimensional, 2-dimensional, or 3-dimensional.)

21. Prove part (*a*) of Theorem 9.

22. Prove part (*b*) of Theorem 9.

23. Prove part (*c*) of Theorem 9.

4.6 ROW AND COLUMN SPACE OF A MATRIX; RANK; APPLICATIONS TO FINDING BASES

In this section we shall study certain vector spaces associated with matrices. Our results will provide a simple procedure for finding bases by reducing an appropriate matrix to row-echelon form.

Definition. Consider the $m \times n$ matrix

$$A = \begin{bmatrix} a_{11} & a_{12} & \cdots & a_{1n} \\ a_{21} & a_{22} & \cdots & a_{2n} \\ \vdots & \vdots & & \vdots \\ a_{m1} & a_{m2} & \cdots & a_{mn} \end{bmatrix}$$

The vectors

$$\begin{aligned}\mathbf{r}_1 &= (a_{11}, a_{12}, \ldots, a_{1n})\\ \mathbf{r}_2 &= (a_{21}, a_{22}, \ldots, a_{2n})\\ &\vdots\\ \mathbf{r}_m &= (a_{m1}, a_{m2}, \ldots, a_{mn})\end{aligned}$$

formed from the rows of A, are called the ***row vectors*** of A and the vectors

$$\mathbf{c}_1 = \begin{bmatrix} a_{11} \\ a_{21} \\ \vdots \\ a_{m1} \end{bmatrix}, \mathbf{c}_2 = \begin{bmatrix} a_{12} \\ a_{22} \\ \vdots \\ a_{m2} \end{bmatrix}, \ldots, \mathbf{c}_n = \begin{bmatrix} a_{1n} \\ a_{2n} \\ \vdots \\ a_{mn} \end{bmatrix}$$

formed from the columns of A are called the ***column vectors*** of A. The subspace of R^n spanned by the row vectors is called the ***row space*** of A and the subspace of R^m spanned by the column vectors is called the ***column space*** of A.

Example 37

Let

$$A = \begin{bmatrix} 2 & 1 & 0 \\ 3 & -1 & 4 \end{bmatrix}$$

The row vectors of A are

$$\mathbf{r}_1 = (2, 1, 0) \quad \text{and} \quad \mathbf{r}_2 = (3, -1, 4)$$

and the column vectors of A are

$$\mathbf{c}_1 = \begin{bmatrix} 2 \\ 3 \end{bmatrix} \quad \mathbf{c}_2 = \begin{bmatrix} 1 \\ -1 \end{bmatrix} \quad \text{and} \quad \mathbf{c}_3 = \begin{bmatrix} 0 \\ 4 \end{bmatrix}$$

The next theorem will help us to find bases for vector spaces. We shall defer its proof to the end of the section.

Theorem 10. *Elementary row operations do not change the row space of a matrix.*

It follows from this theorem that the row space of a matrix A is not changed by reducing the matrix to row-echelon form. However, the nonzero row vectors of a matrix in row-echelon form are always linearly independent (Exercise 14) so that these nonzero row vectors form a basis for the row space. Thus we have the following result.

Theorem 11. *The nonzero row vectors in a row-echelon form of a matrix A form a basis for the row space of A.*

Example 38

Find a basis for the space spanned by the vectors

$$\mathbf{v}_1 = (1, -2, 0, 0, 3) \qquad \mathbf{v}_2 = (2, -5, -3, -2, 6) \qquad \mathbf{v}_3 = (0, 5, 15, 10, 0)$$
$$\mathbf{v}_4 = (2, 6, 18, 8, 6)$$

Solution. The space spanned by these vectors is the row space of the matrix

$$\begin{bmatrix} 1 & -2 & 0 & 0 & 3 \\ 2 & -5 & -3 & -2 & 6 \\ 0 & 5 & 15 & 10 & 0 \\ 2 & 6 & 18 & 8 & 6 \end{bmatrix}$$

Putting this matrix in row-echelon form we obtain (verify):

$$\begin{bmatrix} 1 & -2 & 0 & 0 & 3 \\ 0 & 1 & 3 & 2 & 0 \\ 0 & 0 & 1 & 1 & 0 \\ 0 & 0 & 0 & 0 & 0 \end{bmatrix}$$

The nonzero row vectors in this matrix are

$$\mathbf{w}_1 = (1, -2, 0, 0, 3) \qquad \mathbf{w}_2 = (0, 1, 3, 2, 0) \qquad \mathbf{w}_3 = (0, 0, 1, 1, 0)$$

These form a basis for the row space and consequently a basis for the space spanned by $\mathbf{v}_1$, $\mathbf{v}_2$, $\mathbf{v}_3$. and $\mathbf{v}_4$.

REMARK. We have been writing the row vectors of a matrix in horizontal notation and the column vectors in vertical (matrix) notation because it seems natural to do so. However, there is no reason why the row vectors cannot be written in vertical notation and the column vectors in horizontal notation if it is convenient.

In light of this remark, it is clear that, except for a change from vertical to horizontal notation, the column space of a matrix is the same as the row space of its transpose. Thus we can find a basis for the column space of a matrix A by finding a basis for the row space of A^t and then converting back to vertical notation, if desired.

Example 39

Find a basis for the column space of

$$A = \begin{bmatrix} 1 & 0 & 1 & 1 \\ 3 & 2 & 5 & 1 \\ 0 & 4 & 4 & -4 \end{bmatrix}$$

Solution. Transposing we obtain

$$A^t = \begin{bmatrix} 1 & 3 & 0 \\ 0 & 2 & 4 \\ 1 & 5 & 4 \\ 1 & 1 & -4 \end{bmatrix}$$

and reducing to row-echelon form yields (verify)

$$\begin{bmatrix} 1 & 3 & 0 \\ 0 & 1 & 2 \\ 0 & 0 & 0 \\ 0 & 0 & 0 \end{bmatrix}$$

Thus the vectors (1, 3, 0) and (0, 1, 2) form a basis for the row space of A^t or equivalently

$$\mathbf{w}_1 = \begin{bmatrix} 1 \\ 3 \\ 0 \end{bmatrix} \quad \text{and} \quad \mathbf{w}_2 = \begin{bmatrix} 0 \\ 1 \\ 2 \end{bmatrix}$$

form a basis for the column space of A.

Our next theorem is one of the most fundamental results in linear algebra. The proof is deferred to the end of this section.

Theorem 12. *If A is any matrix, then the row space and column space of A have the same dimension.*

Example 40

In Example 39 we saw that the matrix

$$A = \begin{bmatrix} 1 & 0 & 1 & 1 \\ 3 & 2 & 5 & 1 \\ 0 & 4 & 4 & -4 \end{bmatrix}$$

has a two dimensional column space. Thus Theorem 12 asserts that the row space is also two dimensional. To see that this is in fact the case we reduce A to row-echelon form, obtaining (verify)

$$\begin{bmatrix} 1 & 0 & 1 & 1 \\ 0 & 1 & 1 & -1 \\ 0 & 0 & 0 & 0 \end{bmatrix}$$

Since this matrix has two nonzero rows, the row space of A is two dimensional.

Definition. The dimension of the row and column space of a matrix A is called the ***rank*** of A.

Example 41

The matrix A in Examples 39 and 40 has rank 2.

The next theorem adds three more results to those in Theorem 13 of Section 1.7 and Theorem 6 of Section 2.3.

Theorem 13. *If A is an $n \times n$ matrix, then the following statements are equivalent.*

(a) *A is invertible.*
(b) *$A\mathbf{x} = \mathbf{0}$ has only the trivial solution.*
(c) *A is row equivalent to I_n.*
(d) *$A\mathbf{x} = \mathbf{b}$ is consistent for every $n \times 1$ matrix $\mathbf{b}$.*
(e) *$\det(A) \neq 0$*
(f) *A has rank n*
(g) *The row vectors of A are linearly independent.*
(h) *The column vectors of A are linearly independent.*

Proof. We shall show that (c), (f), (g) and (h) are equivalent by proving the sequence of implications (c) $\Rightarrow$ (f) $\Rightarrow$ (g) $\Rightarrow$ (h) $\Rightarrow$ (c). This will complete the proof since we already know that (c) is equivalent to (a), (b), (d), and (e).

(c) $\Rightarrow$ (f) Since A is row equivalent to I_n, and I_n has n nonzero rows, the row space of A is n-dimensional by Theorem 11. Thus A has rank n.

(f) $\Rightarrow$ (g) Since A has rank n, the row space of A is n dimensional. Since the n row vectors of A span the row space of A, it follows from Theorem 9 in Section 4.5 that the row vectors of A are linearly independent.

(g) $\Rightarrow$ (h) Assume the row vectors of A are linearly independent. Thus the row space of A is n-dimensional. By Theorem 12 the column space of A is also n-dimensional. Since the column vectors of A span the column space, the column vectors of A are linearly independent by Theorem 9 in Section 4.5.

(h) $\Rightarrow$ (c) Assume the column vectors of A are linearly independent. Thus the column space of A is n-dimensional and consequently the row space of A is n-dimensional by Theorem 12. This means that the reduced row-echelon form of A has n nonzero rows, that is all rows are nonzero. As noted in Example 24 of Section 2.3 this implies that the reduced row-echelon form of A is I_n. Thus A is row equivalent to I_n. ∎

It is interesting to note that Theorem 13 ties together all the major topics we have studied so far—matrices, systems of equations, determinants, and vector spaces.

We conclude this section with one additional result about systems of linear equations. Consider a system of linear equations

$$A\mathbf{x} = \mathbf{b}$$

or equivalently

$$\begin{bmatrix} a_{11} & a_{12} & \cdots & a_{1n} \\ a_{21} & a_{22} & \cdots & a_{2n} \\ \vdots & \vdots & & \vdots \\ a_{m1} & a_{m2} & \cdots & a_{mn} \end{bmatrix} \begin{bmatrix} x_1 \\ x_2 \\ \vdots \\ x_n \end{bmatrix} = \begin{bmatrix} b_1 \\ b_2 \\ \vdots \\ b_m \end{bmatrix}$$

By multiplying out the matrices on the left side, this system can be rewritten

$$\begin{bmatrix} a_{11}x_1 + a_{12}x_2 + \cdots + a_{1n}x_n \\ a_{21}x_1 + a_{22}x_2 + \cdots + a_{2n}x_n \\ \vdots \qquad\qquad \vdots \qquad\qquad\qquad \vdots \\ a_{m1}x_1 + a_{m2}x_2 + \cdots + a_{mn}x_n \end{bmatrix} = \begin{bmatrix} b_1 \\ b_2 \\ \vdots \\ b_m \end{bmatrix}$$

or

$$x_1 \begin{bmatrix} a_{11} \\ a_{21} \\ \vdots \\ a_{m1} \end{bmatrix} + x_2 \begin{bmatrix} a_{12} \\ a_{22} \\ \vdots \\ a_{m2} \end{bmatrix} + \cdots + x_n \begin{bmatrix} a_{1n} \\ a_{2n} \\ \vdots \\ a_{mn} \end{bmatrix} = \begin{bmatrix} b_1 \\ b_2 \\ \vdots \\ b_m \end{bmatrix}$$

Since the left side of this equation is a linear combination of the column vectors of A, it follows that the system $A\mathbf{x} = \mathbf{b}$ is consistent if and only if $\mathbf{b}$ is a linear combination of the column vectors of A. Thus we have the following useful theorem.

Theorem 14. *A system of linear equations $A\mathbf{x} = \mathbf{b}$ is consistent if and only if* $\mathbf{b}$ *is in the column space of A.*

OPTIONAL

Proof of Theorem 10. Suppose that the row vectors of a matrix A are $\mathbf{r}_1, \mathbf{r}_2, \ldots, \mathbf{r}_m$ and let B be obtained from A by performing an elementary row operation. We shall show that every vector in the row space of B is also in the row space of A, and conversely that every vector in the row space of A is in the row space of B. We can then conclude that A and B have the same row space.

Consider the possibilities. If the row operation is a row interchange, then B and A have the same row vectors, and consequently the same row space. If the row operation is multiplication of a row by a scalar or addition of a multiple of one row to another, then the row vectors $\mathbf{r}'_1, \mathbf{r}'_2, \ldots, \mathbf{r}'_m$ of B are linear combinations of $\mathbf{r}_1, \mathbf{r}_2, \ldots, \mathbf{r}_m$; thus they lie in the row space of A. Since a vector space is closed under addition and scalar multiplication, all linear combinations of $\mathbf{r}'_1, \mathbf{r}'_2, \ldots, \mathbf{r}'_m$ will also lie in the row space of A. Therefore each vector in the row space of B is in the row space of A.

Since B is obtained from A by performing a row operation, A can be obtained from B by performing the inverse operation (Section 1.7). Thus the argument above shows that row space of A is contained in the row space of B.

OPTIONAL

Proof of Theorem 12. Denote the row vectors of

$$A = \begin{bmatrix} a_{11} & a_{12} & \cdots & a_{1n} \\ a_{21} & a_{22} & \cdots & a_{2n} \\ \vdots & \vdots & & \vdots \\ a_{m1} & a_{m2} & \cdots & a_{mn} \end{bmatrix}$$

by

$$\mathbf{r}_1, \mathbf{r}_2, \ldots, \mathbf{r}_m$$

Suppose the row space of A has dimension k and that $S = \{\mathbf{b}_1, \mathbf{b}_2, \ldots, \mathbf{b}_k\}$ is a basis for the row space, where $\mathbf{b}_i = (b_{i1}, b_{i2}, \ldots, b_{in})$. Since S is a basis for the row space, each row vector is expressible as a linear combination of $\mathbf{b}_1, \mathbf{b}_2, \ldots, \mathbf{b}_k$; thus

$$\begin{aligned} \mathbf{r}_1 &= c_{11}\mathbf{b}_1 + c_{12}\mathbf{b}_2 + \cdots + c_{1k}\mathbf{b}_k \\ \mathbf{r}_2 &= c_{21}\mathbf{b}_1 + c_{22}\mathbf{b}_2 + \cdots + c_{2k}\mathbf{b}_k \\ &\ \vdots \\ \mathbf{r}_m &= c_{m1}\mathbf{b}_1 + c_{m2}\mathbf{b}_2 + \cdots + c_{mk}\mathbf{b}_k \end{aligned} \tag{4.11}$$

Since two vectors in R^n are equal if and only if corresponding components are equal, we can equate the jth component on each side of (4.11) to obtain

$$\begin{aligned} a_{1j} &= c_{11}b_{1j} + c_{12}b_{2j} + \cdots + c_{1k}b_{kj} \\ a_{2j} &= c_{21}b_{1j} + c_{22}b_{2j} + \cdots + c_{2k}b_{kj} \\ &\ \vdots \\ a_{mj} &= c_{m1}b_{1j} + c_{m2}b_{2j} + \cdots + c_{mk}b_{kj} \end{aligned}$$

or equivalently

$$\begin{bmatrix} a_{1j} \\ a_{2j} \\ \vdots \\ a_{mj} \end{bmatrix} = b_{1j}\begin{bmatrix} c_{11} \\ c_{21} \\ \vdots \\ c_{m1} \end{bmatrix} + b_{2j}\begin{bmatrix} c_{12} \\ c_{22} \\ \vdots \\ c_{m2} \end{bmatrix} + \cdots + b_{kj}\begin{bmatrix} c_{1k} \\ c_{2k} \\ \vdots \\ c_{mk} \end{bmatrix} \tag{4.12}$$

The left side of this equation is the jth column vector of A, and $j = 1, 2, \ldots, n$ is arbitrary; therefore each column vector of A lies in the space spanned by the k vectors on the right side of (4.12). Thus the column space of A has dimension $\leq k$.

Since

$$k = \dim(\text{row space of } A)$$

we have

$$\dim(\text{column space of } A) \leq \dim(\text{row space of } A). \tag{4.13}$$

Since the matrix A is completely arbitrary, this same conclusion applies to A^t, that is

$$\dim(\text{column space of } A^t) \leq \dim(\text{row space of } A^t) \tag{4.14}$$

But transposing a matrix converts columns to rows and rows to columns so that

$$\text{column space of } A^t = \text{row space of } A$$

and

$$\text{row space of } A^t = \text{column space of } A$$

Thus (4.14) can be rewritten as

$$\dim(\text{row space of } A) \leq \dim(\text{column space of } A).$$

From this result and (4.13) we conclude that

$$\dim(\text{row space of } A) = \dim(\text{column space of } A). \quad \blacksquare$$

EXERCISE SET 4.6

1. List the row vectors and column vectors of the matrix

$$\begin{bmatrix} 2 & -1 & 0 & 1 \\ 3 & 5 & 7 & -1 \\ 1 & 4 & 2 & 7 \end{bmatrix}$$

In Exercises 2–5, find: (a) a basis for the row space; (b) a basis for the column space; (c) the rank of the matrix.

2. $\begin{bmatrix} 1 & -3 \\ 2 & -6 \end{bmatrix}$ **3.** $\begin{bmatrix} 1 & 2 & -1 \\ 2 & 4 & 6 \\ 0 & 0 & -8 \end{bmatrix}$ **4.** $\begin{bmatrix} 1 & 1 & 2 & 1 \\ 1 & 0 & 1 & 2 \\ 2 & 1 & 3 & 4 \end{bmatrix}$

5. $\begin{bmatrix} 1 & -3 & 2 & 2 & 1 \\ 0 & 3 & 6 & 0 & -2 \\ 2 & -3 & -2 & 4 & 4 \\ 3 & -3 & 6 & 6 & 3 \\ 5 & -3 & 10 & 10 & 5 \end{bmatrix}$

6. Find a basis for the subspace of R^4 spanned by the given vectors.
(a) $(1, 1, -4, -3)$, $(2, 0, 2, -2)$, $(2, -1, 3, 2)$
(b) $(-1, 1, -2, 0)$, $(3, 3, 6, 0)$, $(9, 0, 0, 3)$
(c) $(1, 1, 0, 0)$, $(0, 0, 1, 1)$, $(-2, 0, 2, 2)$, $(0, -3, 0, 3)$

7. Verify that the row space and column space have the same dimension (as guaranteed by Theorem 12).

(a) $\begin{bmatrix} 2 & 0 & 2 & 2 \\ 3 & -4 & -1 & -9 \\ 1 & 2 & 3 & 7 \\ -3 & 1 & -2 & 0 \end{bmatrix}$ (b) $\begin{bmatrix} 2 & 3 & 5 & 7 & 4 \\ -1 & 2 & 1 & 0 & -2 \\ 4 & 1 & 5 & 9 & 8 \end{bmatrix}$

8. (a) If A is a 3×5 matrix, what is the largest possible value for the rank of A?
(b) If A is an $m \times n$ matrix, what is the largest possible value for the rank of A?

9. In each part determine if **b** lies in the column space of A. If it does, express **b** as a linear combination of the column vectors.

(a) $A = \begin{bmatrix} 1 & 3 \\ 4 & -6 \end{bmatrix}$; $\mathbf{b} = \begin{bmatrix} -2 \\ 10 \end{bmatrix}$

(b) $A = \begin{bmatrix} 1 & -4 \\ 2 & -8 \end{bmatrix}$; $\mathbf{b} = \begin{bmatrix} 0 \\ 1 \end{bmatrix}$

(c) $A = \begin{bmatrix} 1 & -1 & 1 \\ 1 & 1 & -1 \\ -1 & -1 & 1 \end{bmatrix}$; $\mathbf{b} = \begin{bmatrix} 2 \\ 0 \\ 0 \end{bmatrix}$

10. (a) Prove: If A is a 3×5 matrix then the column vectors of A are linearly dependent.
(b) Prove: If A is a 5×3 matrix then the row vectors of A are linearly dependent.

11. Prove: If A is a matrix that is not square, then either the row vectors of A or the column vectors of A are linearly dependent.

12. Let

$$A = \begin{bmatrix} a_{11} & a_{12} & a_{13} \\ a_{21} & a_{22} & a_{23} \end{bmatrix}$$

Prove: A has rank 2 if and only if one or more of the determinants

$$\begin{vmatrix} a_{11} & a_{12} \\ a_{21} & a_{22} \end{vmatrix}, \begin{vmatrix} a_{11} & a_{13} \\ a_{21} & a_{23} \end{vmatrix}, \begin{vmatrix} a_{12} & a_{13} \\ a_{22} & a_{23} \end{vmatrix}$$

is nonzero.

13. Prove that a system of equations $A\mathbf{x} = \mathbf{b}$ is consistent if and only if the rank of its augmented matrix is equal to the rank of A.

14. Prove Theorem 11.

15. Prove that the row vectors in an $n \times n$ invertible matrix A form a basis for R^n.

4.7 INNER PRODUCT SPACES

In Section 4.1 we studied the Euclidean inner product on the vector space R^n. In this section we introduce the notion of an inner product on an arbitrary vector space. As a consequence of our work, we will be able to define meaningfully notions of angle, length, and distance in more general vector spaces.

In Theorem 2 of Section 4.1, we collected the most important properties of the Euclidean inner product. In a general vector space, an inner product is defined axiomatically using these properties as axioms.

Definition. An ***inner product*** on a vector space V is a function that associates a real number $\langle \mathbf{u}, \mathbf{v} \rangle$ with each pair of vectors $\mathbf{u}$ and $\mathbf{v}$ in V in such a way that the following axioms are satisfied for all vectors $\mathbf{u}$, $\mathbf{v}$, and $\mathbf{w}$ in V and all scalars k.

(1) $\langle \mathbf{u}, \mathbf{v} \rangle = \langle \mathbf{v}, \mathbf{u} \rangle$ (symmetry axiom)
(2) $\langle \mathbf{u} + \mathbf{v}, \mathbf{w} \rangle = \langle \mathbf{u}, \mathbf{w} \rangle + \langle \mathbf{v}, \mathbf{w} \rangle$ (additivity axiom)
(3) $\langle k\mathbf{u}, \mathbf{v} \rangle = k\langle \mathbf{u}, \mathbf{v} \rangle$ (homogeneity axiom)
(4) $\langle \mathbf{v}, \mathbf{v} \rangle \geq 0$ and $\langle \mathbf{v}, \mathbf{v} \rangle = 0$ if and only if $\mathbf{v} = \mathbf{0}$ (positivity axiom)

A vector space with an inner product is called an ***inner product space***.

The following additional properties follow immediately from the four inner product axioms:

(i) $\langle \mathbf{0}, \mathbf{v} \rangle = \langle \mathbf{v}, \mathbf{0} \rangle = 0$
(ii) $\langle \mathbf{u}, \mathbf{v} + \mathbf{w} \rangle = \langle \mathbf{u}, \mathbf{v} \rangle + \langle \mathbf{u}, \mathbf{w} \rangle$
(iii) $\langle \mathbf{u}, k\mathbf{v} \rangle = k\langle \mathbf{u}, \mathbf{v} \rangle$

We prove (ii) and leave (i) and (iii) as exercises.

$$\begin{aligned} \langle \mathbf{u}, \mathbf{v} + \mathbf{w} \rangle &= \langle \mathbf{v} + \mathbf{w}, \mathbf{u} \rangle && \text{(by symmetry)} \\ &= \langle \mathbf{v}, \mathbf{u} \rangle + \langle \mathbf{w}, \mathbf{u} \rangle && \text{(by additivity)} \\ &= \langle \mathbf{u}, \mathbf{v} \rangle + \langle \mathbf{u}, \mathbf{w} \rangle && \text{(by symmetry)} \end{aligned}$$

Example 42

Let $\mathbf{u} = (u_1, u_2, \ldots, u_n)$ and $\mathbf{v} = (v_1, v_2, \ldots, v_n)$; the Euclidean inner product $\langle \mathbf{u}, \mathbf{v} \rangle = \mathbf{u} \cdot \mathbf{v} = u_1v_1 + u_2v_2 + \cdots + u_nv_n$ satisfies all the inner product axioms by Theorem 2 of Section 4.1.

Example 43

If $\mathbf{u} = (u_1, u_2)$ and $\mathbf{v} = (v_1, v_2)$ are vectors in R^2, then

$$\langle \mathbf{u}, \mathbf{v} \rangle = 3u_1v_1 + 2u_2v_2$$

defines an inner product. To see this, note first that if $\mathbf{u}$ and $\mathbf{v}$ are interchanged in this equation, then the right side remains the same. Therefore

$$\langle \mathbf{u}, \mathbf{v} \rangle = \langle \mathbf{v}, \mathbf{u} \rangle$$

If $\mathbf{w} = (w_1, w_2)$, then

$$\begin{aligned}\langle \mathbf{u} + \mathbf{v}, \mathbf{w}\rangle &= 3(u_1 + v_1)w_1 + 2(u_2 + v_2)w_2 \\ &= (3u_1w_1 + 2u_2w_2) + (3v_1w_1 + 2v_2w_2) \\ &= \langle \mathbf{u}, \mathbf{w}\rangle + \langle \mathbf{v}, \mathbf{w}\rangle\end{aligned}$$

which establishes the second axiom.

Next

$$\begin{aligned}\langle k\mathbf{u}, \mathbf{v}\rangle &= 3(ku_1)v_1 + 2(ku_2)v_2 \\ &= k(3u_1v_1 + 2u_2v_2) \\ &= k\langle \mathbf{u}, \mathbf{v}\rangle\end{aligned}$$

which establishes the third axiom.

Finally

$$\langle \mathbf{v}, \mathbf{v}\rangle = 3v_1v_1 + 2v_2v_2 = 3v_1{}^2 + 2v_2{}^2$$

Obviously $\langle \mathbf{v}, \mathbf{v}\rangle = 3v_1{}^2 + 2v_2{}^2 \geq 0$. Further $\langle \mathbf{v}, \mathbf{v}\rangle = 3v_1{}^2 + 2v_2{}^2 = 0$ if and only if $v_1 = v_2 = 0$; that is, if and only if $\mathbf{v} = (v_1, v_2) = \mathbf{0}$. Thus the fourth axiom is satisfied.

The inner product in this example is different from the Euclidean inner product on R^2; this shows that a vector space can have more than one inner product.

Example 44

If

$$U = \begin{bmatrix} u_1 & u_2 \\ u_3 & u_4 \end{bmatrix} \quad \text{and} \quad V = \begin{bmatrix} v_1 & v_2 \\ v_3 & v_4 \end{bmatrix}$$

are any two 2×2 matrices, then the following formula defines an inner product on M_{22} (verify):

$$\langle U, V\rangle = u_1v_1 + u_2v_2 + u_3v_3 + u_4v_4$$

For example, if

$$U = \begin{bmatrix} 1 & 2 \\ 3 & 4 \end{bmatrix} \quad \text{and} \quad V = \begin{bmatrix} -1 & 0 \\ 3 & 2 \end{bmatrix}$$

then

$$\langle U, V\rangle = 1(-1) + 2(0) + 3(3) + 4(2) = 16$$

Example 45

If

$$\mathbf{p} = a_0 + a_1x + a_2x^2 \quad \text{and} \quad \mathbf{q} = b_0 + b_1x + b_2x^2$$

are any two vectors in P_2, then the following formula defines an inner product on P_2 (verify):

$$\langle \mathbf{p}, \mathbf{q}\rangle = a_0b_0 + a_1b_1 + a_2b_2$$

Example 46

(For readers who have studied calculus.)

Let $\mathbf{p} = p(x)$ and $\mathbf{q} = q(x)$ be two polynomials in P_n, and define

$$\langle \mathbf{p}, \mathbf{q} \rangle = \int_a^b p(x)q(x)\,dx \tag{4.15}$$

where a and b are any fixed real numbers such that $a < b$. We shall show that (4.15) defines an inner product on P_n.

(1) $\langle \mathbf{p}, \mathbf{q} \rangle = \int_a^b p(x)q(x)\,dx = \int_a^b q(x)p(x)\,dx = \langle \mathbf{q}, \mathbf{p} \rangle$

which proves that Axiom 1 holds.

(2) $$\begin{aligned} \langle \mathbf{p} + \mathbf{q}, \mathbf{s} \rangle &= \int_a^b (p(x) + q(x))s(x)\,dx \\ &= \int_a^b p(x)s(x)\,dx + \int_a^b q(x)s(x)\,dx \\ &= \langle \mathbf{p}, \mathbf{s} \rangle + \langle \mathbf{q}, \mathbf{s} \rangle \end{aligned}$$

which proves that Axiom 2 holds.

(3) $\langle k\mathbf{p}, \mathbf{q} \rangle = \int_a^b kp(x)q(x)\,dx = k\int_a^b p(x)q(x)\,dx = k\langle \mathbf{p}, \mathbf{q} \rangle$

which proves that Axiom 3 holds.

(4) If $\mathbf{p} = p(x)$ is any polynomial in P_n then $p^2(x) \geq 0$ for all x; therefore

$$\langle \mathbf{p}, \mathbf{p} \rangle = \int_a^b p^2(x)\,dx \geq 0$$

Further, since $p^2(x) \geq 0$ and polynomials are continuous functions, $\int_a^b p^2(x)\,dx = 0$ if and only if $p(x) = 0$ for all x satisfying $a \leq x \leq b$. Therefore, $\langle \mathbf{p}, \mathbf{p} \rangle = \int_a^b p^2(x)\,dx = 0$ if and only if $\mathbf{p} = \mathbf{0}$. This establishes Axiom 4.

We note that the arguments given here can also be used to show that the vector space $C[a, b]$ discussed in Example 14 is an inner product space under the inner product

$$\langle \mathbf{f}, \mathbf{g} \rangle = \int_a^b f(x)g(x)\,dx$$

If $\mathbf{u}$ and $\mathbf{v}$ are nonzero vectors in R^3, then $\mathbf{u} \cdot \mathbf{v} = \|\mathbf{u}\|\,\|\mathbf{v}\| \cos\theta$, where θ is the angle between $\mathbf{u}$ and $\mathbf{v}$ (Section 3.3). If we square both sides of this inequality and use the relationships $\|\mathbf{u}\|^2 = \mathbf{u} \cdot \mathbf{u}$, $\|\mathbf{v}\|^2 = \mathbf{v} \cdot \mathbf{v}$, and $\cos^2\theta \leq 1$, we obtain the inequality

$$(\mathbf{u} \cdot \mathbf{v})^2 \leq (\mathbf{u} \cdot \mathbf{u})(\mathbf{v} \cdot \mathbf{v})$$

The following theorem shows that this inequality can be generalized to any inner product space. The resulting inequality, called the *Cauchy-Schwarz inequality*, will enable us to introduce notions of length and angle in any inner product space.

Theorem 15. (Cauchy-Schwarz* Inequality). *If* **u** *and* **v** *are vectors in an inner product space V, then*

$$\langle \mathbf{u}, \mathbf{v} \rangle^2 \le \langle \mathbf{u}, \mathbf{u} \rangle \langle \mathbf{v}, \mathbf{v} \rangle$$

Proof. We warn the reader in advance that the proof presented here depends on a clever but highly unmotivated trick. If $\mathbf{u} = \mathbf{0}$, then $\langle \mathbf{u}, \mathbf{v} \rangle = \langle \mathbf{u}, \mathbf{u} \rangle = 0$, so that the equality clearly holds. Assume now that $\mathbf{u} \neq \mathbf{0}$. Let $a = \langle \mathbf{u}, \mathbf{u} \rangle$, $b = 2\langle \mathbf{u}, \mathbf{v} \rangle$, $c = \langle \mathbf{v}, \mathbf{v} \rangle$, and let t be any real number. By the positivity axiom, the inner product of any vector with itself is always nonnegative. Therefore

$$\begin{aligned} 0 \le \langle (t\mathbf{u} + \mathbf{v}), (t\mathbf{u} + \mathbf{v}) \rangle &= \langle \mathbf{u}, \mathbf{u} \rangle t^2 + 2\langle \mathbf{u}, \mathbf{v} \rangle t + \langle \mathbf{v}, \mathbf{v} \rangle \\ &= at^2 + bt + c \end{aligned}$$

This inequality implies that the quadratic polynomial $at^2 + bt + c$ has either no real roots or a repeated real root. Therefore its discriminant must satisfy $b^2 - 4ac \le 0$. Expressing a, b, and c in terms of **u** and **v** gives $4\langle \mathbf{u}, \mathbf{v} \rangle^2 - 4\langle \mathbf{u}, \mathbf{u} \rangle \langle \mathbf{v}, \mathbf{v} \rangle \le 0$; or equivalently, $\langle \mathbf{u}, \mathbf{v} \rangle^2 \le \langle \mathbf{u}, \mathbf{u} \rangle \langle \mathbf{v}, \mathbf{v} \rangle$. ∎

Example 47

If $\mathbf{u} = (u_1, u_2, \ldots, u_n)$ and $\mathbf{v} = (v_1, v_2, \ldots, v_n)$ are any two vectors in R^n, then the Cauchy-Schwarz inequality applied to **u** and **v** yields

$$(u_1 v_1 + u_2 v_2 + \cdots + u_n v_n)^2 \le (u_1^2 + u_2^2 + \cdots + u_n^2)(v_1^2 + v_2^2 + \cdots + v_n^2)$$

which is called ***Cauchy's inequality***.

EXERCISE SET 4.7

1. Compute $\langle \mathbf{u}, \mathbf{v} \rangle$ using the inner product in Example 43.
   (a) $\mathbf{u} = (2, -1)$, $\mathbf{v} = (-1, 3)$ (b) $\mathbf{u} = (0, 0)$, $\mathbf{v} = (7, 2)$
   (c) $\mathbf{u} = (3, 1)$, $\mathbf{v} = (-2, 9)$ (d) $\mathbf{u} = (4, 6)$, $\mathbf{v} = (4, 6)$
2. Repeat Exercise 1 using the Euclidean inner product on R^2.
3. Compute $\langle \mathbf{u}, \mathbf{v} \rangle$ using the inner product in Example 44.
   (a) $\mathbf{u} = \begin{bmatrix} 2 & -1 \\ 3 & 7 \end{bmatrix}$ $\mathbf{v} = \begin{bmatrix} 0 & 4 \\ 2 & 2 \end{bmatrix}$ (b) $\mathbf{u} = \begin{bmatrix} 1 & 2 \\ -3 & 5 \end{bmatrix}$ $\mathbf{v} = \begin{bmatrix} 4 & 6 \\ 0 & 8 \end{bmatrix}$
4. Compute $\langle \mathbf{p}, \mathbf{q} \rangle$ using the inner product in Example 45.
   (a) $\mathbf{p} = -1 + 2x + x^2$ $\mathbf{q} = 2 - 4x^2$
   (b) $\mathbf{p} = -3 + 2x + x^2$ $\mathbf{q} = 2 + 4x - 2x^2$

* *Augustin Louis (Baron de) Cauchy* (1789–1857). Sometimes called the father of modern analysis, Cauchy helped put calculus on a firm mathematical footing. He was a partisan of the Bourbons and spent several years in exile for his political involvements.

Hermann Amandus Schwarz (1843–1921). German mathematician.

5. Let $\mathbf{u} = (u_1, u_2)$ and $\mathbf{v} = (v_1, v_2)$. Show that the following are inner products on R^2.
(a) $\langle \mathbf{u}, \mathbf{v} \rangle = 6u_1v_1 + 2u_2v_2$
(b) $\langle \mathbf{u}, \mathbf{v} \rangle = 2u_1v_1 + u_2v_1 + u_1v_2 + 2u_2v_2$

6. Let $\mathbf{u} = (u_1, u_2, u_3)$ and $\mathbf{v} = (v_1, v_2, v_3)$. Determine which of the following are inner products on R^3. For those that are not, list the axioms which do not hold.
(a) $\langle \mathbf{u}, \mathbf{v} \rangle = u_1v_1 + u_3v_3$
(b) $\langle \mathbf{u}, \mathbf{v} \rangle = u_1^2v_1^2 + u_2^2v_2^2 + u_3^2v_3^2$
(c) $\langle \mathbf{u}, \mathbf{v} \rangle = 2u_1v_1 + u_2v_2 + 4u_3v_3$
(d) $\langle \mathbf{u}, \mathbf{v} \rangle = u_1v_1 - u_2v_2 + u_3v_3$

7. Let $U = \begin{bmatrix} u_1 & u_2 \\ u_3 & u_4 \end{bmatrix}$ and $V = \begin{bmatrix} v_1 & v_2 \\ v_3 & v_4 \end{bmatrix}$.

Show that $\langle U, V \rangle = u_1v_1 + u_2v_3 + u_3v_2 + u_4v_4$ is an inner product on M_{22}.

8. Let $\mathbf{p} = p(x)$ and $\mathbf{q} = q(x)$ be polynomials in P_2. Show that $\langle \mathbf{p}, \mathbf{q} \rangle = p(0)q(0) + p(1/2)q(1/2) + p(1)q(1)$ is an inner product on P_2.

9. Verify the Cauchy-Schwarz inequality for
(a) $\mathbf{u} = (2, 1)$ and $\mathbf{v} = (1, -3)$, using the inner product in Example 43
(b) $\mathbf{u} = (2, 1, 5)$ and $\mathbf{v} = (1, -3, 4)$, using the Euclidean inner product

(c) $U = \begin{bmatrix} -1 & 2 \\ 6 & 1 \end{bmatrix}$ and $V = \begin{bmatrix} 1 & 0 \\ 3 & 3 \end{bmatrix}$

using the inner product in Example 44
(d) $\mathbf{p} = -1 + 2x + x^2$ and $\mathbf{q} = 2 - 4x^2$, using the inner product in Example 45.

10. Let R^2 have the Euclidean inner product. Apply the Cauchy-Schwarz inequality to the vectors $\mathbf{u} = (a, b)$, $\mathbf{v} = (\cos\theta, \sin\theta)$ to show $|a \cos\theta + b \sin\theta|^2 \le a^2 + b^2$.

11. Prove that if $\langle \mathbf{u}, \mathbf{v} \rangle$ is any inner product, then $\langle \mathbf{0}, \mathbf{v} \rangle = \langle \mathbf{v}, \mathbf{0} \rangle = 0$.

12. Prove that if $\langle \mathbf{u}, \mathbf{v} \rangle$ is any inner product and k is any scalar, then $\langle \mathbf{u}, k\mathbf{v} \rangle = k\langle \mathbf{u}, \mathbf{v} \rangle$.

13. Show that equality holds in the Cauchy-Schwarz inequality if and only if $\mathbf{u}$ and $\mathbf{v}$ are linearly dependent.

14. Let c_1, c_2, and c_3 be positive real numbers and let $\mathbf{u} = (u_1, u_2, u_3)$ and $\mathbf{v} = (v_1, v_2, v_3)$. Show that $\langle \mathbf{u}, \mathbf{v} \rangle = c_1u_1v_1 + c_2u_2v_2 + c_3u_3v_3$ is an inner product on R^3.

15. Let $c_1, c_2, \ldots, c_n$ be positive real numbers and let $\mathbf{u} = (u_1, u_2, \ldots, u_n)$ and $\mathbf{v} = (v_1, v_2, \ldots, v_n)$. Show that $\langle \mathbf{u}, \mathbf{v} \rangle = c_1u_1v_1 + c_2u_2v_2 + \cdots + c_nu_nv_n$ is an inner product on R_n.

16. **(For readers who have studied calculus.)** Use the inner product

$$\langle \mathbf{p}, \mathbf{q} \rangle = \int_{-1}^{1} p(x)q(x)\,dx$$

to compute $\langle \mathbf{p}, \mathbf{q} \rangle$ for the vectors $\mathbf{p} = p(x)$ and $\mathbf{q} = q(x)$ in P_3.
(a) $\mathbf{p} = 1 - x + x^2 + 5x^3$ $\quad \mathbf{q} = x - 3x^2$
(b) $\mathbf{p} = x - 5x^3$ $\quad \mathbf{q} = 2 + 8x^2$

17. (For readers who have studied calculus.) Use the inner product

$$\langle \mathbf{f}, \mathbf{g} \rangle = \int_0^1 f(x)g(x)\,dx$$

to compute $\langle \mathbf{f}, \mathbf{g} \rangle$ for the vectors $\mathbf{f} = f(x)$ and $\mathbf{g} = g(x)$ in $C[0, 1]$.

(a) $\mathbf{f} = \cos 2\pi x \quad \mathbf{g} = \sin 2\pi x$
(b) $\mathbf{f} = x \quad \mathbf{g} = e^x$
(c) $\mathbf{f} = \tan \frac{\pi}{4} x \quad \mathbf{g} = 1$

18. (For readers who have studied calculus.) Let $f(x)$ and $g(x)$ be continuous functions on $[0, 1]$. Prove:

(a) $\left[\int_0^1 f(x)g(x)\,dx\right]^2 \leq \left[\int_0^1 f^2(x)\,dx\right]\left[\int_0^1 g^2(x)\,dx\right]$

(b) $\left[\int_0^1 [f(x) + g(x)]^2\,dx\right]^{1/2} \leq \left[\int_0^1 f^2(x)\,dx\right]^{1/2} + \left[\int_0^1 g^2(x)\,dx\right]^{1/2}$

(***Hint.*** Use the Cauchy-Schwarz inequality and the inner product of Exercise 17.)

4.8 LENGTH AND ANGLE IN INNER PRODUCT SPACES

In this section we use the Cauchy-Schwarz inequality to develop notions of length, distance, and angle in general inner product spaces.

Definition. If V is an inner product space, then the ***norm*** (or length) of a vector $\mathbf{u}$ is denoted by $\|\mathbf{u}\|$ and defined by

$$\|\mathbf{u}\| = \langle \mathbf{u}, \mathbf{u} \rangle^{1/2}$$

Further, the ***distance*** between two points (vectors) $\mathbf{u}$ and $\mathbf{v}$ is denoted by $d(\mathbf{u}, \mathbf{v})$ and defined by

$$d(\mathbf{u}, \mathbf{v}) = \|\mathbf{u} - \mathbf{v}\|$$

Example 48

If $\mathbf{u} = (u_1, u_2, \ldots, u_n)$ and $\mathbf{v} = (v_1, v_2, \ldots, v_n)$ are vectors in R^n with the Euclidean inner product, then

$$\|\mathbf{u}\| = \langle \mathbf{u}, \mathbf{u} \rangle^{1/2} = \sqrt{u_1^2 + u_2^2 + \cdots + u_n^2}$$

and

$$\begin{aligned} d(\mathbf{u}, \mathbf{v}) &= \|\mathbf{u} - \mathbf{v}\| = \langle \mathbf{u} - \mathbf{v}, \mathbf{u} - \mathbf{v} \rangle^{1/2} \\ &= \sqrt{(u_1 - v_1)^2 + (u_2 - v_2)^2 + \cdots + (u_n - v_n)^2} \end{aligned}$$

Observe that these are just the formulas for the Euclidean norm and distance discussed in Section 4.1.

Example 49

Suppose R^2 has the inner product $\langle \mathbf{u}, \mathbf{v} \rangle = 3u_1v_1 + 2u_2v_2$ discussed in Example 43. If $\mathbf{u} = (1, 0)$ and $\mathbf{v} = (0, 1)$ then

$$\|\mathbf{u}\| = \langle \mathbf{u}, \mathbf{u} \rangle^{1/2} = [3(1)(1) + 2(0)(0)]^{1/2} = \sqrt{3}$$

and

$$d(\mathbf{u}, \mathbf{v}) = \|\mathbf{u} - \mathbf{v}\| = \langle (1, -1), (1, -1) \rangle^{1/2}$$
$$= [3(1)(1) + 2(-1)(-1)]^{1/2} = \sqrt{5}$$

It is important to keep in mind that norm and distance depend on the inner product being used. If the inner product is changed, then the norms and distances between vectors change. For example if R^2 has the Euclidean inner product, then the norm of the vector $\mathbf{u}$ in the preceding example is 1, and the distance between $\mathbf{u}$ and $\mathbf{v}$ is $\sqrt{2}$.

The reader might, at this point, object to our use of the terms length and distance for the quantities $\langle \mathbf{u}, \mathbf{u} \rangle^{1/2}$ and $\|\mathbf{u} - \mathbf{v}\|$. Although these defining formulas arise by imitating the formulas in R^2 and R^3, the odd results obtained in Example 49 cast some doubt on the wisdom of these definitions. It does, after all, require a stretch of the imagination to state that the length of the vector $\mathbf{u} = (1, 0)$ is $\sqrt{3}$! We shall now give some arguments in support of these definitions.

Through the years mathematicians have isolated what are considered to be the most important properties of Euclidean length and distance in R^2 and R^3. They are listed in Figure 4.8.

Basic properties of length	*Basic properties of distance*
*L*1. $\|\mathbf{u}\| \geq 0$	*D*1. $d(\mathbf{u}, \mathbf{v}) \geq 0$
*L*2. $\|\mathbf{u}\| = 0$ if and only if $\mathbf{u} = 0$	*D*2. $d(\mathbf{u}, \mathbf{v}) = 0$ if and only if $\mathbf{u} = \mathbf{v}$
*L*3. $\|k\mathbf{u}\| = \lvert k\rvert \, \|\mathbf{u}\|$	*D*3. $d(\mathbf{u}, \mathbf{v}) = d(\mathbf{v}, \mathbf{u})$
*L*4. $\|\mathbf{u} + \mathbf{v}\| \leq \|\mathbf{u}\| + \|\mathbf{v}\|$ (triangle inequality)	*D*4. $d(\mathbf{u}, \mathbf{v}) \leq d(\mathbf{u}, \mathbf{w}) + d(\mathbf{w}, \mathbf{v})$ (triangle inequality)

Figure 4.8

The next theorem justifies our definitions of norm and distance in inner product spaces.

Theorem 16. *If V is an inner product space, then the norm $\|\mathbf{u}\| = \langle \mathbf{u}, \mathbf{u} \rangle^{1/2}$ and the distance $d(\mathbf{u}, \mathbf{v}) = \|\mathbf{u} - \mathbf{v}\|$ satisfy all the properties listed in Figure 4.8.*

We shall prove property $L4$ and leave the proofs of the remaining parts as exercises. Before beginning the proof, we note that the Cauchy-Schwarz inequality

$$\langle \mathbf{u}, \mathbf{v} \rangle^2 \leq \langle \mathbf{u}, \mathbf{u} \rangle \langle \mathbf{v}, \mathbf{v} \rangle$$

can be written in alternative forms. Since $\|\mathbf{u}\|^2 = \langle \mathbf{u}, \mathbf{u} \rangle$ and $\|\mathbf{v}\|^2 = \langle \mathbf{v}, \mathbf{v} \rangle$ it can be written

$$\langle \mathbf{u}, \mathbf{v} \rangle^2 \leq \|\mathbf{u}\|^2 \, \|\mathbf{v}\|^2 \tag{4.16}$$

or, upon taking square roots, as

$$|\langle \mathbf{u}, \mathbf{v} \rangle| \leq \|\mathbf{u}\| \, \|\mathbf{v}\| \tag{4.17}$$

Proof of Property L4. By definition

$$\begin{aligned}
\|\mathbf{u} + \mathbf{v}\|^2 &= \langle \mathbf{u} + \mathbf{v}, \mathbf{u} + \mathbf{v} \rangle \\
&= \langle \mathbf{u}, \mathbf{u} \rangle + 2\langle \mathbf{u}, \mathbf{v} \rangle + \langle \mathbf{v}, \mathbf{v} \rangle \\
&\leq \langle \mathbf{u}, \mathbf{u} \rangle + 2|\langle \mathbf{u}, \mathbf{v} \rangle| + \langle \mathbf{v}, \mathbf{v} \rangle \\
&\leq \langle \mathbf{u}, \mathbf{u} \rangle + 2\|\mathbf{u}\| \, \|\mathbf{v}\| + \langle \mathbf{v}, \mathbf{v} \rangle && \text{(by 4.17)} \\
&= \|\mathbf{u}\|^2 + 2\|\mathbf{u}\| \, \|\mathbf{v}\| + \|\mathbf{v}\|^2 \\
&= (\|\mathbf{u}\| + \|\mathbf{v}\|)^2.
\end{aligned}$$

Taking square roots gives

$$\|\mathbf{u} + \mathbf{v}\| \leq \|\mathbf{u}\| + \|\mathbf{v}\|. \quad \blacksquare$$

In R^2 or R^3 the result just proved states the familiar geometric fact that the sum of the lengths of two sides of a triangle is at least as large as the length of the third side (Figure 4.9).

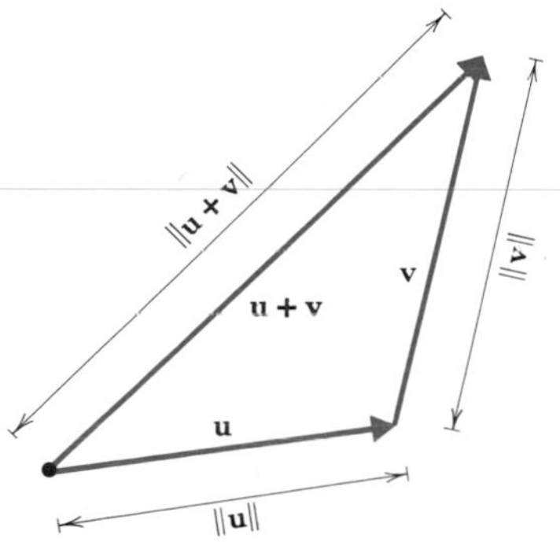

Figure 4.9

Suppose that **u** and **v** are nonzero vectors in an inner product space V. The Cauchy-Schwarz inequality as given in (4.16) can be rewritten as

$$\left(\frac{\langle \mathbf{u}, \mathbf{v}\rangle}{\|\mathbf{u}\|\,\|\mathbf{v}\|}\right)^2 \leq 1$$

or equivalently

$$-1 \leq \frac{\langle \mathbf{u}, \mathbf{v}\rangle}{\|\mathbf{u}\|\,\|\mathbf{v}\|} \leq 1$$

As a consequence of this result, there is a unique angle θ such that

$$\cos\theta = \frac{\langle \mathbf{u}, \mathbf{v}\rangle}{\|\mathbf{u}\|\,\|\mathbf{v}\|} \quad \text{and} \quad 0 \leq \theta \leq \pi \tag{4.18}$$

We define θ to be the ***angle between*** the vectors **u** and **v**. Observe that in R^2 or R^3 with the Euclidean inner product, (4.18) agrees with the usual formula for the cosine of the angle between two nonzero vectors (Section 3.3 of Chapter 3).

Example 50

Find the cosine of the angle θ between the vectors

$$\mathbf{u} = (4, 3, 1, -2) \quad \text{and} \quad \mathbf{v} = (-2, 1, 2, 3)$$

where the vector space is R^4 with the Eulidean inner product.

Solution.

$$\|\mathbf{u}\| = \sqrt{30} \qquad \|\mathbf{v}\| = \sqrt{18} \qquad \text{and} \qquad \langle \mathbf{u}, \mathbf{v}\rangle = -9$$

so that

$$\cos\theta = -\frac{9}{\sqrt{30}\sqrt{18}} = -\frac{3}{\sqrt{30}\sqrt{2}}$$

Example 51

If M_{22} has the inner product given in Example 44, then the angle between the matrices

$$U = \begin{bmatrix} 1 & 0 \\ 1 & 1 \end{bmatrix} \quad \text{and} \quad V = \begin{bmatrix} 0 & 2 \\ 0 & 0 \end{bmatrix}$$

is $\pi/2$ since

$$\cos\theta = \frac{U \cdot V}{\|U\|\,\|V\|} = \frac{1(0) + 0(2) + 1(0) + 1(0)}{\|U\|\,\|V\|} = 0$$

If **u** and **v** are nonzero vectors such that $\langle \mathbf{u}, \mathbf{v}\rangle = 0$, then from (4.18) it follows that $\cos\theta = 0$ and $\theta = \pi/2$. This suggests the following terminology.

Definition. In an inner product space, two vectors **u** and **v** are called ***orthogonal*** if $\langle \mathbf{u}, \mathbf{v}\rangle = 0$. Further, if **u** is orthogonal to each vector in a set W, we say that **u** is ***orthogonal to W***.

We emphasize that orthogonality depends on the selection of the inner product. Two vectors can be othogonal with respect to one inner product but not another.

Example 52

(For readers who have studied calculus.)
Let P_2 have the inner product

$$\langle \mathbf{p}, \mathbf{q} \rangle = \int_{-1}^{1} p(x)q(x)\, dx$$

discussed in Example 46, and let

$$\mathbf{p} = x, \qquad \mathbf{q} = x^2$$

Then

$$\|\mathbf{p}\| = \langle \mathbf{p}, \mathbf{p} \rangle^{1/2} = \left[\int_{-1}^{1} xx\, dx\right]^{1/2} = \left[\int_{-1}^{1} x^2\, dx\right]^{1/2} = \sqrt{\frac{2}{3}}$$

$$\|\mathbf{q}\| = \langle \mathbf{q}, \mathbf{q} \rangle^{1/2} = \left[\int_{-1}^{1} x^2x^2\, dx\right]^{1/2} = \left[\int_{-1}^{1} x^4\, dx\right]^{1/2} = \sqrt{\frac{2}{5}}$$

$$\langle \mathbf{p}, \mathbf{q} \rangle = \int_{-1}^{1} xx^2\, dx = \int_{-1}^{1} x^3\, dx = 0$$

Because $\langle \mathbf{p}, \mathbf{q} \rangle = 0$ the vectors $\mathbf{p} = x$ and $\mathbf{q} = x^2$ are orthogonal relative to the given inner product.

We conclude this section with an interesting and useful generalization of a familiar result.

Theorem 17. (*Generalized Theorem of Pythagoras*). *If* **u** *and* **v** *are orthogonal vectors in an inner product space, then*

$$\|\mathbf{u} + \mathbf{v}\|^2 = \|\mathbf{u}\|^2 + \|\mathbf{v}\|^2.$$

Proof.

$$\|\mathbf{u} + \mathbf{v}\|^2 = \langle (\mathbf{u} + \mathbf{v}), (\mathbf{u} + \mathbf{v}) \rangle = \|\mathbf{u}\|^2 + 2\langle \mathbf{u}, \mathbf{v} \rangle + \|\mathbf{v}\|^2$$
$$= \|\mathbf{u}\|^2 + \|\mathbf{v}\|^2. \quad ■$$

Note that in R^2 or R^3 with the Euclidean inner product this theorem reduces to the ordinary Pythagorean theorem (Figure 4.10).

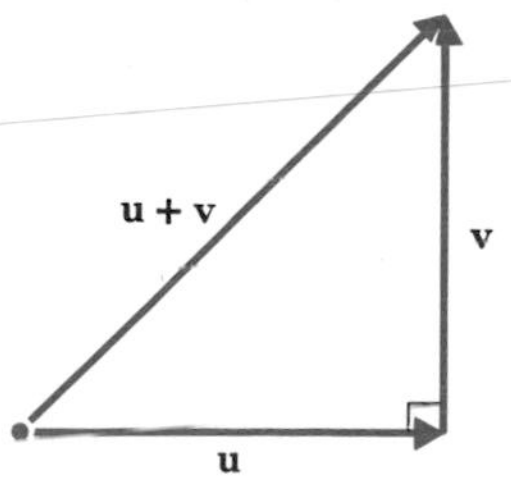

Figure 4.10

EXERCISE SET 4.8

1. Let R^2 have the inner product $\langle \mathbf{u}, \mathbf{v} \rangle = 3u_1v_1 + 2u_2v_2$, where $\mathbf{u} = (u_1, u_2)$ and $\mathbf{v} = (v_1, v_2)$. Find $\|\mathbf{w}\|$ when
(a) $\mathbf{w} = (-1, 3)$ (b) $\mathbf{w} = (6, 7)$ (c) $\mathbf{w} = (0, 1)$ (d) $\mathbf{w} = (0, 0)$

2. Repeat Exercise 1 using the Euclidean inner product on R^2.

3. Let P_2 have the inner product in Example 45. Find $\|\mathbf{p}\|$ when
(a) $\mathbf{p} = -1 + 2x + x^2$ (b) $\mathbf{p} = 3 - 4x^2$

4. Let M_{22} have the inner product in Example 44. Find $\|A\|$ when

(a) $A = \begin{bmatrix} -1 & 7 \\ 6 & 2 \end{bmatrix}$ (b) $A = \begin{bmatrix} 0 & 0 \\ 0 & 0 \end{bmatrix}$

5. Let R^2 have the inner product in Exercise 1. Find $d(\mathbf{x}, \mathbf{y})$ when
(a) $\mathbf{x} = (-1, 2), \mathbf{y} = (2, 5)$ (b) $\mathbf{x} = (3, 9), \mathbf{y} = (3, 9)$

6. Repeat Exercise 5 using the Euclidean inner product on R^2.

7. Let P_2 have the inner product in Example 45. Find $d(\mathbf{p}, \mathbf{q})$ when

$$\mathbf{p} = 2 - x + x^2, \qquad \mathbf{q} = 1 + 5x^2$$

8. Let M_{22} have the inner product in Example 44. Find $d(A, B)$ when

(a) $A = \begin{bmatrix} 1 & 5 \\ 8 & 3 \end{bmatrix}$ $B = \begin{bmatrix} -5 & 0 \\ 7 & -3 \end{bmatrix}$

(b) $A = \begin{bmatrix} 6 & 3 \\ 2 & 1 \end{bmatrix}$ $B = \begin{bmatrix} 6 & 3 \\ 2 & 1 \end{bmatrix}$

9. Let R^2, R^3, and R^4 have the Euclidean inner product. In each part find the cosine of the angle between $\mathbf{u}$ and $\mathbf{v}$.
(a) $\mathbf{u} = (1, -3), \mathbf{v} = (2, 4)$ (b) $\mathbf{u} = (-1, 0), \mathbf{v} = (3, 8)$
(c) $\mathbf{u} = (-1, 5, 2), \mathbf{v} = (2, 4, -9)$ (d) $\mathbf{u} = (4, 1, 8), \mathbf{v} = (1, 0, -3)$
(e) $\mathbf{u} = (1, 0, 1, 0), \mathbf{v} = (-3, -3, -3, -3)$ (f) $\mathbf{u} = (2, 1, 7, -1), \mathbf{v} = (4, 0, 0, 0)$

10. Let P_2 have the inner product in Example 45. Find the cosine of the angle between $\mathbf{p}$ and $\mathbf{q}$.
(a) $\mathbf{p} = -1 + 5x + 2x^2$ $\mathbf{q} = 2 + 4x - 9x^2$
(b) $\mathbf{p} = x - x^2$ $\mathbf{q} = 7 + 3x + 3x^2$

11. Let M_{22} have the inner product in Example 44. Find the cosine of the angle between A and B

(a) $A = \begin{bmatrix} 2 & 6 \\ 1 & -3 \end{bmatrix}$ $B = \begin{bmatrix} 3 & 2 \\ 1 & 0 \end{bmatrix}$

(b) $A = \begin{bmatrix} 2 & 4 \\ -1 & 3 \end{bmatrix}$ $B = \begin{bmatrix} -3 & 1 \\ 4 & 2 \end{bmatrix}$

12. Let R^3 have the Euclidean inner product. For which values of k are $\mathbf{u}$ and $\mathbf{v}$ orthogonal?
(a) $\mathbf{u} = (2, 1, 3)$ $\mathbf{v} = (1, 7, k)$
(b) $\mathbf{u} = (k, k, 1)$ $\mathbf{v} = (k, 5, 6)$

13. Let P_2 have the inner product in Example 45. Show that $\mathbf{p} = 1 - x + 2x^2$ and $\mathbf{q} = 2x + x^2$ are orthogonal.

14. Let M_{22} have the inner product in Example 44. Determine which of the following are orthogonal to

$$A = \begin{bmatrix} 2 & 1 \\ -1 & 3 \end{bmatrix}$$

(a) $\begin{bmatrix} -3 & 0 \\ 0 & 2 \end{bmatrix}$ (b) $\begin{bmatrix} 1 & 1 \\ 0 & -1 \end{bmatrix}$ (c) $\begin{bmatrix} 0 & 0 \\ 0 & 0 \end{bmatrix}$ (d) $\begin{bmatrix} 2 & 1 \\ 5 & 2 \end{bmatrix}$

15. Let R^4 have the Euclidean inner product. Find two vectors of norm 1 orthogonal to all of the vectors $\mathbf{u} = (2, 1, -4, 0)$, $\mathbf{v} = (-1, -1, 2, 2)$, and $\mathbf{w} = (3, 2, 5, 4)$.

16. Let V be an inner product space. Show that if $\mathbf{w}$ is orthogonal to both $\mathbf{u}_1$ and $\mathbf{u}_2$, it is orthogonal to $k_1\mathbf{u}_1 + k_2\mathbf{u}_2$ for all scalars k_1 and k_2. Interpret this result geometrically in R^3 with the Euclidean inner product.

17. Let V be an inner product space. Show that if $\mathbf{w}$ is orthogonal to each of the vectors $\mathbf{u}_1, \mathbf{u}_2, \ldots, \mathbf{u}_r$, then it is orthogonal to every vector in $\text{lin}\{\mathbf{u}_1, \mathbf{u}_2, \ldots, \mathbf{u}_r\}$.

18. Let V be an inner product space. Show that if $\mathbf{u}$ and $\mathbf{v}$ are orthogonal vectors in V such that $\|\mathbf{u}\| = \|\mathbf{v}\| = 1$, then $\|\mathbf{u} - \mathbf{v}\| = \sqrt{2}$.

19. Let V be an inner product space. Establish the identity

$$\|\mathbf{u} + \mathbf{v}\|^2 + \|\mathbf{u} - \mathbf{v}\|^2 = 2\|\mathbf{u}\|^2 + 2\|\mathbf{v}\|^2$$

for vectors in V.

20. Let V be an inner product space. Establish the identity

$$\langle \mathbf{u}, \mathbf{v} \rangle = \tfrac{1}{4}\|\mathbf{u} + \mathbf{v}\|^2 - \tfrac{1}{4}\|\mathbf{u} - \mathbf{v}\|^2$$

for vectors in V.

21. Let $\{\mathbf{v}_1, \mathbf{v}_2, \ldots, \mathbf{v}_r\}$ be a basis for an inner product space V. Show that the zero vector is the only vector in V that is orthogonal to all of the basis vectors.

22. Let $\mathbf{v}$ be a vector in an inner product space V.
(a) Show that the set of all vectors in V orthogonal to $\mathbf{v}$ forms a subspace of V.
(b) Describe the subspace geometrically in R^2 and R^3 with the Euclidean inner product.

23. Prove the following generalization of Theorem 17. If $\mathbf{v}_1, \mathbf{v}_2, \ldots, \mathbf{v}_r$ are pairwise orthogonal vectors in an inner product space V, then

$$\|\mathbf{v}_1 + \mathbf{v}_2 + \cdots + \mathbf{v}_r\|^2 = \|\mathbf{v}_1\|^2 + \|\mathbf{v}_2\|^2 + \cdots + \|\mathbf{v}_r\|^2$$

24. Prove the following parts of Theorem 16.
(a) part $L1$ (b) part $L2$ (c) part $L3$ (d) part $D1$
(e) part $D2$ (f) part $D3$ (g) part $D4$

25. Use vector methods to prove that a triangle inscribed in a circle and having a diameter for a side must be a right triangle. (***Hint.*** Express the vectors $\overrightarrow{AB}$ and $\overrightarrow{BC}$ in the following figure in terms of $\mathbf{u}$ and $\mathbf{v}$.)

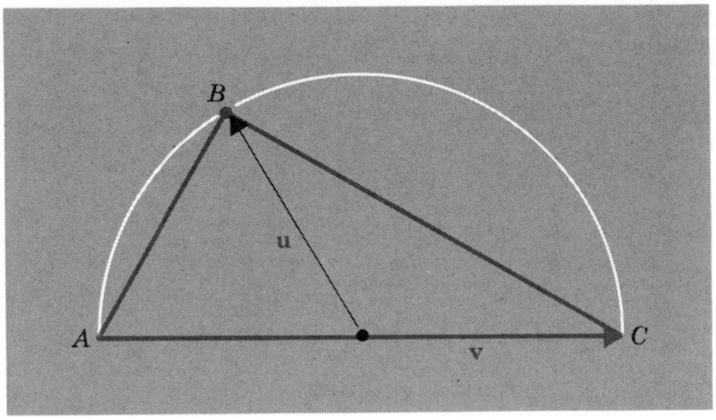

26. **(For readers who have studied calculus.)** Let $C[0, \pi]$ have the inner product

$$\langle \mathbf{f}, \mathbf{g} \rangle = \int_0^{\pi} f(x)g(x)\,dx$$

and let $\mathbf{f}_n = \cos nx (n = 0, 1, 2, \ldots)$. Show that if $k \neq l$, then $\mathbf{f}_k$ and $\mathbf{f}_l$ are orthogonal with respect to the given inner product.

4.9 ORTHONORMAL BASES; GRAM-SCHMIDT PROCESS

In many problems concerned with vector spaces, the selection of a basis for the space is at the discretion of the problem solver. Naturally, the best strategy is to choose the basis to simplify the solution of the problem at hand. In inner product spaces, it is often the case that the best choice is a basis in which all the vectors are orthogonal to one another. In this section we show how such bases can be constructed.

Definition. A set of vectors in an inner product space is called an ***orthogonal set*** if any two distinct vectors in the set are orthogonal. An orthogonal set in which each vector has norm 1 is called ***orthonormal***.

Example 53

Let

$$\mathbf{v}_1 = (0, 1, 0), \quad \mathbf{v}_2 = \left(\frac{1}{\sqrt{2}}, 0, \frac{1}{\sqrt{2}}\right), \quad \mathbf{v}_3 = \left(\frac{1}{\sqrt{2}}, 0, -\frac{1}{\sqrt{2}}\right).$$

The set $S = \{\mathbf{v}_1, \mathbf{v}_2, \mathbf{v}_3\}$ is orthonormal if R^3 has the Euclidean inner product, since

$$\langle \mathbf{v}_1, \mathbf{v}_2 \rangle = \langle \mathbf{v}_2, \mathbf{v}_3 \rangle = \langle \mathbf{v}_1, \mathbf{v}_3 \rangle = 0$$

and

$$\|\mathbf{v}_1\| = \|\mathbf{v}_2\| = \|\mathbf{v}_3\| = 1$$

Example 54

If $\mathbf{v}$ is a nonzero vector in an inner product space, then by property $L3$ in Figure 4.8 the vector

$$\frac{1}{\|\mathbf{v}\|}\mathbf{v}$$

has norm 1, since

$$\left\|\frac{1}{\|\mathbf{v}\|}\mathbf{v}\right\| = \frac{1}{\|\mathbf{v}\|}\|\mathbf{v}\| = 1$$

This process of multiplying a nonzero vector $\mathbf{v}$ by the reciprocal of its length to obtain a vector of norm 1 is called ***normalizing*** $\mathbf{v}$.

The interest in finding orthonormal bases for inner product spaces is, in part, motivated by the next theorem which shows that it is exceptionally simple to express a vector in terms of an orthonormal basis.

Theorem 18. *If $S = \{\mathbf{v}_1, \mathbf{v}_2, \ldots, \mathbf{v}_n\}$ is an orthonormal basis for an inner product space V, and $\mathbf{u}$ is any vector in V, then*

$$\mathbf{u} = \langle \mathbf{u}, \mathbf{v}_1\rangle \mathbf{v}_1 + \langle \mathbf{u}, \mathbf{v}_2\rangle \mathbf{v}_2 + \cdots + \langle \mathbf{u}, \mathbf{v}_n\rangle \mathbf{v}_n$$

Proof. Since $S = \{\mathbf{v}_1, \mathbf{v}_2, \ldots, \mathbf{v}_n\}$ is a basis, a vector $\mathbf{u}$ can be expressed in the form

$$\mathbf{u} = k_1\mathbf{v}_1 + k_2\mathbf{v}_2 + \cdots + k_n\mathbf{v}_n$$

We shall complete the proof by showing $k_i = \langle \mathbf{u}, \mathbf{v}_i\rangle$ for $i = 1, 2, \ldots, n$. For each vector $\mathbf{v}_i$ in S we have

$$\begin{aligned}\langle \mathbf{u}, \mathbf{v}_i\rangle &= \langle k_1\mathbf{v}_1 + k_2\mathbf{v}_2 + \cdots + k_n\mathbf{v}_n, \mathbf{v}_i\rangle \\ &= k_1\langle \mathbf{v}_1, \mathbf{v}_i\rangle + k_2\langle \mathbf{v}_2, \mathbf{v}_i\rangle + \cdots + k_n\langle \mathbf{v}_n, \mathbf{v}_i\rangle\end{aligned}$$

Since $S = \{\mathbf{v}_1, \mathbf{v}_2, \ldots, \mathbf{v}_n\}$ is an orthonormal set, we have

$$\langle \mathbf{v}_i, \mathbf{v}_i\rangle = \|\mathbf{v}_i\|^2 = 1 \qquad \text{and} \qquad \langle \mathbf{v}_i, \mathbf{v}_j\rangle = 0 \qquad \text{if } j \neq i$$

Therefore the above equation simplifies to

$$\langle \mathbf{u}, \mathbf{v}_i\rangle = k_i \quad \blacksquare$$

Example 55

Let

$$\mathbf{v}_1 = (0, 1, 0), \quad \mathbf{v}_2 = (-\tfrac{4}{5}, 0, \tfrac{3}{5}), \quad \mathbf{v}_3 = (\tfrac{3}{5}, 0, \tfrac{4}{5})$$

It is easy to check that $S = \{\mathbf{v}_1, \mathbf{v}_2, \mathbf{v}_3\}$ is an orthonormal basis for R^3 with the Euclidean inner product. Express the vector $\mathbf{u} = (1, 1, 1)$ as a linear combination of the vectors in S.

Solution.

$$\langle \mathbf{u}, \mathbf{v}_1 \rangle = 1 \qquad \langle \mathbf{u}, \mathbf{v}_2 \rangle = -\tfrac{1}{5} \qquad \text{and} \qquad \langle \mathbf{u}, \mathbf{v}_3 \rangle = \tfrac{7}{5}$$

Therefore by Theorem 18

$$\mathbf{u} = \mathbf{v}_1 - \tfrac{1}{5}\mathbf{v}_2 + \tfrac{7}{5}\mathbf{v}_3$$

that is,

$$(1, 1, 1) = (0, 1, 0) - \tfrac{1}{5}(-\tfrac{4}{5}, 0, \tfrac{3}{5}) + \tfrac{7}{5}(\tfrac{3}{5}, 0, \tfrac{4}{5})$$

The usefulness of Theorem 18 should be evident from this example if it is kept in mind that for nonorthonormal bases, it is usually necessary to solve a system of equations in order to express a vector in terms of a basis.

Theorem 19. *If $S = \{\mathbf{v}_1, \mathbf{v}_2, \ldots, \mathbf{v}_n\}$ is an orthogonal set of nonzero vectors in an inner product space, then S is linearly independent.*

Proof. Assume

$$k_1\mathbf{v}_1 + k_2\mathbf{v}_2 + \cdots + k_n\mathbf{v}_n = \mathbf{0} \tag{4.19}$$

To demonstrate that $S = \{\mathbf{v}_1, \mathbf{v}_2, \ldots, \mathbf{v}_n\}$ is linearly independent, we must prove that $k_1 = k_2 = \cdots = k_n = 0$.

For each $\mathbf{v}_i$ in S, it follows from (4.19) that

$$\langle k_1\mathbf{v}_1 + k_2\mathbf{v}_2 + \cdots + k_n\mathbf{v}_n, \mathbf{v}_i \rangle = \langle \mathbf{0}, \mathbf{v}_i \rangle = 0$$

or equivalently

$$k_1\langle \mathbf{v}_1, \mathbf{v}_i \rangle + k_2\langle \mathbf{v}_2, \mathbf{v}_i \rangle + \cdots + k_n\langle \mathbf{v}_n, \mathbf{v}_i \rangle = 0$$

From the orthogonality of S, $\langle \mathbf{v}_j, \mathbf{v}_i \rangle = 0$ when $j \neq i$, so that this equation reduces to

$$k_i\langle \mathbf{v}_i, \mathbf{v}_i \rangle = 0$$

Since the vectors in S are assumed to be nonzero, $\langle \mathbf{v}_i, \mathbf{v}_i \rangle \neq 0$ by the positivity axiom for inner products. Therefore $k_i = 0$. Since the subscript i is arbitrary, we have $k_1 = k_2 = \cdots = k_n = 0$; thus S is linearly independent. ∎

Example 56

In Example 53 we showed that

$$\mathbf{v}_1 = (0, 1, 0), \mathbf{v}_2 = \left(\frac{1}{\sqrt{2}}, 0, \frac{1}{\sqrt{2}}\right), \text{and } \mathbf{v}_3 = \left(\frac{1}{\sqrt{2}}, 0, -\frac{1}{\sqrt{2}}\right)$$

form an orthonormal set with respect to the Euclidean inner product on R^3. By Theorem 19, these vectors form a linearly independent set. Therefore, since R^3 is three dimensional, $S = \{\mathbf{v}_1, \mathbf{v}_2, \mathbf{v}_3\}$ is an orthonormal basis for R^3.

We now turn to the problem of constructing orthonormal bases for inner product spaces. The proof of the following preliminary result is discussed in the exercises at the end of this section.

Theorem 20. *Let V be an inner product space and $\{\mathbf{v}_1, \mathbf{v}_2, \ldots, \mathbf{v}_r\}$ an orthonormal set of vectors in V. If W denotes the space spanned by $\mathbf{v}_1, \mathbf{v}_2, \ldots, \mathbf{v}_r$, then every vector $\mathbf{u}$ in V can be expressed in the form*

$$\mathbf{u} = \mathbf{w}_1 + \mathbf{w}_2$$

where $\mathbf{w}_1$ is in W and $\mathbf{w}_2$ is orthogonal to W by letting

$$\mathbf{w}_1 = \langle \mathbf{u}, \mathbf{v}_1\rangle \mathbf{v}_1 + \langle \mathbf{u}, \mathbf{v}_2\rangle \mathbf{v}_2 + \cdots + \langle \mathbf{u}, \mathbf{v}_r\rangle \mathbf{v}_r \tag{4.20}$$

and

$$\mathbf{w}_2 = \mathbf{u} - \langle \mathbf{u}, \mathbf{v}_1\rangle \mathbf{v}_1 - \langle \mathbf{u}, \mathbf{v}_2\rangle \mathbf{v}_2 - \cdots - \langle \mathbf{u}, \mathbf{v}_r\rangle \mathbf{v}_r \tag{4.21}$$

(See Figure 4.11 for an illustration in R^3.)

Motivated by Figure 4.11, we call $\mathbf{w}_1$ the ***orthogonal projection of* u *on* W** and denote it by $\text{proj}_W \mathbf{u}$. The vector $\mathbf{w}_2 = \mathbf{u} - \text{proj}_W \mathbf{u}$ is called the ***component of* u *orthogonal to* W**.

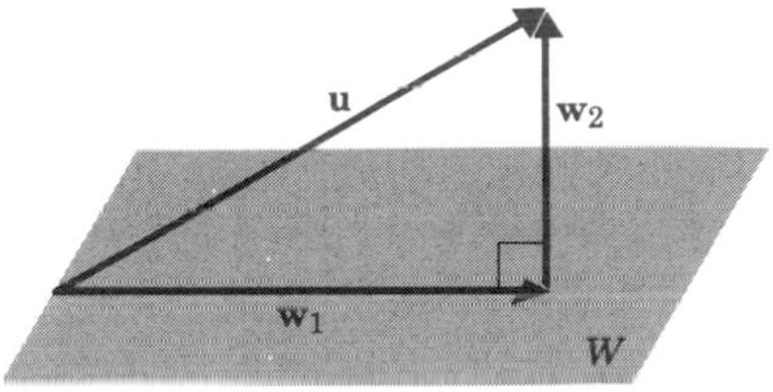

Figure 4.11

Example 57

Let R^3 have the Euclidean inner product, and let W be the subspace spanned by the orthonormal vectors $\mathbf{v}_1 = (0, 1, 0)$ and $\mathbf{v}_2 = (-\frac{4}{5}, 0, \frac{3}{5})$. The orthogonal projection of $\mathbf{u} = (1, 1, 1)$ on W is

$$\begin{aligned}\text{proj}_W \mathbf{u} &= \langle \mathbf{u}, \mathbf{v}_1\rangle \mathbf{v}_1 + \langle \mathbf{u}, \mathbf{v}_2\rangle \mathbf{v}_2 \\ &= (1)(0, 1, 0) + (-\tfrac{1}{5})(-\tfrac{4}{5}, 0, \tfrac{3}{5}) \\ &= (\tfrac{4}{25}, 1, -\tfrac{3}{25})\end{aligned}$$

The component of $\mathbf{u}$ orthogonal to W is

$$\mathbf{u} - \text{proj}_W \mathbf{u} = (1, 1, 1) - (\tfrac{4}{25}, 1, -\tfrac{3}{25}) = (\tfrac{21}{25}, 0, \tfrac{28}{25})$$

Observe that $\mathbf{u} - \text{proj}_W \mathbf{u}$ is orthogonal to both $\mathbf{v}_1$ and $\mathbf{v}_2$ so that this vector is orthogonal to each vector in the space W spanned by $\mathbf{v}_1$ and $\mathbf{v}_2$ as it should be.

We are now in a position to prove the main result of this section.

Theorem 21. *Every nonzero finite dimensional inner product space has an orthonormal basis.*

Proof. Let V be any nonzero, n-dimensional inner product space, and let $S = \{\mathbf{u}_1, \mathbf{u}_2, \ldots, \mathbf{u}_n\}$ be any basis for V. The following sequence of steps will produce an orthonormal basis $\{\mathbf{v}_1, \mathbf{v}_2, \ldots, \mathbf{v}_n\}$ for V.

Step 1. Let $\mathbf{v}_1 = \mathbf{u}_1/\|\mathbf{u}_1\|$. The vector $\mathbf{v}_1$ has norm 1.

Step 2. To construct a vector $\mathbf{v}_2$ of norm 1 that is orthogonal to $\mathbf{v}_1$, we compute the component of $\mathbf{u}_2$ orthogonal to the space W_1 spanned by $\mathbf{v}_1$ and then normalize it; that is,

$$\mathbf{v}_2 = \frac{\mathbf{u}_2 - \text{proj}_{W_1}\mathbf{u}_2}{\|\mathbf{u}_2 - \text{proj}_{W_1}\mathbf{u}_2\|} = \frac{\mathbf{u}_2 - \langle \mathbf{u}_2, \mathbf{v}_1\rangle \mathbf{v}_1}{\|\mathbf{u}_2 - \langle \mathbf{u}_2, \mathbf{v}_1\rangle \mathbf{v}_1\|}$$

(Figure 4.12). Of course, if $\mathbf{u}_2 - \langle \mathbf{u}_2, \mathbf{v}_1\rangle \mathbf{v}_1 = \mathbf{0}$, then we cannot carry out the normalization. But this cannot happen since we would then have

$$\mathbf{u}_2 = \langle \mathbf{u}_2, \mathbf{v}_1\rangle \mathbf{v}_1 = \frac{\langle \mathbf{u}_2, \mathbf{v}_1\rangle}{\|\mathbf{u}_1\|}\mathbf{u}_1$$

which says that $\mathbf{u}_2$ is a multiple of $\mathbf{u}_1$, contradicting the linear independence of the basis $S = \{\mathbf{u}_1, \mathbf{u}_2, \ldots, \mathbf{u}_n\}$.

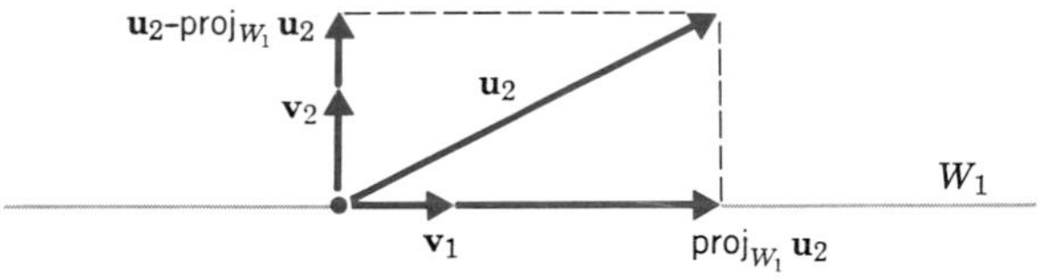

Figure 4.12

Step 3. To construct a vector $\mathbf{v}_3$ of norm 1 that is orthogonal to both $\mathbf{v}_1$ and $\mathbf{v}_2$, we compute the component of $\mathbf{u}_3$ orthogonal to the space W_2 spanned by $\mathbf{v}_1$ and $\mathbf{v}_2$ and normalize it (Figure 4.13); that is

$$\mathbf{v}_3 = \frac{\mathbf{u}_3 - \text{proj}_{W_2}\mathbf{u}_3}{\|\mathbf{u}_3 - \text{proj}_{W_2}\mathbf{u}_3\|} = \frac{\mathbf{u}_3 - \langle \mathbf{u}_3, \mathbf{v}_1\rangle \mathbf{v}_1 - \langle \mathbf{u}_3, \mathbf{v}_2\rangle \mathbf{v}_2}{\|\mathbf{u}_3 - \langle \mathbf{u}_3, \mathbf{v}_1\rangle \mathbf{v}_1 - \langle \mathbf{u}_3, \mathbf{v}_2\rangle \mathbf{v}_2\|}$$

As in Step 2, the linear independence of $\{\mathbf{u}_1, \mathbf{u}_2, \ldots, \mathbf{u}_n\}$ assures that

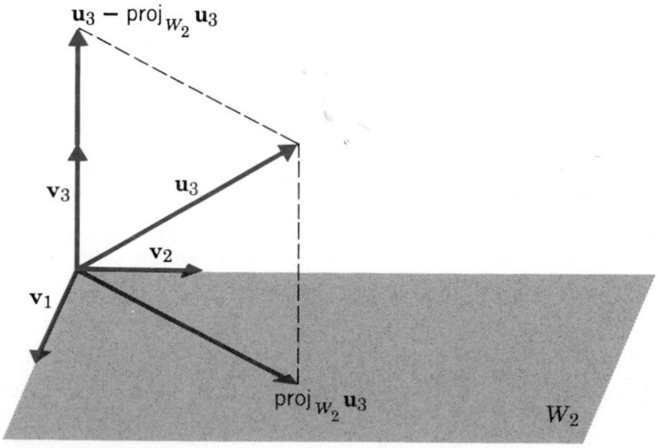

Figure 4.13

$\mathbf{u}_3 - \langle \mathbf{u}_3, \mathbf{v}_1 \rangle \mathbf{v}_1 - \langle \mathbf{u}_3, \mathbf{v}_2 \rangle \mathbf{v}_2 \neq \mathbf{0}$ so that the normalization can always be carried out. We leave the details as an exercise.

Step 4. To determine a vector $\mathbf{v}_4$ of norm 1 that is orthogonal to $\mathbf{v}_1$, $\mathbf{v}_2$, and $\mathbf{v}_3$, we compute the component of $\mathbf{u}_4$ orthogonal to the space W_3 spanned by $\mathbf{v}_1$, $\mathbf{v}_2$ and $\mathbf{v}_3$ and normalize it. Thus

$$\mathbf{v}_4 = \frac{\mathbf{u}_4 - \text{proj}_{W_4} \mathbf{u}_4}{\|\mathbf{u}_4 - \text{proj}_{W_4} \mathbf{u}_4\|} = \frac{\mathbf{u}_4 - \langle \mathbf{u}_4, \mathbf{v}_1 \rangle \mathbf{v}_1 - \langle \mathbf{u}_4, \mathbf{v}_2 \rangle \mathbf{v}_2 - \langle \mathbf{u}_4, \mathbf{v}_3 \rangle \mathbf{v}_3}{\|\mathbf{u}_4 - \langle \mathbf{u}_4, \mathbf{v}_1 \rangle \mathbf{v}_1 - \langle \mathbf{u}_4, \mathbf{v}_2 \rangle \mathbf{v}_2 - \langle \mathbf{u}_4, \mathbf{v}_3 \rangle \mathbf{v}_3\|}$$

Continuing in this way, we will obtain an orthonormal set of vectors, $\{\mathbf{v}_1, \mathbf{v}_2, \ldots, \mathbf{v}_n\}$. Since V is n-dimensional and every orthonormal set is linearly independent, the set $\{\mathbf{v}_1, \mathbf{v}_2, \ldots, \mathbf{v}_n\}$ will be an orthonormal basis for V. ∎

The above step by step construction for converting an arbitrary basis into an orthonormal basis is called the ***Gram-Schmidt* process***.

Example 58

Consider the vector space R^3 with the Euclidean inner product. Apply the Gram-Schmidt process to transform the basis $\mathbf{u}_1 = (1, 1, 1)$, $\mathbf{u}_2 = (0, 1, 1)$, $\mathbf{u}_3 = (0, 0, 1)$ into an orthonormal basis.

Solution.

Step 1.
$$\mathbf{v}_1 = \frac{\mathbf{u}_1}{\|\mathbf{u}_1\|} = \frac{(1, 1, 1)}{\sqrt{3}} = \left(\frac{1}{\sqrt{3}}, \frac{1}{\sqrt{3}}, \frac{1}{\sqrt{3}}\right)$$

Step 2.
$$\begin{aligned}\mathbf{u}_2 - \text{proj}_{W_1} \mathbf{u}_2 &= \mathbf{u}_2 - \langle \mathbf{u}_2, \mathbf{v}_1 \rangle \mathbf{v}_1 \\ &= (0, 1, 1) - \frac{2}{\sqrt{3}}\left(\frac{1}{\sqrt{3}}, \frac{1}{\sqrt{3}}, \frac{1}{\sqrt{3}}\right) \\ &= \left(-\frac{2}{3}, \frac{1}{3}, \frac{1}{3}\right)\end{aligned}$$

Therefore

$$\mathbf{v}_2 = \frac{\mathbf{u}_2 - \text{proj}_{W_1} \mathbf{u}_2}{\|\mathbf{u}_2 - \text{proj}_{W_1} \mathbf{u}_2\|} = \frac{3}{\sqrt{6}}\left(-\frac{2}{3}, \frac{1}{3}, \frac{1}{3}\right) = \left(-\frac{2}{\sqrt{6}}, \frac{1}{\sqrt{6}}, \frac{1}{\sqrt{6}}\right)$$

Step 3.
$$\begin{aligned}\mathbf{u}_3 - \text{proj}_{W_2} \mathbf{u}_3 &= \mathbf{u}_3 - \langle \mathbf{u}_3, \mathbf{v}_1 \rangle \mathbf{v}_1 - \langle \mathbf{u}_3, \mathbf{v}_2 \rangle \mathbf{v}_2 \\ &= (0, 0, 1) - \frac{1}{\sqrt{3}}\left(\frac{1}{\sqrt{3}}, \frac{1}{\sqrt{3}}, \frac{1}{\sqrt{3}}\right) - \frac{1}{\sqrt{6}}\left(-\frac{2}{\sqrt{6}}, \frac{1}{\sqrt{6}}, \frac{1}{\sqrt{6}}\right) \\ &= \left(0, -\frac{1}{2}, \frac{1}{2}\right)\end{aligned}$$

* *Jörgen Pederson Gram* (1850–1916). Danish actuary.
Erhardt Schmidt (1876–1959). German mathematician.

Therefore

$$\mathbf{v}_3 = \frac{\mathbf{u}_3 - \text{proj}_{W_2}\,\mathbf{u}_3}{\|\mathbf{u}_3 - \text{proj}_{W_2}\,\mathbf{u}_3\|} = \sqrt{2}\left(0, -\frac{1}{2}, \frac{1}{2}\right) = \left(0, -\frac{1}{\sqrt{2}}, \frac{1}{\sqrt{2}}\right)$$

Thus

$$\mathbf{v}_1 = \left(\frac{1}{\sqrt{3}}, \frac{1}{\sqrt{3}}, \frac{1}{\sqrt{3}}\right), \mathbf{v}_2 = \left(-\frac{2}{\sqrt{6}}, \frac{1}{\sqrt{6}}, \frac{1}{\sqrt{6}}\right), \mathbf{v}_3 = \left(0, -\frac{1}{\sqrt{2}}, \frac{1}{\sqrt{2}}\right)$$

form an orthonormal basis for R^3.

OPTIONAL

The following consequences of the Gram-Schmidt process have numerous applications, some of which are discussed in Section 7.2. The reader will have the background to read these applications after completing this optional section.

Theorem 22. (Projection Theorem). *If W is a finite dimensional subspace of an inner product space V then every vector $\mathbf{u}$ in V can be expressed in exactly one way as*

$$\mathbf{u} = \mathbf{w}_1 + \mathbf{w}_2$$

where $\mathbf{w}_1$ is in W and $\mathbf{w}_2$ is orthogonal to W.

Proof. There are two parts to the proof. First we must find vectors $\mathbf{w}_1$ and $\mathbf{w}_2$ with the stated properties, and then we must show that these are the only such vectors.

By the Gram-Schmidt process there is an orthonormal basis $\{\mathbf{v}_1, \mathbf{v}_2, \ldots, \mathbf{v}_r\}$ for W, so that $W = \text{lin}\{\mathbf{v}_1, \mathbf{v}_2, \ldots, \mathbf{v}_r\}$. Therefore, by Theorem 20, the vectors

$$\mathbf{w}_1 = \text{proj}_W\,\mathbf{u} \quad \text{and} \quad \mathbf{w}_2 = \mathbf{u} - \text{proj}_W\,\mathbf{u}$$

will have the properties stated in this theorem. To see that these are the only vectors with these properties suppose that we can also write

$$\mathbf{u} = \mathbf{w}'_1 + \mathbf{w}'_2 \tag{4.22}$$

where $\mathbf{w}'_1$ is in W and $\mathbf{w}'_2$ is orthogonal to W. If we subtract from (4.22) the equation

$$\mathbf{u} = \mathbf{w}_1 + \mathbf{w}_2$$

we obtain

$$\mathbf{0} = (\mathbf{w}'_1 - \mathbf{w}_1) + (\mathbf{w}'_2 - \mathbf{w}_2)$$

or

$$\mathbf{w}_1 - \mathbf{w}'_1 = \mathbf{w}'_2 - \mathbf{w}_2 \tag{4.23}$$

Since $\mathbf{w}_2$ and $\mathbf{w}'_2$ are orthogonal to W, their difference will also be orthogonal to W, since for any vector $\mathbf{w}$ in W we can write

$$\langle \mathbf{w}, \mathbf{w}'_2 - \mathbf{w}_2 \rangle = \langle \mathbf{w}, \mathbf{w}'_2 \rangle - \langle \mathbf{w}, \mathbf{w}_2 \rangle = 0 - 0 = 0$$

But $\mathbf{w}_2' - \mathbf{w}_2$ is itself a vector in W, since from (4.23) it is a difference of two vectors in the subspace W. Thus $\mathbf{w}_2' - \mathbf{w}_2$ must be orthogonal to itself, that is

$$\langle \mathbf{w}_2' - \mathbf{w}_2, \mathbf{w}_2' - \mathbf{w}_2 \rangle = 0$$

But this implies that $\mathbf{w}_2' - \mathbf{w}_2 = 0$ by Axiom 4 for inner products. Thus $\mathbf{w}_2' = \mathbf{w}_2$ and by (4.23) $\mathbf{w}_1' = \mathbf{w}_1$ ▮

If P is a point in ordinary 3-space and W is a plane through the origin, then the point Q in W, closest to P, is obtained by dropping a perpendicular from P to W (Figure 4.14*a*). Therefore, if we let $\mathbf{u} = \overrightarrow{OP}$ then the distance between P and W is given by

$$\|\mathbf{u} - \text{proj}_W \mathbf{u}\|$$

In other words, among all vectors $\mathbf{w}$ in W, the vector $\mathbf{w} = \text{proj}_W \mathbf{u}$ minimizes the distance $\|\mathbf{u} - \mathbf{w}\|$ (Figure 4.14*b*).

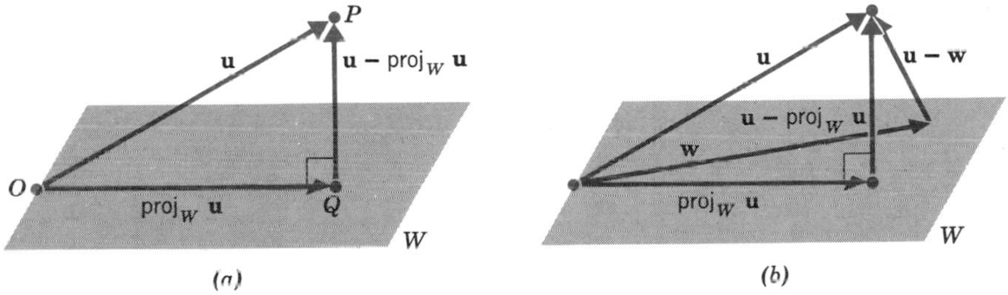

Figure 4.14

There is another way of thinking about this idea. View $\mathbf{u}$ as a fixed vector that we would like to approximate by a vector in W. Any such approximation $\mathbf{w}$ will result in an "error vector"

$$\mathbf{u} - \mathbf{w}$$

which, unless $\mathbf{u}$ is in W, cannot be made equal to $\mathbf{0}$. However, by choosing

$$\mathbf{w} = \text{proj}_W \mathbf{u}$$

we can make the length of the error vector

$$\|\mathbf{u} - \mathbf{w}\| = \|\mathbf{u} - \text{proj}_W \mathbf{u}\|$$

as small as possible. Thus we can describe $\text{proj}_W \mathbf{u}$ as the "best approximation" to $\mathbf{u}$ by vectors in W. The following theorem will make these intuitive ideas precise.

Theorem 23. (Best Approximation Theorem). If W is a finite dimensional subspace of an inner product space V, and if $\mathbf{u}$ is a vector in V, then $\text{proj}_W \mathbf{u}$ is the

best approximation to **u** from W in the sense that

$$\|\mathbf{u} - \text{proj}_W \mathbf{u}\| < \|\mathbf{u} - \mathbf{w}\|$$

for every vector **w** in W different from proj_W **u**.

Proof. For any vector **w** in W we can write

$$\mathbf{u} - \mathbf{w} = (\mathbf{u} - \text{proj}_W \mathbf{u}) + (\text{proj}_W \mathbf{u} - \mathbf{w}) \tag{4.24}$$

But $\text{proj}_W \mathbf{u} - \mathbf{w}$, being a difference of vectors in W, is in W; and $\mathbf{u} - \text{proj}_W \mathbf{u}$ is orthogonal to W, so that the two terms on the right side of (4.24) are orthogonal. Thus by the Theorem by Pythagoras (Theorem 17 of Section 4.8)

$$\|\mathbf{u} - \mathbf{w}\|^2 = \|\mathbf{u} - \text{proj}_W \mathbf{u}\|^2 + \|\text{proj}_W \mathbf{u} - \mathbf{w}\|^2$$

If $\mathbf{w} \neq \text{proj}_W \mathbf{u}$ then the second term in this sum will be positive, so that

$$\|\mathbf{u} - \mathbf{w}\|^2 > \|\mathbf{u} - \text{proj}_W \mathbf{u}\|^2$$

or equivalently

$$\|\mathbf{u} - \mathbf{w}\| > \|\mathbf{u} - \text{proj}_W \mathbf{u}\| \quad \blacksquare$$

Applications of the last two theorems are given in Section 7.2.

EXERCISE SET 4.9

1. Let R^2 have the Euclidean inner product. Which of the following form orthonormal sets?

(a) $(1, 0), (0, 2)$

(b) $\left(\frac{1}{\sqrt{2}}, -\frac{1}{\sqrt{2}}\right), \left(\frac{1}{\sqrt{2}}, \frac{1}{\sqrt{2}}\right)$

(c) $\left(\frac{1}{\sqrt{2}}, \frac{1}{\sqrt{2}}\right), \left(-\frac{1}{\sqrt{2}}, -\frac{1}{\sqrt{2}}\right)$

(d) $(1, 0), (0, 0)$

2. Let R^3 have the Euclidean inner product. Which of the following form orthonormal sets?

(a) $\left(\frac{1}{\sqrt{2}}, 0, \frac{1}{\sqrt{2}}\right), \left(\frac{1}{\sqrt{3}}, \frac{1}{\sqrt{3}}, -\frac{1}{\sqrt{3}}\right), \left(-\frac{1}{\sqrt{2}}, 0, \frac{1}{\sqrt{2}}\right)$

(b) $\left(\frac{2}{3}, -\frac{2}{3}, \frac{1}{3}\right), \quad \left(\frac{2}{3}, \frac{1}{3}, -\frac{2}{3}\right), \quad \left(\frac{1}{3}, \frac{2}{3}, \frac{2}{3}\right)$

(c) $(1, 0, 0), \quad \left(0, \frac{1}{\sqrt{2}}, \frac{1}{\sqrt{2}}\right), \quad (0, 0, 1)$

(d) $\left(\frac{1}{\sqrt{6}}, \frac{1}{\sqrt{6}}, -\frac{2}{\sqrt{6}}\right), \left(\frac{1}{\sqrt{2}}, -\frac{1}{\sqrt{2}}, 0\right)$

3. Let P_2 have the inner product in Example 45. Which of the following form orthonormal sets?

(a) $\frac{2}{3} - \frac{2}{3}x + \frac{1}{3}x^2,\ \frac{2}{3} + \frac{1}{3}x - \frac{2}{3}x^2,\ \frac{1}{3} + \frac{2}{3}x + \frac{2}{3}x^2$ (b) $1,\ \frac{1}{\sqrt{2}}x + \frac{1}{\sqrt{2}}x^2,\ x^2$

4. Let M_{22} have the inner product in Example 42. Which of the following form orthonormal sets?

(a) $\begin{bmatrix} 1 & 0 \\ 0 & 0 \end{bmatrix}\ \begin{bmatrix} 0 & \frac{2}{3} \\ \frac{1}{3} & -\frac{2}{3} \end{bmatrix}\ \begin{bmatrix} 0 & \frac{2}{3} \\ -\frac{2}{3} & \frac{1}{3} \end{bmatrix}\ \begin{bmatrix} 0 & \frac{1}{3} \\ \frac{2}{3} & \frac{2}{3} \end{bmatrix}$

(b) $\begin{bmatrix} 1 & 0 \\ 0 & 0 \end{bmatrix}\ \begin{bmatrix} 0 & 1 \\ 0 & 0 \end{bmatrix}\ \begin{bmatrix} 0 & 0 \\ 1 & 1 \end{bmatrix}\ \begin{bmatrix} 0 & 0 \\ 1 & -1 \end{bmatrix}$

5. Let $\mathbf{x} = \left(\frac{1}{\sqrt{5}}, -\frac{1}{\sqrt{5}}\right)$ and $\mathbf{y} = \left(\frac{2}{\sqrt{30}}, \frac{3}{\sqrt{30}}\right)$.

Show that $\{\mathbf{x}, \mathbf{y}\}$ is orthonormal if R^2 has the inner product $\langle \mathbf{u}, \mathbf{v} \rangle = 3u_1v_1 + 2u_2v_2$, but is not orthonormal if R^2 has the Euclidean inner product.

6. Show that

$$\mathbf{u}_1 = (1, 0, 0, 1),\ \mathbf{u}_2 = (-1, 0, 2, 1),\ \mathbf{u}_3 = (2, 3, 2, -2),\ \mathbf{u}_4 = (-1, 2, -1, 1)$$

is an orthogonal set in R^4 with the Euclidean inner product. By normalizing each of these vectors, obtain an orthonormal set.

7. Let R^2 have the Euclidean inner product. Use the Gram-Schmidt process to transform the basis $\{\mathbf{u}_1, \mathbf{u}_2\}$ into an orthonormal basis.
(a) $\mathbf{u}_1 = (1, -3), \mathbf{u}_2 = (2, 2)$ (b) $\mathbf{u}_1 = (1, 0), \mathbf{u}_2 = (3, -5)$

8. Let R^3 have the Euclidean inner product. Use the Gram-Schmidt process to transform the basis $\{\mathbf{u}_1, \mathbf{u}_2, \mathbf{u}_3\}$ into an orthonormal basis.
(a) $\mathbf{u}_1 = (1, 1, 1), \mathbf{u}_2 = (-1, 1, 0), \mathbf{u}_3 = (1, 2, 1)$
(b) $\mathbf{u}_1 = (1, 0, 0), \mathbf{u}_2 = (3, 7, -2), \mathbf{u}_3 = (0, 4, 1)$

9. Let R^4 have the Euclidean inner product. Use the Gram-Schmidt process to transform the basis $\{\mathbf{u}_1, \mathbf{u}_2, \mathbf{u}_3, \mathbf{u}_4\}$ into an orthonormal basis.

$$\mathbf{u}_1 = (0, 2, 1, 0),\ \mathbf{u}_2 = (1, -1, 0, 0),\ \mathbf{u}_3 = (1, 2, 0, -1),\ \mathbf{u}_4 = (1, 0, 0, 1)$$

10. Let R^3 have the Euclidean inner product. Find an orthonormal basis for the subspace spanned by $(0, 1, 2), (-1, 0, 1)$.

11. Let R^3 have the inner product $\langle \mathbf{u}, \mathbf{v} \rangle = u_1v_1 + 2u_2v_2 + 3u_3v_3$. Use the Gram-Schmidt process to transform

$$\mathbf{u}_1 = (1, 1, 1) \qquad \mathbf{u}_2 = (1, 1, 0) \qquad \mathbf{u}_3 = (1, 0, 0)$$

into an orthonormal basis.

12. The subspace of R^3 spanned by the vectors $\mathbf{u}_1 = (\frac{4}{5}, 0, -\frac{3}{5})$ and $\mathbf{u}_2 = (0, 1, 0)$ is a plane passing through the origin. Express $\mathbf{w} = (1, 2, 3)$ in the form $\mathbf{w} = \mathbf{w}_1 + \mathbf{w}_2$, where $\mathbf{w}_1$ lies in the plane and $\mathbf{w}_2$ is perpendicular to the plane.

13. Repeat Exercise 12 with $\mathbf{u}_1 = (1, 1, 1)$ and $\mathbf{u}_2 = (2, 0, -1)$.

14. Let R^4 have the Euclidean inner product. Express $\mathbf{w} = (-1, 2, 6, 0)$ in the form $\mathbf{w} = \mathbf{w}_1 + \mathbf{w}_2$, where $\mathbf{w}_1$ is in the space W spanned by $\mathbf{u}_1 = (-1, 0, 1, 2)$ and $\mathbf{u}_2 = (0, 1, 0, 1)$, and $\mathbf{w}_2$ is orthogonal to W.

15. Let $\{\mathbf{v}_1, \mathbf{v}_2, \mathbf{v}_3\}$ be an orthonormal basis for an inner product space V. Show that if $\mathbf{w}$ is a vector in V, then $\|\mathbf{w}\|^2 = \langle \mathbf{w}, \mathbf{v}_1 \rangle^2 + \langle \mathbf{w}, \mathbf{v}_2 \rangle^2 + \langle \mathbf{w}, \mathbf{v}_3 \rangle^2$.

16. Let $\{\mathbf{v}_1, \mathbf{v}_2, \ldots, \mathbf{v}_n\}$ be an orthonormal basis for an inner product space V. Show that if $\mathbf{w}$ is a vector in V, then

$$\|\mathbf{w}\|^2 = \langle \mathbf{w}, \mathbf{v}_1 \rangle^2 + \langle \mathbf{w}, \mathbf{v}_2 \rangle^2 + \cdots + \langle \mathbf{w}, \mathbf{v}_n \rangle^2$$

17. In Step 3 of the proof of Theorem 21, it was stated that, "the linear independence of $\{\mathbf{u}_1, \mathbf{u}_2, \ldots, \mathbf{u}_n\}$ assures that

$$\mathbf{u}_3 - \langle \mathbf{u}_3, \mathbf{v}_1 \rangle \mathbf{v}_1 - \langle \mathbf{u}_3, \mathbf{v}_2 \rangle \mathbf{v}_2 \neq \mathbf{0}."$$

Prove this statement.

18. Prove Theorem 20.
(***Hint.*** Show that the vector $\mathbf{w}_1$ in (4.20) lies in W, the vector $\mathbf{w}_2$ in (4.21) is orthogonal to W, and that $\mathbf{u} = \mathbf{w}_1 + \mathbf{w}_2$.

19. **(For readers who have studied calculus.)** Let the vector space P_2 have the inner product

$$\langle \mathbf{p}, \mathbf{q} \rangle = \int_{-1}^{1} p(x)q(x)\, dx$$

Apply the Gram-Schmidt process to transform the standard basis $S = \{1, x, x^2\}$ into an orthonormal basis. (The polynomials in the resulting basis are called the first three ***normalized Legendre polynomials***.)

20. **(For readers who have studied calculus.)** Use Theorem 16 to express the following as linear combinations of the first three normalized Legendre polynomials (Exercise 19).
(a) $1 + x + 4x^2$ (b) $2 - 7x^2$ (c) $4 + 3x$

21. **(For readers who have studied calculus.)** Let P_2 have the inner product

$$\langle \mathbf{p}, \mathbf{q} \rangle = \int_{0}^{1} p(x)q(x)\, dx$$

Apply the Gram-Schmidt process to transform the standard basis $S = \{1, x, x^2\}$ into an orthonormal basis.

22. **(For readers who have studied the optional material in this section.)** Find the point Q in the plane $5x - 3y + z = 0$ closest to $P(1, -2, 4)$, and determine the distance between the point P and the plane. (***Hint.*** View the plane as a subspace W of R^3 with the Euclidean inner product and apply Theorem 23.)

23. **(For readers who have studied the optional material in this section.)** Find the point Q on the line

$$\begin{aligned} x &= 2t \\ y &= -t \qquad -\infty < t < +\infty \\ z &= 4t \end{aligned}$$

closest to $P(-4, 8, 1)$. (***Hint.*** See the hint in the previous exercise.)

4.10 COORDINATES; CHANGE OF BASIS

There is a close relationship between the notion of a basis and the notion of a coordinate system. In this section we develop this idea and also discuss results about changing bases for vector spaces.

In plane analytic geometry we associate a pair of coordinates (a, b) with a point P in the plane by using two perpendicular coordinate axes. However, coordinates can also be introduced without reference to coordinate axes by using vectors. For example, instead of introducing coordinate axes as in Figure 4.15a, consider two perpendicular vectors $\mathbf{v}_1$ and $\mathbf{v}_2$, each of length 1, and having the same initial point O. (These vectors form a basis for R^2.) By dropping perpendiculars from a point P onto the lines determined by $\mathbf{v}_1$ and $\mathbf{v}_2$ we obtain vectors $a\mathbf{v}_1$ and $b\mathbf{v}_2$ such that

$$\overrightarrow{OP} = a\mathbf{v}_1 + b\mathbf{v}_2$$

(Figure 4.15b). Clearly, the numbers a and b just obtained are the same as the coordinates of P relative to the coordinate system in Figure 4.15a. Thus we can view the coordinates of P as the numbers needed to express the vector $\overrightarrow{OP}$ in terms of the basis vectors $\mathbf{v}_1$ and $\mathbf{v}_2$.

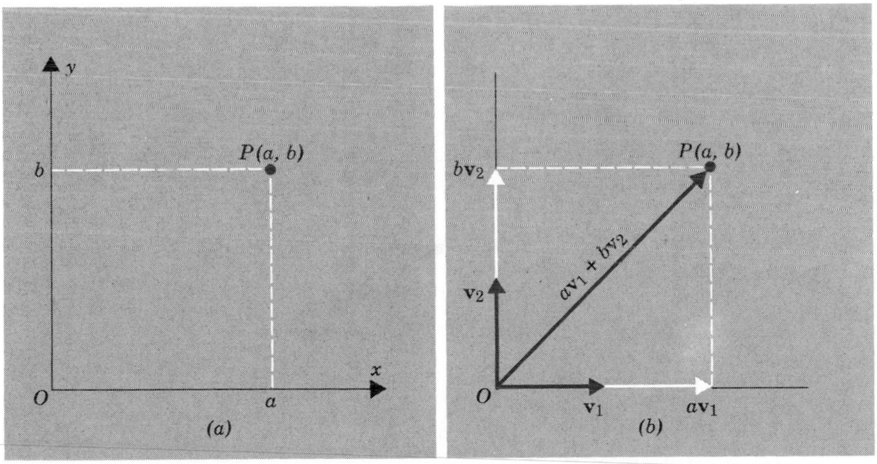

Figure 4.15

For the purposes of attaching coordinates to points in the plane it is not essential that the basis vectors $\mathbf{v}_1$ and $\mathbf{v}_2$ be perpendicular or have length 1; any basis for R^2 will do. For example, using the basis vectors $\mathbf{v}_1$ and $\mathbf{v}_2$ in Figure 4.16, we can attach a unique pair of coordinates to a point P by projecting P parallel

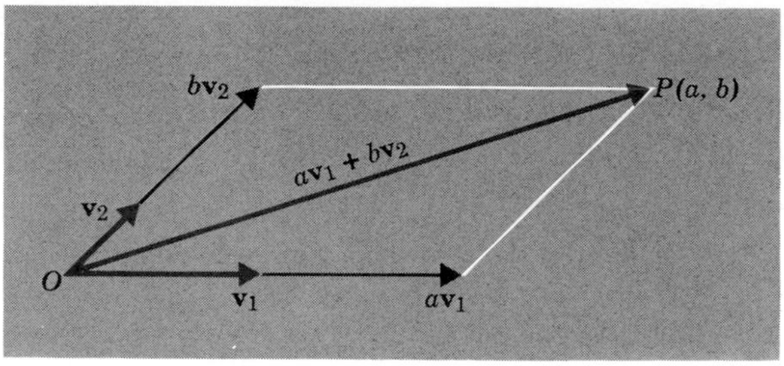

Figure 4.16

to the basis vectors in order to make $\overrightarrow{OP}$ the diagonal of a parallelogram determined by vectors $a\mathbf{v}_1$ and $b\mathbf{v}_2$; thus

$$\overrightarrow{OP} = a\mathbf{v}_1 + b\mathbf{v}_2$$

We can regard (a, b) as the coordinates of P relative to the basis $\{\mathbf{v}_1, \mathbf{v}_2\}$. This generalized notion of coordinates is important because it can be extended to more general vector spaces. But first we will need some preliminary results.

Suppose $S = \{\mathbf{v}_1, \mathbf{v}_2, \ldots, \mathbf{v}_n\}$ is a basis for a finite dimensional vector space V. Since S spans V, every vector in V is expressible as a linear combination of vectors in S. Moreover, the linear independence of S assures that there is only *one* way to express a vector as a linear combination of vectors in S. To see why, suppose a vector $\mathbf{v}$ can be written as

$$\mathbf{v} = c_1\mathbf{v}_1 + c_2\mathbf{v}_2 + \cdots + c_n\mathbf{v}_n$$

and also

$$\mathbf{v} = k_1\mathbf{v}_1 + k_2\mathbf{v}_2 + \cdots + k_n\mathbf{v}_n$$

Subtracting the second equation from the first gives

$$\mathbf{0} = (c_1 - k_1)\mathbf{v}_1 + (c_2 - k_2)\mathbf{v}_2 + \cdots + (c_n - k_n)\mathbf{v}_n$$

Since the right side of this equation is a linear combination of vectors in S, the linear independence of S implies that

$$c_1 - k_1 = 0, \qquad c_2 - k_2 = 0, \ldots, c_n - k_n = 0$$

that is,

$$c_1 = k_1, \qquad c_2 = k_2, \ldots, c_n = k_n$$

In summary, we have the following result.

Theorem 24. *If $S = \{\mathbf{v}_1, \mathbf{v}_2, \ldots, \mathbf{v}_n\}$ is a basis for a vector space V, then every vector $\mathbf{v}$ in V can be expressed in the form $\mathbf{v} = c_1\mathbf{v}_1 + c_2\mathbf{v}_2 + \cdots + c_n\mathbf{v}_n$ in exactly one way.*

If $S = \{\mathbf{v}_1, \mathbf{v}_2, \ldots, \mathbf{v}_n\}$ is a basis for a finite dimensional vector space V, and

$$\mathbf{v} = c_1\mathbf{v}_1 + c_2\mathbf{v}_2 + \cdots + c_n\mathbf{v}_n$$

is the expression for $\mathbf{v}$ in terms of the basis S, then the scalars $c_1, c_2, \ldots, c_n$ are called the ***coordinates*** of $\mathbf{v}$ relative to the basis S. The ***coordinate vector*** of $\mathbf{v}$ relative to S is denoted by $(\mathbf{v})_S$ and is the vector in R^n defined by

$$(\mathbf{v})_S = (c_1, c_2, \ldots, c_n)$$

The ***coordinate matrix*** of $\mathbf{v}$ relative to S is denoted by $[\mathbf{v}]_S$ and is the $n \times 1$ matrix defined by

$$\begin{bmatrix} c_1 \\ c_2 \\ \vdots \\ c_n \end{bmatrix}$$

Example 59

In Example 28 of Section 4.5 we showed $S = \{\mathbf{v}_1, \mathbf{v}_2, \mathbf{v}_3\}$ is a basis for R^3, where $\mathbf{v}_1 = (1, 2, 1)$, $\mathbf{v}_2 = (2, 9, 0)$, and $\mathbf{v}_3 = (3, 3, 4)$.

(a) Find the coordinate vector and coordinate matrix of $\mathbf{v} = (5, -1, 9)$ with respect to S.
(b) Find the vector $\mathbf{v}$ in R^3 whose coordinate vector with respect to S is $(\mathbf{v})_S = (-1, 3, 2)$.

Solution (*a*). We must find scalars c_1, c_2, c_3 such that

$$\mathbf{v} = c_1\mathbf{v}_1 + c_2\mathbf{v}_2 + c_3\mathbf{v}_3$$

or in terms of components

$$(5, -1, 9) = c_1(1, 2, 1) + c_2(2, 9, 0) + c_3(3, 3, 4)$$

Equating corresponding components gives

$$\begin{aligned} c_1 + 2c_2 + 3c_3 &= 5 \\ 2c_1 + 9c_2 + 3c_3 &= -1 \\ c_1 \qquad\quad + 4c_3 &= 9 \end{aligned}$$

Solving this system, we obtain $c_1 = 1$, $c_2 = -1$, $c_3 = 2$. Therefore

$$(\mathbf{v})_S = (1, -1, 2) \qquad \text{and} \qquad [\mathbf{v}]_S = \begin{bmatrix} 1 \\ -1 \\ 2 \end{bmatrix}$$

Solution (*b*). Using the definition of the coordinate vector $(\mathbf{v})_S$, we obtain

$$\mathbf{v} = (-1)\mathbf{v}_1 + 3\mathbf{v}_2 + 2\mathbf{v}_3 = (11, 31, 7)$$

Coordinate vectors and matrices depend on the order in which the basis vectors are written; a change in the order of the basis vectors results in a corresponding change of order for the entries in the coordinate matrices and coordinate vectors.

Example 60

Consider the basis $S = \{1, x, x^2\}$ for P_2. By inspection the coordinate vector and coordinate matrix with respect to S for a polynomial $\mathbf{p} = a_0 + a_1x + a_2x^2$ are

$$(\mathbf{p})_S = (a_0, a_1, a_2) \qquad \text{and} \qquad [\mathbf{p}]_S = \begin{bmatrix} a_0 \\ a_1 \\ a_2 \end{bmatrix}$$

Example 61

Let a rectangular xyz-coordinate system be introduced into 3-space and consider the standard basis $S = \{\mathbf{i}, \mathbf{j}, \mathbf{k}\}$ where

$$\mathbf{i} = (1, 0, 0), \quad \mathbf{j} = (0, 1, 0), \text{ and } \mathbf{k} = (0, 0, 1)$$

If, as in Figure 4.17, $\mathbf{v} = (a, b, c)$ is any vector in R^3 then

$$\mathbf{v} = (a, b, c) = a(1, 0, 0) + b(0, 1, 0) + c(0, 0, 1) = a\mathbf{i} + b\mathbf{j} + c\mathbf{k}$$

which means

$$\mathbf{v} = (a, b, c) = (\mathbf{v})_S$$

In other words the components of a vector $\mathbf{v}$ relative to a rectangular xyz-coordinate system are the same as the coordinates of $\mathbf{v}$ relative to the standard basis $\{\mathbf{i}, \mathbf{j}, \mathbf{k}\}$.

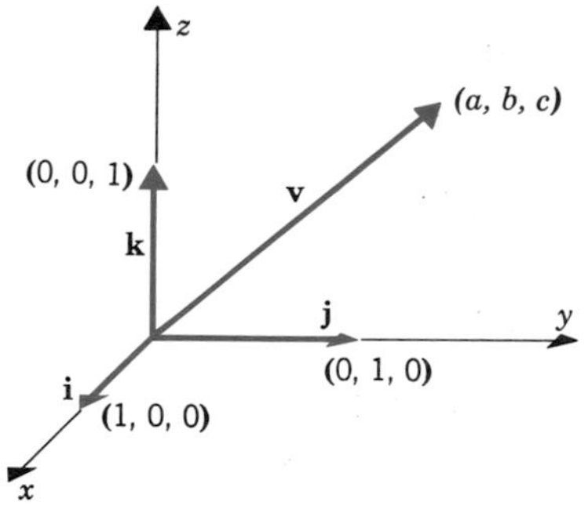

Figure 4.17

Example 62

If $S = \{\mathbf{v}_1, \mathbf{v}_2, \ldots, \mathbf{v}_n\}$ is an orthonormal basis for an inner product space V, then by Theorem 18 of Section 4.9 the expression for a vector $\mathbf{u}$ in terms of the basis S is

$$\mathbf{u} = \langle \mathbf{u}, \mathbf{v}_1 \rangle \mathbf{v}_1 + \langle \mathbf{u}, \mathbf{v}_2 \rangle \mathbf{v}_2 + \cdots + \langle \mathbf{u}, \mathbf{v}_n \rangle \mathbf{v}_n$$

which means that

$$(\mathbf{u})_S = (\langle \mathbf{u}, \mathbf{v}_1 \rangle, \langle \mathbf{u}, \mathbf{v}_2 \rangle, \ldots, \langle \mathbf{u}, \mathbf{v}_n \rangle)$$

and

$$[\mathbf{u}]_S = \begin{bmatrix} \langle \mathbf{u}, \mathbf{v}_1 \rangle \\ \langle \mathbf{u}, \mathbf{v}_2 \rangle \\ \vdots \\ \langle \mathbf{u}, \mathbf{v}_n \rangle \end{bmatrix}$$

For example, if

$$\mathbf{v}_1 = (0, 1, 0), \mathbf{v}_2 = (-\tfrac{4}{5}, 0, \tfrac{3}{5}), \mathbf{v}_3 = (\tfrac{3}{5}, 0, \tfrac{4}{5})$$

then, as observed in Example 55 of Section 4.9, $S = \{\mathbf{v}_1, \mathbf{v}_2, \mathbf{v}_3\}$ is an orthonormal basis for R^3 with the Euclidean inner product. If $\mathbf{u} = (2, -1, 4)$ then

$$\langle \mathbf{u}, \mathbf{v}_1 \rangle = -1, \langle \mathbf{u}, \mathbf{v}_2 \rangle = \tfrac{4}{5}, \langle \mathbf{u}, \mathbf{v}_3 \rangle = \tfrac{22}{5}$$

so that

$$(\mathbf{u})_S = (-1, \tfrac{4}{5}, \tfrac{22}{5}) \quad \text{and} \quad [\mathbf{u}]_S = \begin{bmatrix} -1 \\ \tfrac{4}{5} \\ \tfrac{22}{5} \end{bmatrix}$$

Orthonormal bases for inner product spaces are convenient because, as the following theorem shows, many familiar formulas hold in such spaces.

Theorem 25. *If S is an orthonormal basis for an n-dimensional inner product space and if*

$$(\mathbf{u})_S = (u_1, u_2, \ldots, u_n) \quad \text{and} \quad (\mathbf{v})_S = (v_1, v_2, \ldots, v_n)$$

then

(*a*) $\|\mathbf{u}\| = \sqrt{u_1^2 + u_2^2 + \cdots + u_n^2}$

(*b*) $d(\mathbf{u}, \mathbf{v}) = \sqrt{(u_1 - v_1)^2 + (u_2 - v_2)^2 + \cdots + (u_n - v_n)^2}$

(*c*) $\langle \mathbf{u}, \mathbf{v} \rangle = u_1 v_1 + u_2 v_2 + \cdots + u_n v_n$

The proofs and some numerical examples are discussed in the exercises.

We now turn to the main problem in this section.

Change of Basis Problem. If we change the basis for a vector space from some old basis B to some new basis B', how is the old coordinate matrix $[\mathbf{v}]_B$ of a vector $\mathbf{v}$ related to the new coordinate matrix $[\mathbf{v}]_{B'}$?

For simplicity, we will solve this problem for two dimensional spaces. The solution for n-dimensional spaces is similar, and will be left as an exercise. Let

$$B = \{\mathbf{u}_1, \mathbf{u}_2\} \quad \text{and} \quad B' = \{\mathbf{u}_1', \mathbf{u}_2'\}$$

be the old and new bases, respectively. We will need the coordinate matrices for the old basis vectors relative to the new basis. Suppose they are

$$[\mathbf{u}_1]_{B'} = \begin{bmatrix} a \\ b \end{bmatrix} \quad \text{and} \quad [\mathbf{u}_2]_{B'} = \begin{bmatrix} c \\ d \end{bmatrix}; \tag{4.22}$$

that is

$$\begin{aligned} \mathbf{u}_1 &= a\mathbf{u}'_1 + b\mathbf{u}'_2 \\ \mathbf{u}_2 &= c\mathbf{u}'_1 + d\mathbf{u}'_2 \end{aligned} \tag{4.23}$$

Now let $\mathbf{v}$ be any vector in V and let

$$[\mathbf{v}]_B = \begin{bmatrix} k_1 \\ k_2 \end{bmatrix} \tag{4.24}$$

be the old coordinate matrix, so that

$$\mathbf{v} = k_1\mathbf{u}_1 + k_2\mathbf{u}_2 \tag{4.25}$$

In order to find the new coordinates of $\mathbf{v}$ we must express $\mathbf{v}$ in terms of the new basis B'. To do this, we substitute (4.23) into (4.25). This yields

$$\mathbf{v} = k_1(a\mathbf{u}'_1 + b\mathbf{u}'_2) + k_2(c\mathbf{u}'_1 + d\mathbf{u}'_2)$$

or

$$\mathbf{v} = (k_1 a + k_2 c)\mathbf{u}'_1 + (k_1 b + k_2 d)\mathbf{u}'_2$$

Thus the new coordinate matrix for $\mathbf{v}$ is

$$[\mathbf{v}]_{B'} = \begin{bmatrix} k_1 a + k_2 c \\ k_1 b + k_2 d \end{bmatrix}$$

which can be rewritten

$$[\mathbf{v}]_{B'} = \begin{bmatrix} a & c \\ b & d \end{bmatrix} \begin{bmatrix} k_1 \\ k_2 \end{bmatrix}$$

or from (4.24)

$$[\mathbf{v}]_{B'} = \begin{bmatrix} a & c \\ b & d \end{bmatrix} [\mathbf{v}]_B$$

This equation states that the new coordinate matrix $[\mathbf{v}]_{B'}$ can be obtained by multiplying the old coordinate matrix $[\mathbf{v}]_B$ on the left by the matrix

$$P = \begin{bmatrix} a & c \\ b & d \end{bmatrix}$$

whose columns are the coordinates of the old basis vectors relative to the new basis (see 4.22). Thus we have the following solution of the change of basis problem:

Solution of the Change of Basis Problem. If we change the basis for a vector space V from some old basis $B = \{\mathbf{u}_1, \mathbf{u}_2, \ldots, \mathbf{u}_n\}$ to some new basis $B' = \{\mathbf{u}'_1, \mathbf{u}'_2, \ldots, \mathbf{u}'_n\}$ then the old coordinate matrix $[\mathbf{v}]_B$ of a vector $\mathbf{v}$ is related to the new coordinate matrix $[\mathbf{v}]_{B'}$ by the equation

$$[\mathbf{v}]_{B'} = P[\mathbf{v}]_B \tag{4.26}$$

where the columns of P are the coordinate matrices of the old basis vectors relative to the new basis, that is the column vectors of P are

$$[\mathbf{u}_1]_{B'}, [\mathbf{u}_2]_{B'}, \ldots, [\mathbf{u}_n]_{B'}$$

Symbolically, the matrix P can be written

$$P = \left[[\mathbf{u}_1]_{B'} \;\middle|\; [\mathbf{u}_2]_{B'} \;\middle|\; \cdots \;\middle|\; [\mathbf{u}_n]_{B'} \right];$$

it is called the ***transition matrix*** from B to B'.

Example 63

Consider the bases

$$B = \{\mathbf{u}_1, \mathbf{u}_2\} \quad \text{and} \quad B' = \{\mathbf{u}'_1, \mathbf{u}'_2\}$$

for R^2 where

$$\mathbf{u}_1 = \begin{bmatrix} 1 \\ 0 \end{bmatrix}, \mathbf{u}_2 = \begin{bmatrix} 0 \\ 1 \end{bmatrix}; \mathbf{u}'_1 = \begin{bmatrix} 1 \\ 1 \end{bmatrix}, \mathbf{u}'_2 = \begin{bmatrix} 2 \\ 1 \end{bmatrix}$$

(a) Find the transition matrix from B to B'.
(b) Use (4.26) to find $[\mathbf{v}]_{B'}$ if

$$\mathbf{v} = \begin{bmatrix} 7 \\ 2 \end{bmatrix}$$

Solution (*a*). First we must find the coordinate matrices for the old basis vectors $\mathbf{u}_1$ and $\mathbf{u}_2$ relative to the new basis B'. Following the procedure of Example 59*a* the reader should be able to show that

$$\begin{aligned} \mathbf{u}_1 &= -\mathbf{u}'_1 + \mathbf{u}'_2 \\ \mathbf{u}_2 &= 2\mathbf{u}'_1 - \mathbf{u}'_2 \end{aligned}$$

(verify), so that

$$[\mathbf{u}_1]_{B'} = \begin{bmatrix} -1 \\ 1 \end{bmatrix} \quad \text{and} \quad [\mathbf{u}_2]_{B'} = \begin{bmatrix} 2 \\ -1 \end{bmatrix}$$

Thus the transition matrix from B to B' is

$$P = \begin{bmatrix} -1 & 2 \\ 1 & -1 \end{bmatrix}$$

Solution (*b*). By inspection

$$[\mathbf{v}]_B = \begin{bmatrix} 7 \\ 2 \end{bmatrix}$$

so that using (4.26) and the transition matrix in part (*a*)

$$[\mathbf{v}]_{B'} = \begin{bmatrix} -1 & 2 \\ 1 & -1 \end{bmatrix} \begin{bmatrix} 7 \\ 2 \end{bmatrix} = \begin{bmatrix} -3 \\ 5 \end{bmatrix}$$

The reader may wish to check this result by verifying that $\mathbf{v} = -3\mathbf{u}'_1 + 5\mathbf{u}'_2$.

Example 64 (Application to Rotation of Coordinate Axes.)

In many problems a rectangular xy-coordinate system is given and a new $x'y'$-coordinate system is obtained by rotating the xy-system counterclockwise about the origin through an angle θ. When this is done each point Q in the plane has two sets of coordinates: coordinates (x, y) relative to the xy-system and coordinates (x', y') relative to the $x'y'$-system (Figure 4.18*a*).

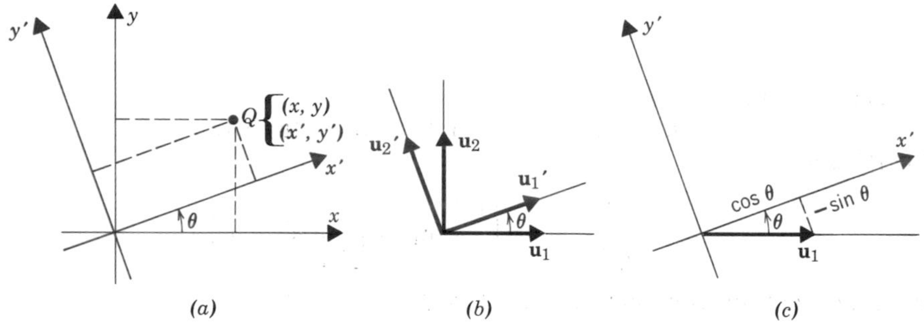

Figure 4.18

By introducing unit vectors $\mathbf{u}_1$ and $\mathbf{u}_2$ along the positive x and y axes and unit vectors $\mathbf{u}_1'$ and $\mathbf{u}_2'$ along the positive x' and y' axes, we can regard this rotation as a change from an old basis $B = \{\mathbf{u}_1, \mathbf{u}_2\}$ to a new basis $B' = \{\mathbf{u}_1', \mathbf{u}_2'\}$ (Figure 4.18*b*). Thus the new coordinates (x', y') and the old coordinates (x, y) of a point Q will be related by

$$\begin{bmatrix} x' \\ y' \end{bmatrix} = P \begin{bmatrix} x \\ y \end{bmatrix} \tag{4.27}$$

where P is the transition from B to B'. To find P we must determine the coordinate matrices of the old basis vectors $\mathbf{u}_1$ and $\mathbf{u}_2$ relative to the new basis. As indicated in Figure 4.18*c* the components of $\mathbf{u}_1$ in the new basis are $\cos\theta$ and $-\sin\theta$ so that

$$[\mathbf{u}_1]_{B'} = \begin{bmatrix} \cos\theta \\ -\sin\theta \end{bmatrix}$$

while, as indicated in Figure 4.18*d* below, the components of $\mathbf{u}_2$ in the new basis are $\cos(\pi/2 - \theta) = \sin\theta$ and $\sin(\pi/2 - \theta) = \cos\theta$, so that

$$[\mathbf{u}_2]_{B'} = \begin{bmatrix} \sin\theta \\ \cos\theta \end{bmatrix}$$

Thus the transition matrix from B to B' is

$$P = \begin{bmatrix} \cos\theta & \sin\theta \\ -\sin\theta & \cos\theta \end{bmatrix}$$

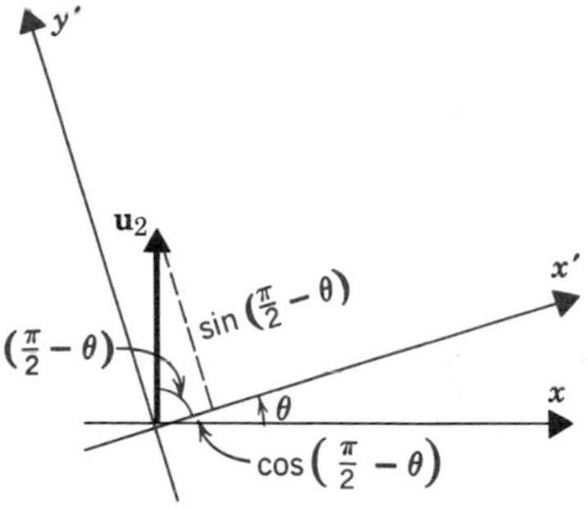

Figure 4.18 (*d*)

and (4.27) becomes

$$\begin{bmatrix} x' \\ y' \end{bmatrix} = \begin{bmatrix} \cos\theta & \sin\theta \\ -\sin\theta & \cos\theta \end{bmatrix} \begin{bmatrix} x \\ y \end{bmatrix} \tag{4.28}$$

or equivalently

$$\begin{aligned} x' &= x\cos\theta + y\sin\theta \\ y' &= -x\sin\theta + y\cos\theta \end{aligned}$$

For example, if the axes are rotated $\theta = 45°$ then since

$$\sin 45° = \cos 45° = \frac{1}{\sqrt{2}}$$

(4.28) becomes

$$\begin{bmatrix} x' \\ y' \end{bmatrix} = \begin{bmatrix} \frac{1}{\sqrt{2}} & \frac{1}{\sqrt{2}} \\ -\frac{1}{\sqrt{2}} & \frac{1}{\sqrt{2}} \end{bmatrix} \begin{bmatrix} x \\ y \end{bmatrix}$$

Thus if the old coordinates of a point Q are $(x, y) = (2, -1)$ then

$$\begin{bmatrix} x' \\ y' \end{bmatrix} = \begin{bmatrix} \frac{1}{\sqrt{2}} & \frac{1}{\sqrt{2}} \\ -\frac{1}{\sqrt{2}} & \frac{1}{\sqrt{2}} \end{bmatrix} \begin{bmatrix} 2 \\ -1 \end{bmatrix} = \begin{bmatrix} \frac{1}{\sqrt{2}} \\ -\frac{3}{\sqrt{2}} \end{bmatrix}$$

so the new coordinates of Q are $(x', y') = (1/\sqrt{2}, -3/\sqrt{2})$.

Example 65 (Application to Rotation of Axes in 3-space.)
Suppose a rectangular xyz-coordinate system is rotated around its z-axis counterclockwise (looking down the positive z-axis) through an angle θ (Figure 4.19). If we introduce unit vectors $\mathbf{u}_1$, $\mathbf{u}_2$, and $\mathbf{u}_3$ along the positive x, y, and z axes and

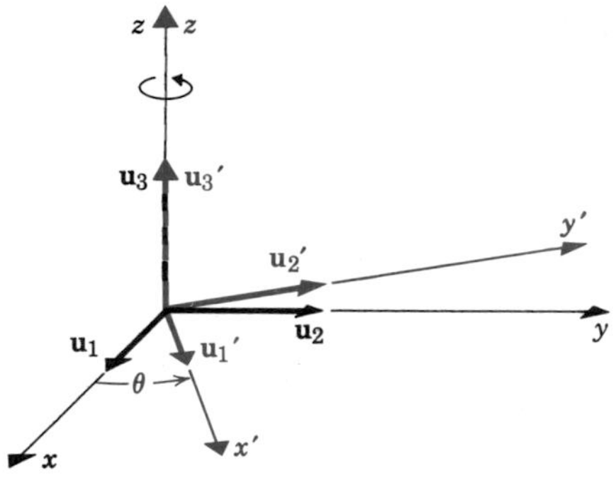

Figure 4.19

unit vectors $\mathbf{u}_1'$, $\mathbf{u}_2'$, and $\mathbf{u}_3'$ along the x', y', and z' axes, we can regard the rotation as a change from the old basis $B = \{\mathbf{u}_1, \mathbf{u}_2, \mathbf{u}_3\}$ to the new basis $B' = \{\mathbf{u}_1', \mathbf{u}_2', \mathbf{u}_3'\}$. In light of Example 64 it should be evident that

$$[\mathbf{u}_1]_{B'} = \begin{bmatrix} \cos\theta \\ -\sin\theta \\ 0 \end{bmatrix} \quad \text{and} \quad [\mathbf{u}_2]_{B'} = \begin{bmatrix} \sin\theta \\ \cos\theta \\ 0 \end{bmatrix}$$

Moreover, since $\mathbf{u}_3$ extends 1 unit up the positive z' axis

$$[\mathbf{u}_3]_{B'} = \begin{bmatrix} 0 \\ 0 \\ 1 \end{bmatrix}$$

Thus the transition matrix from B to B' is

$$\begin{bmatrix} \cos\theta & \sin\theta & 0 \\ -\sin\theta & \cos\theta & 0 \\ 0 & 0 & 1 \end{bmatrix}$$

and the old coordinates (x, y, z) of a point Q are related to its new coordinates (x', y', z') by

$$\begin{bmatrix} x' \\ y' \\ z' \end{bmatrix} = \begin{bmatrix} \cos\theta & \sin\theta & 0 \\ -\sin\theta & \cos\theta & 0 \\ 0 & 0 & 1 \end{bmatrix} \begin{bmatrix} x \\ y \\ z \end{bmatrix}$$

Example 66

Consider the vectors

$$\mathbf{u}_1 = \begin{bmatrix} 1 \\ 0 \end{bmatrix}, \mathbf{u}_2 = \begin{bmatrix} 0 \\ 1 \end{bmatrix}, \mathbf{u}_1' = \begin{bmatrix} 1 \\ 1 \end{bmatrix}, \mathbf{u}_2' = \begin{bmatrix} 2 \\ 1 \end{bmatrix}$$

In Example 63 we found the transition matrix from the basis $B = \{\mathbf{u}_1, \mathbf{u}_2\}$ for R^2 to the basis $B' = \{\mathbf{u}_1', \mathbf{u}_2'\}$. However, we can just as well ask for the transition matrix from B' to B. To obtain this matrix, we simply change our point of view and regard B' as the old basis and B as the new basis. As usual, the columns of the transition matrix will be the coordinates of the old basis vectors relative to the new basis.

By inspection

$$\mathbf{u}'_1 = \mathbf{u}_1 + \mathbf{u}_2$$
$$\mathbf{u}'_2 = 2\mathbf{u}_1 + \mathbf{u}_2$$

so that

$$[\mathbf{u}'_1]_B = \begin{bmatrix} 1 \\ 1 \end{bmatrix} \quad \text{and} \quad [\mathbf{u}'_2]_B = \begin{bmatrix} 2 \\ 1 \end{bmatrix}$$

Thus the transition matrix from B' to B is

$$Q = \begin{bmatrix} 1 & 2 \\ 1 & 1 \end{bmatrix}$$

If we multiply the transition matrix from B to B' obtained in Example 63 and the transition matrix from B' to B obtained in this example we find

$$PQ = \begin{bmatrix} -1 & 2 \\ 1 & -1 \end{bmatrix}\begin{bmatrix} 1 & 2 \\ 1 & 1 \end{bmatrix} = \begin{bmatrix} 1 & 0 \\ 0 & 1 \end{bmatrix} = I$$

which shows that $Q = P^{-1}$. As the following theorem shows, this is not accidental.

Theorem 26. *If P is the transition matrix from a basis B to a basis B' then*

(a) P is invertible
(b) P^{-1} is the transition matrix from B to B'.

(The proof is deferred to the end of the section.) To summarize, if P is the transition matrix from a basis B to a basis B' then for every vector $\mathbf{v}$:

$$[\mathbf{v}]_{B'} = P[\mathbf{v}]_B$$
$$[\mathbf{v}]_B = P^{-1}[\mathbf{v}]_{B'}$$

The next theorem shows that if P is the transition matrix from one *orthonormal* basis to another, then the inverse of P is especially easy to find.

Theorem 27. *If P is the transition matrix from one orthonormal basis to another orthonormal basis for an inner product space then*

$$P^{-1} = P^t$$

(We omit the proof.)

To illustrate this result, consider the transition matrix

$$P = \begin{bmatrix} \cos\theta & \sin\theta \\ -\sin\theta & \cos\theta \end{bmatrix}$$

obtained in Example 64 when we rotated coordinate axes (thereby changing the orthonormal basis $\{\mathbf{u}_1, \mathbf{u}_2\}$ of Figure 4.18*b* into the orthonormal basis $\{\mathbf{u}_1', \mathbf{u}_2'\}$). It is easy to verify that

$$P^{-1} = \begin{bmatrix} \cos\theta & -\sin\theta \\ \sin\theta & \cos\theta \end{bmatrix}$$

so that $P^{-1} = P^t$.

Definition. A square matrix A with the property

$$A^{-1} = A^t$$

is said to be an ***orthogonal matrix***.

Thus Theorem 27 states that a transition matrix from one orthonormal basis to another is always orthogonal.

The following result, whose proof is discussed in the exercises, makes it easy to determine when an $n \times n$ matrix A is orthogonal.

Theorem 28. *The following are equivalent:*

(*a*) *A is orthogonal*
(*b*) *The row vectors of A form an orthonormal set in R^n with the Euclidean inner product*
(*c*) *The column vectors of A form an orthonormal set in R^n with the Euclidean inner product.*

Example 67

Consider the matrix

$$A = \begin{bmatrix} \frac{1}{\sqrt{2}} & \frac{1}{\sqrt{2}} & 0 \\ 0 & 0 & 1 \\ \frac{1}{\sqrt{2}} & -\frac{1}{\sqrt{2}} & 0 \end{bmatrix}$$

The row vectors of A are

$$\mathbf{r}_1 = \left(\frac{1}{\sqrt{2}}, \frac{1}{\sqrt{2}}, 0\right), \quad \mathbf{r}_2 = (0, 0, 1), \quad \mathbf{r}_3 = \left(\frac{1}{\sqrt{2}}, -\frac{1}{\sqrt{2}}, 0\right)$$

Relative to the Euclidean inner product we have

$$\|\mathbf{r}_1\| = \|\mathbf{r}_2\| = \|\mathbf{r}_3\| = 1$$

and

$$\mathbf{r}_1 \cdot \mathbf{r}_2 = \mathbf{r}_2 \cdot \mathbf{r}_3 = \mathbf{r}_1 \cdot \mathbf{r}_3 = 0$$

so that the row vectors of A form an orthonormal set in R^3. Thus A is orthogonal and

$$A^{-1} = A^t = \begin{bmatrix} \frac{1}{\sqrt{2}} & 0 & \frac{1}{\sqrt{2}} \\ \frac{1}{\sqrt{2}} & 0 & -\frac{1}{\sqrt{2}} \\ 0 & 1 & 0 \end{bmatrix}$$

(The reader will find it instructive to check that the column vectors of A also form an orthonormal set.)

In Examples 64 and 65 we considered the problem of relating old and new coordinates when a geometric change (rotation) was made in the coordinate axes. Sometimes the following converse problem arises. A relationship

$$\begin{bmatrix} x \\ y \end{bmatrix} = \begin{bmatrix} a_1 & a_2 \\ b_1 & b_2 \end{bmatrix} \begin{bmatrix} x' \\ y' \end{bmatrix} \tag{4.29}$$

between old and new coordinates is known, where the 2×2 matrix is orthogonal; and it is of interest to determine how the xy-coordinate system and $x'y'$-coordinate system are related geometrically. Equation (4.29) is called an ***orthogonal coordinate transformation*** on R^2. To study the effect of an orthogonal coordinate transformation, consider the vectors

$$\mathbf{u}_1 = \begin{bmatrix} 1 \\ 0 \end{bmatrix}, \mathbf{u}_2 = \begin{bmatrix} 0 \\ 1 \end{bmatrix}, \mathbf{u}_1' = \begin{bmatrix} a_1 \\ b_1 \end{bmatrix}, \mathbf{u}_2' = \begin{bmatrix} a_2 \\ b_2 \end{bmatrix}$$

in the xy-coordinate system and introduce an $x'y'$-coordinate system with positive x'-axis along $\mathbf{u}_1'$ and positive y'-axis along $\mathbf{u}_2'$ (Figure 4.20). Because the 2×2 matrix in (4.29) is assumed orthogonal, the vectors $\mathbf{u}_1'$ and $\mathbf{u}_2'$ are orthogonal, which assures

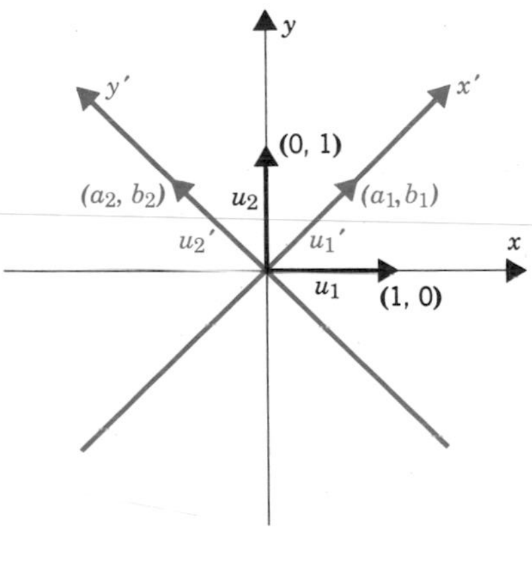

Figure 4.20

that the x' and y' axes are perpendicular. Since

$$\mathbf{u}_1' = a_1\mathbf{u}_1 + b_1\mathbf{u}_2$$

and

$$\mathbf{u}_2' = a_2\mathbf{u}_1 + b_2\mathbf{u}_2$$

the matrix

$$Q = \begin{bmatrix} a_1 & a_2 \\ b_1 & b_2 \end{bmatrix}$$

in (4.29) is the transition matrix from the basis $\{\mathbf{u}_1', \mathbf{u}_2'\}$ to the basis $\{\mathbf{u}_1, \mathbf{u}_2\}$. Clearly, there are two possibilities; either the $x'y'$-coordinate system can be obtained by rotating the xy-coordinate system (Figure 4.21*a*) or the $x'y'$-coordinate system can be obtained by first reflecting the xy-coordinate system about the x-axis and then rotating the reflected coordinate system (Figure 4.21*b*). It is shown in the exercises that the determinant of an orthogonal matrix is always $+1$ or -1. Moreover, it can be proved that the orthogonal coordinate transformation (4.29) is a rotation if

$$\begin{vmatrix} a_1 & a_2 \\ b_1 & b_2 \end{vmatrix} = 1$$

and is a reflection following by a rotation if this determinant is -1.

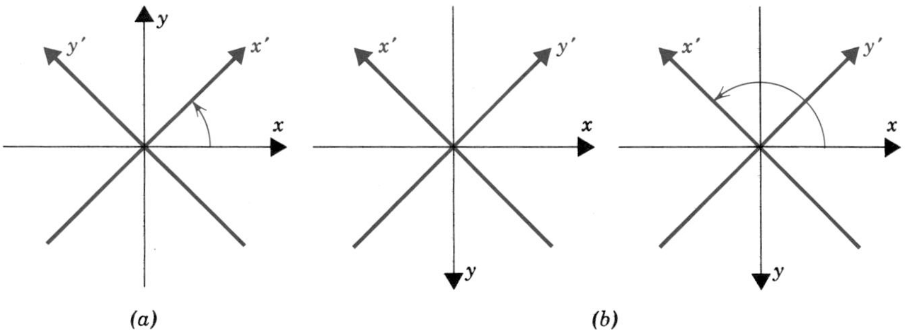

Figure 4.21

Similarly, an orthogonal coordinate transformation

$$\begin{bmatrix} x \\ y \\ z \end{bmatrix} = \begin{bmatrix} a_1 & a_2 & a_3 \\ b_1 & b_2 & b_3 \\ c_1 & c_2 & c_3 \end{bmatrix} \begin{bmatrix} x' \\ y' \\ z' \end{bmatrix}$$

on R^3 is a rotation if

$$\begin{vmatrix} a_1 & a_2 & a_3 \\ b_1 & b_2 & b_3 \\ c_1 & c_2 & c_3 \end{vmatrix} = 1$$

and is a rotation combined with a reflection in one of the coordinate planes if this determinant is -1.

Example 68

The orthogonal coordinate transformation

$$\begin{bmatrix} x \\ y \end{bmatrix} = \begin{bmatrix} \frac{1}{\sqrt{2}} & -\frac{1}{\sqrt{2}} \\ \frac{1}{\sqrt{2}} & \frac{1}{\sqrt{2}} \end{bmatrix} \begin{bmatrix} x' \\ y' \end{bmatrix}$$

is a rotation since

$$\begin{vmatrix} \frac{1}{\sqrt{2}} & -\frac{1}{\sqrt{2}} \\ \frac{1}{\sqrt{2}} & \frac{1}{\sqrt{2}} \end{vmatrix} = 1$$

The positive x' and y' axes are along the column vectors

$$\mathbf{u}_1' = \begin{bmatrix} \frac{1}{\sqrt{2}} \\ \frac{1}{\sqrt{2}} \end{bmatrix} \quad \text{and} \quad \mathbf{u}_2' = \begin{bmatrix} -\frac{1}{\sqrt{2}} \\ \frac{1}{\sqrt{2}} \end{bmatrix}$$

(Figure 4.21).

OPTIONAL

Proof of Theorem 26. Let Q be the transition matrix from B' to B. We shall show that $QP = I$ and thus conclude that $Q = P^{-1}$ to complete the proof.

Assume that $B = \{\mathbf{u}_1, \mathbf{u}_2, \ldots, \mathbf{u}_n\}$ and suppose

$$QP = \begin{bmatrix} c_{11} & c_{12} & \cdots & c_{1n} \\ c_{21} & c_{22} & \cdots & c_{2n} \\ \vdots & \vdots & & \vdots \\ c_{n1} & c_{n2} & \cdots & c_{nn} \end{bmatrix}$$

From (4.26)

$$[\mathbf{x}]_{B'} = P[\mathbf{x}]_B$$

and

$$[\mathbf{x}]_B = Q[\mathbf{x}]_{B'}$$

for all $\mathbf{x}$ in V. Multiplying the top equation through on the left by Q and substituting the second equation gives

$$[\mathbf{x}]_B = QP[\mathbf{x}]_B \tag{4.30}$$

for all $\mathbf{x}$ in V. Letting $\mathbf{x} = \mathbf{u}_1$ in (4.30) gives

$$\begin{bmatrix} 1 \\ 0 \\ 0 \\ \vdots \\ 0 \end{bmatrix} = \begin{bmatrix} c_{11} & c_{12} & \cdots & c_{1n} \\ c_{21} & c_{22} & \cdots & c_{2n} \\ \vdots & \vdots & & \vdots \\ c_{n1} & c_{n2} & \cdots & c_{nn} \end{bmatrix} \begin{bmatrix} 1 \\ 0 \\ 0 \\ \vdots \\ 0 \end{bmatrix}$$

or

$$\begin{bmatrix} 1 \\ 0 \\ 0 \\ \vdots \\ 0 \end{bmatrix} = \begin{bmatrix} c_{11} \\ c_{21} \\ \vdots \\ c_{n1} \end{bmatrix}$$

Similarly, successively substituting $\mathbf{x} = \mathbf{u}_2, \ldots, \mathbf{u}_n$ in (4.30) yields

$$\begin{bmatrix} c_{12} \\ c_{22} \\ \vdots \\ c_{n2} \end{bmatrix} = \begin{bmatrix} 0 \\ 1 \\ 0 \\ \vdots \\ 0 \end{bmatrix}, \ldots, \begin{bmatrix} c_{1n} \\ c_{2n} \\ \vdots \\ c_{nn} \end{bmatrix} = \begin{bmatrix} 0 \\ 0 \\ \vdots \\ 1 \end{bmatrix}$$

Therefore $QP = I$. ▮

EXERCISE SET 4.10

1. Find the coordinate matrix and coordinate vector for $\mathbf{w}$ relative to the basis $S = \{\mathbf{u}_1, \mathbf{u}_2\}$
   (a) $\mathbf{u}_1 = (1, 0)$, $\mathbf{u}_2 = (0, 1)$; $\mathbf{w} = (3, -7)$
   (b) $\mathbf{u}_1 = (2, -4)$, $\mathbf{u}_2 = (3, 8)$; $\mathbf{w} = (1, 1)$
   (c) $\mathbf{u}_1 = (1, 1)$, $\mathbf{u}_2 = (0, 2)$; $\mathbf{w} = (a, b)$

2. Find the coordinate vector and coordinate matrix for $\mathbf{v}$ relative to $S = \{\mathbf{v}_1, \mathbf{v}_2, \mathbf{v}_3\}$
   (a) $\mathbf{v} = (2, -1, 3)$, $\mathbf{v}_1 = (1, 0, 0)$, $\mathbf{v}_2 = (2, 2, 0)$, $\mathbf{v}_3 = (3, 3, 3)$
   (b) $\mathbf{v} = (5, -12, 3)$, $\mathbf{v}_1 = (1, 2, 3)$, $\mathbf{v}_2 = (-4, 5, 6)$, $\mathbf{v}_3 = (7, -8, 9)$

3. Find the coordinate vector and coordinate matrix for $\mathbf{p}$ relative to $S = \{\mathbf{p}, \mathbf{p}_2, \mathbf{p}_3\}$.
   (a) $\mathbf{p} = 4 - 3x + x^2$, $\mathbf{p}_1 = 1$, $\mathbf{p}_2 = x$, $\mathbf{p}_3 = x^2$
   (b) $\mathbf{p} = 2 - x + x^2$, $\mathbf{p}_1 = 1 + x$, $\mathbf{p}_2 = 1 + x^2$, $\mathbf{p}_3 = x + x^2$

4. Find the coordinate vector and coordinate matrix for A relative to $S = \{A_1, A_2, A_3, A_4\}$.

$$A = \begin{bmatrix} 2 & 0 \\ -1 & 3 \end{bmatrix} \quad A_1 = \begin{bmatrix} -1 & 1 \\ 0 & 0 \end{bmatrix} \quad A_2 = \begin{bmatrix} 1 & 1 \\ 0 & 0 \end{bmatrix} \quad A_3 = \begin{bmatrix} 0 & 0 \\ 1 & 0 \end{bmatrix} \quad A_4 = \begin{bmatrix} 0 & 0 \\ 0 & 1 \end{bmatrix}$$

5. In each part an orthonormal basis relative to the Euclidean inner product is given. Use the method of Example 62 to find the coordinate vector and coordinate matrix of $\mathbf{w}$.

   (a) $\mathbf{w} = (3, 7)$; $\mathbf{u}_1 = \left(\frac{1}{\sqrt{2}}, -\frac{1}{\sqrt{2}}\right)$, $\mathbf{u}_2 = \left(\frac{1}{\sqrt{2}}, \frac{1}{\sqrt{2}}\right)$

(b) $\mathbf{w} = (-1, 0, 2)$; $\mathbf{u}_1 = (\frac{2}{3}, -\frac{2}{3}, \frac{1}{3})$, $\mathbf{u}_2 = (\frac{2}{3}, \frac{1}{3}, -\frac{2}{3})$, $\mathbf{u}_3 = (\frac{1}{3}, \frac{2}{3}, \frac{2}{3})$

6. (a) Find $\mathbf{w}$ if $(\mathbf{w})_S = (6, -1, 4)$ and S is the basis in Exercise 2a.
(b) Find $\mathbf{q}$ if $(\mathbf{q})_S = (3, 0, 4)$ and S is the basis in Exercise 3a.
(c) Find B if $(B)_S = (-8, 7, 6, 3)$ and S is the basis in Exercise 4.

7. Let R^2 have the Euclidean inner product and let $S = \{\mathbf{w}_1, \mathbf{w}_2\}$ be the orthonormal basis with $\mathbf{w}_1 = (\frac{3}{5}, -\frac{4}{5})$, $\mathbf{w}_2 = (\frac{4}{5}, \frac{3}{5})$. Let $\mathbf{u}$, $\mathbf{v}$ be the vectors in R^2 for which $(\mathbf{u})_S = (1, 1)$ and $(\mathbf{v})_S = (-1, 4)$.
(a) Compute $\|\mathbf{u}\|$, $d(\mathbf{u}, \mathbf{v})$, and $\langle \mathbf{u}, \mathbf{v} \rangle$ using Theorem 25.
(b) Find $\mathbf{u}$ and $\mathbf{v}$ and check the results of part (a) by computing $\|\mathbf{u}\|$, $d(\mathbf{u}, \mathbf{v})$, and $\langle \mathbf{u}, \mathbf{v} \rangle$ directly.

8. Consider the bases $B = \{\mathbf{u}_1, \mathbf{u}_2\}$ and $B' = \{\mathbf{v}_1, \mathbf{v}_2\}$ for R^2, where

$$\mathbf{u}_1 = \begin{bmatrix} 1 \\ 0 \end{bmatrix} \quad \mathbf{u}_2 = \begin{bmatrix} 0 \\ 1 \end{bmatrix} \quad \mathbf{v}_1 = \begin{bmatrix} 2 \\ 1 \end{bmatrix} \quad \text{and} \quad \mathbf{v}_2 = \begin{bmatrix} -3 \\ 4 \end{bmatrix}$$

(a) Find the transition matrix from B to B'.
(b) Compute the coordinate matrix $[\mathbf{w}]_B$, where

$$\mathbf{w} = \begin{bmatrix} 3 \\ -5 \end{bmatrix}$$

and use (4.26) to compute $[\mathbf{w}]_{B'}$.
(c) Check your work by computing $[\mathbf{w}]_{B'}$ directly.
(d) Find the transition matrix from B' to B.

9. Repeat the directions of Exercise 8 with

$$\mathbf{u}_1 = \begin{bmatrix} 2 \\ 2 \end{bmatrix} \quad \mathbf{u}_2 = \begin{bmatrix} 4 \\ -1 \end{bmatrix} \quad \mathbf{v}_1 = \begin{bmatrix} 1 \\ 3 \end{bmatrix} \quad \mathbf{v}_2 = \begin{bmatrix} -1 \\ -1 \end{bmatrix}$$

10. Consider the bases $B = \{\mathbf{u}_1, \mathbf{u}_2, \mathbf{u}_3\}$ and $B' = \{\mathbf{v}_1, \mathbf{v}_2, \mathbf{v}_3\}$ for R^3, where

$$\mathbf{u}_1 = \begin{bmatrix} -3 \\ 0 \\ -3 \end{bmatrix} \quad \mathbf{u}_2 = \begin{bmatrix} -3 \\ 2 \\ -1 \end{bmatrix} \quad \mathbf{u}_3 = \begin{bmatrix} 1 \\ 6 \\ -1 \end{bmatrix} \quad \mathbf{v}_1 = \begin{bmatrix} -6 \\ -6 \\ 0 \end{bmatrix}$$

$$\mathbf{v}_2 = \begin{bmatrix} -2 \\ -6 \\ 4 \end{bmatrix} \quad \text{and} \quad \mathbf{v}_3 = \begin{bmatrix} -2 \\ -3 \\ 7 \end{bmatrix}$$

(a) Find the transition matrix from B to B'.
(b) Compute the coordinate matrix $[\mathbf{w}]_B$, where

$$\mathbf{w} = \begin{bmatrix} -5 \\ 8 \\ -5 \end{bmatrix}$$

and use (4.26) to compute $[\mathbf{w}]_{B'}$.
(c) Check your work by computing $[\mathbf{w}]_{B'}$ directly.

11. Repeat the directions of Exercise 10 with

$$\mathbf{u}_1 = \begin{bmatrix} 2 \\ 1 \\ 1 \end{bmatrix} \quad \mathbf{u}_2 = \begin{bmatrix} 2 \\ -1 \\ 1 \end{bmatrix} \quad \mathbf{u}_3 = \begin{bmatrix} 1 \\ 2 \\ 1 \end{bmatrix} \quad \mathbf{v}_1 = \begin{bmatrix} 3 \\ 1 \\ -5 \end{bmatrix} \quad \mathbf{v}_2 = \begin{bmatrix} 1 \\ 1 \\ -3 \end{bmatrix} \quad \mathbf{v}_3 = \begin{bmatrix} -1 \\ 0 \\ 2 \end{bmatrix}$$

12. Consider the bases $B = \{\mathbf{p}_1, \mathbf{p}_2\}$ and $B' = \{\mathbf{q}_1, \mathbf{q}_2\}$ for P_1, where $\mathbf{p}_1 = 6 + 3x$, $\mathbf{p}_2 = 10 + 2x$, $\mathbf{q}_1 = 2$, $\mathbf{q}_2 = 3 + 2x$.
(a) Find the transition matrix from B to B'.
(b) Compute the coordinate matrix $[\mathbf{p}]_B$, where $\mathbf{p} = -4 + x$, and use (4.26) to compute $[\mathbf{p}]_{B'}$.
(c) Check your work by computing $[\mathbf{p}]_{B'}$ directly.
(d) Find the transition matrix from B' to B.

13. Let V be the space spanned by $\mathbf{f}_1 = \sin x$ and $\mathbf{f}_2 = \cos x$.
(a) Show that $\mathbf{g}_1 = 2 \sin x + \cos x$ and $\mathbf{g}_2 = 3 \cos x$ form a basis for V.
(b) Find the transition matrix from $B = \{\mathbf{f}_1, \mathbf{f}_2\}$ to $B' = \{\mathbf{g}_1, \mathbf{g}_2\}$.
(c) Compute the coordinate matrix $[\mathbf{h}]_B$, where $\mathbf{h} = 2 \sin x - 5 \cos x$, and use (4.26) to obtain $[\mathbf{h}]_{B'}$.
(d) Check your work by computing $[\mathbf{h}]_{B'}$ directly.
(e) Find the transition matrix from B' to B.

14. Let a rectangular $x'y'$-coordinate system be obtained by rotating a rectangular xy-coordinate system through an angle $\theta = 3\pi/4$.
(a) Find the $x'y'$-coordinates of the point whose xy-coordinates are $(-2, 6)$.
(b) Find the xy-coordinates of the point whose $x'y'$-coordinates are $(5, 2)$.

15. Repeat Exercise 14 with $\theta = \pi/3$.

16. Let a rectangular $x'y'z'$-coordinate system be obtained by rotating a rectangular xyz-coordinate system counterclockwise about the z-axis (looking down the z-axis) through an angle $\theta = \pi/4$.
(a) Find the $x'y'z'$-coordinates of the point whose xyz-coordinates are $(-1, 2, 5)$.
(b) Find the xyz-coordinates of the point whose $x'y'z'$-coordinates are $(1, 6, -3)$.

17. Repeat Exercise 16 for a rotation of $\theta = \pi/3$ counterclockwise about the y-axis (looking along the positive y-axis toward the origin).

18. Repeat Exercise 16 for a rotation of $\theta = 3\pi/4$ counterclockwise about the x-axis (looking along the positive x-axis toward the origin).

19. Use Theorem 28 to determine which of the following are orthogonal.

(a) $\begin{bmatrix} 1 & 0 \\ 0 & 1 \end{bmatrix}$ (b) $\begin{bmatrix} \frac{1}{\sqrt{2}} & -\frac{1}{\sqrt{2}} \\ \frac{1}{\sqrt{2}} & \frac{1}{\sqrt{2}} \end{bmatrix}$ (c) $\begin{bmatrix} 0 & 1 & \frac{1}{\sqrt{2}} \\ 1 & 0 & 0 \\ 0 & 0 & \frac{1}{\sqrt{2}} \end{bmatrix}$

(d) $\begin{bmatrix} -\frac{1}{\sqrt{2}} & \frac{1}{\sqrt{6}} & \frac{1}{\sqrt{3}} \\ 0 & -\frac{2}{\sqrt{6}} & \frac{1}{\sqrt{3}} \\ \frac{1}{\sqrt{2}} & \frac{1}{\sqrt{6}} & \frac{1}{\sqrt{3}} \end{bmatrix}$ (e) $\begin{bmatrix} \frac{1}{2} & \frac{1}{2} & \frac{1}{2} & \frac{1}{2} \\ \frac{1}{2} & -\frac{5}{6} & \frac{1}{6} & \frac{1}{6} \\ \frac{1}{2} & \frac{1}{6} & \frac{1}{6} & -\frac{5}{6} \\ \frac{1}{2} & \frac{1}{6} & -\frac{5}{6} & \frac{1}{6} \end{bmatrix}$ (f) $\begin{bmatrix} 1 & 0 & 0 & 0 \\ 0 & \frac{1}{\sqrt{3}} & -\frac{1}{2} & 0 \\ 0 & \frac{1}{\sqrt{3}} & 0 & 1 \\ 0 & \frac{1}{\sqrt{3}} & \frac{1}{2} & 0 \end{bmatrix}$

20. Find the inverses of those matrices in Exercise 19 that are orthogonal.

21. Show that the following are orthogonal matrices for every value of θ.

(a) $\begin{bmatrix} \cos\theta & -\sin\theta \\ \sin\theta & \cos\theta \end{bmatrix}$ (b) $\begin{bmatrix} \cos\theta & -\sin\theta & 0 \\ \sin\theta & \cos\theta & 0 \\ 0 & 0 & 1 \end{bmatrix}$

22. Find the inverses of the matrices in Exercise 21.

23. Consider the orthogonal coordinate transformation

$$\begin{bmatrix} x \\ y \end{bmatrix} = \begin{bmatrix} -\frac{3}{5} & -\frac{4}{5} \\ \frac{4}{5} & -\frac{3}{5} \end{bmatrix} \begin{bmatrix} x' \\ y' \end{bmatrix}$$

Find (x', y') for the points with the following (x, y) coordinates.
(a) $(2, -1)$ (b) $(4, 2)$ (c) $(-7, -8)$ (d) $(0, 0)$

24. Sketch the xy-axes and the $x'y'$-axes for the coordinate transformation in Exercise 23.

25. For which of the following is $\mathbf{x} = P\mathbf{x}'$ a rotation?

(a) $P = \begin{bmatrix} \frac{1}{\sqrt{2}} & \frac{1}{\sqrt{2}} \\ -\frac{1}{\sqrt{2}} & \frac{1}{\sqrt{2}} \end{bmatrix}$ (b) $P = \begin{bmatrix} -\frac{1}{\sqrt{2}} & \frac{1}{\sqrt{2}} \\ \frac{1}{\sqrt{2}} & \frac{1}{\sqrt{2}} \end{bmatrix}$

(c) $P = \begin{bmatrix} \frac{3}{5} & \frac{4}{5} \\ -\frac{4}{5} & \frac{3}{5} \end{bmatrix}$ (d) $P = \begin{bmatrix} -\frac{3}{5} & -\frac{4}{5} \\ -\frac{4}{5} & \frac{3}{5} \end{bmatrix}$

26. Sketch the xy-axes and the $x'y'$-axes for the coordinate transformations in Exercise 25.

27. Consider the orthogonal coordinate transformation

$$\begin{bmatrix} x \\ y \\ z \end{bmatrix} = \begin{bmatrix} \frac{4}{5} & -\frac{3}{5} & 0 \\ \frac{3}{5} & \frac{4}{5} & 0 \\ 0 & 0 & 1 \end{bmatrix} \begin{bmatrix} x' \\ y' \\ z' \end{bmatrix}$$

Find (x', y', z') for the points with the following (x, y, z) coordinates.
(a) $(3, 0, -7)$ (b) $(1, 2, 6)$ (c) $(-9, -2, -3)$ (d) $(0, 0, 0)$

28. Sketch the xyz-axes and the $x'y'z'$-axes for the coordinate transformation in Exercise 27.

29. For which of the following is $\mathbf{x} = P\mathbf{x}'$ a rotation?

(a) $P = \begin{bmatrix} \frac{4}{5} & 0 & -\frac{3}{5} \\ \frac{3}{5} & 0 & \frac{4}{5} \\ 0 & 1 & 0 \end{bmatrix}$ (b) $P = \begin{bmatrix} \frac{6}{7} & \frac{2}{7} & \frac{3}{7} \\ \frac{2}{7} & \frac{3}{7} & -\frac{6}{7} \\ -\frac{3}{7} & \frac{6}{7} & \frac{2}{7} \end{bmatrix}$

30. Sketch the xyz-axes and the $x'y'z'$-axes for the coordinate transformations in Exercise 29.

31. (a) A rectangular $x'y'z'$-coordinate system is obtained by rotating an xyz-coordinate system counterclockwise about the y-axis through an angle θ (looking along the positive y-axis toward the origin.) Find a matrix A such that

$$\begin{bmatrix} x' \\ y' \\ z' \end{bmatrix} = A \begin{bmatrix} x \\ y \\ z \end{bmatrix}$$

where (x, y, z) and (x', y', z') are the coordinates of a point in the xyz and $x'y'z'$-systems respectively.

(b) Repeat part (*a*) for a rotation about the x-axis.

32. A rectangular $x''y''z''$-coordinate system is obtained by first rotating a rectangular xyz-coordinate system 60° counterclockwise about the z-axis (looking down the positive z-axis) to obtain an $x'y'z'$-coordinate system, and then rotating the $x'y'z'$-coordinate system 45° counterclockwise about the y'-axis (looking along the positive y'-axis toward the origin). Find a matrix A such that

$$\begin{bmatrix} x'' \\ y'' \\ z'' \end{bmatrix} = A \begin{bmatrix} x \\ y \\ z \end{bmatrix}$$

where (x, y, z) and (x'', y'', z'') are the xyz and $x''y''z''$-coordinates of a point.

33. Show that if A is an orthogonal matrix, then A^t is also orthogonal.

34. Prove that an $n \times n$ matrix is orthogonal if and only if its rows form an orthonormal set in R^n.

35. Use Exercises 33 and 34 to show that an $n \times n$ matrix is orthogonal if and only if its columns form an orthonormal set in R^n.

36. Prove that if P is an orthogonal matrix, then $\det(P) = 1$ or -1.

37. Prove Theorem 25*a*.

38. Prove Theorem 25*b*.

39. Prove Theorem 25*c*.

5 Linear Transformations

5.1 INTRODUCTION TO LINEAR TRANSFORMATIONS

In this section we begin the study of vector-valued functions of a vector variable. That is, functions having the form $\mathbf{w} = F(\mathbf{v})$, where the independent variable $\mathbf{v}$ and the dependent variable $\mathbf{w}$ are both vectors. We shall concentrate on a special class of vector functions called linear transformations. These have many important applications in physics, engineering, social sciences, and various branches of mathematics.

If V and W are vector spaces and F is a function that associates a unique vector in W with each vector in V, we say F ***maps*** V into W, and write $F: V \to W$. Further, if F associates the vector $\mathbf{w}$ with the vector $\mathbf{v}$, we write $\mathbf{w} = F(\mathbf{v})$ and say that $\mathbf{w}$ is the ***image*** of $\mathbf{v}$ under F.

To illustrate, if $\mathbf{v} = (x, y)$ is a vector in R^2, then the formula

$$F(\mathbf{v}) = (x, x + y, x - y) \tag{5.1}$$

defines a function that maps R^2 into R^3. In particular, if $\mathbf{v} = (1, 1)$, then the image of $\mathbf{v}$ under F is $F(\mathbf{v}) = (1, 2, 0)$.

Definition. If $F: V \to W$ is a function from the vector space V into the vector space W, then F is called a ***linear transformation*** if

(i) $F(\mathbf{u} + \mathbf{v}) = F(\mathbf{u}) + F(\mathbf{v})$ for all vectors $\mathbf{u}$ and $\mathbf{v}$ in V.
(ii) $F(k\mathbf{u}) = kF(\mathbf{u})$ for all vectors $\mathbf{u}$ in V and all scalars k.

To illustrate, let $F: R^2 \to R^3$ be the function defined by (5.1). If $\mathbf{u} = (x_1, y_1)$ and $\mathbf{v} = (x_2, y_2)$, then $\mathbf{u} + \mathbf{v} = (x_1 + x_2, y_1 + y_2)$, so that

$$\begin{aligned} F(\mathbf{u} + \mathbf{v}) &= (x_1 + x_2, [x_1 + x_2] + [y_1 + y_2], [x_1 + x_2] - [y_1 + y_2]) \\ &= (x_1, x_1 + y_1, x_1 - y_1) + (x_2, x_2 + y_2, x_2 - y_2) \\ &= F(\mathbf{u}) + F(\mathbf{v}) \end{aligned}$$

Also, if k is a scalar, $k\mathbf{u} = (kx_1, ky_1)$, so that

$$\begin{aligned} F(k\mathbf{u}) &= (kx_1, kx_1 + ky_1, kx_1 - ky_1) \\ &= k(x_1, x_1 + y_1, x_1 - y_1) \\ &= kF(\mathbf{u}) \end{aligned}$$

Thus F is a linear transformation.

If $F: V \to W$ is a linear transformation, then for any $\mathbf{v}_1$ and $\mathbf{v}_2$ in V and any scalars k_1 and k_2, we have

$$F(k_1\mathbf{v}_1 + k_2\mathbf{v}_2) = F(k_1\mathbf{v}_1) + F(k_2\mathbf{v}_2) = k_1F(\mathbf{v}_1) + k_2F(\mathbf{v}_2)$$

Similarly, if $\mathbf{v}_1, \mathbf{v}_2, \ldots, \mathbf{v}_n$ are vectors in V and $k_1, k_2, \ldots, k_n$ are scalars, then

$$F(k_1\mathbf{v}_1 + k_2\mathbf{v}_2 + \cdots + k_n\mathbf{v}_n) = k_1F(\mathbf{v}_1) + k_2F(\mathbf{v}_2) + \cdots + k_nF(\mathbf{v}_n) \quad (5.2)$$

We now give some further examples of linear transformations.

Example 1

Let A be a fixed $m \times n$ matrix. If we use matrix notation for vectors in R^m and R^n, then we can define a function $T: R^n \to R^m$ by

$$T(\mathbf{x}) = A\mathbf{x}$$

Observe that if $\mathbf{x}$ is an $n \times 1$ matrix, then the product $A\mathbf{x}$ is an $m \times 1$ matrix; thus T maps R^n into R^m. Moreover, T is linear; to see this, let $\mathbf{u}$ and $\mathbf{v}$ be $n \times 1$ matrices and let k be a scalar. Using properties of matrix multiplication, we obtain

$$A(\mathbf{u} + \mathbf{v}) = A\mathbf{u} + A\mathbf{v} \qquad \text{and} \qquad A(k\mathbf{u}) = k(A\mathbf{u})$$

or equivalently

$$T(\mathbf{u} + \mathbf{v}) = T(\mathbf{u}) + T(\mathbf{v}) \qquad \text{and} \qquad T(k\mathbf{u}) = kT(\mathbf{u})$$

We shall call the linear transformation in this example ***multiplication by A***. Linear transformations of this kind are called ***matrix transformations***.

Example 2

As a special case of the previous example, let θ be a fixed angle, and let $T: R^2 \to R^2$ be multiplication by the matrix

$$A = \begin{bmatrix} \cos\theta & -\sin\theta \\ \sin\theta & \cos\theta \end{bmatrix}$$

If $\mathbf{v}$ is the vector

$$\mathbf{v} = \begin{bmatrix} x \\ y \end{bmatrix}$$

then

$$T(\mathbf{v}) = A\mathbf{v} = \begin{bmatrix} \cos\theta & -\sin\theta \\ \sin\theta & \cos\theta \end{bmatrix}\begin{bmatrix} x \\ y \end{bmatrix} = \begin{bmatrix} x\cos\theta - y\sin\theta \\ x\sin\theta + y\cos\theta \end{bmatrix}$$

Geometrically, $T(\mathbf{v})$ is the vector that results if $\mathbf{v}$ is rotated through an angle θ.

To see this, let ϕ be the angle between $\mathbf{v}$ and the positive x axis, and let

$$\mathbf{v}' = \begin{bmatrix} x' \\ y' \end{bmatrix}$$

be the vector that results when $\mathbf{v}$ is rotated through an angle θ (Figure 5.1). We shall show $\mathbf{v}' = T(\mathbf{v})$. If r denotes the length of $\mathbf{v}$, then

$$x = r\cos\phi \qquad y = r\sin\phi$$

Similarly, since $\mathbf{v}'$ has the same length as $\mathbf{v}$, we have

$$x' = r\cos(\theta + \phi) \qquad y' = r\sin(\theta + \phi)$$

Therefore

$$\begin{aligned} \mathbf{v}' = \begin{bmatrix} x' \\ y' \end{bmatrix} &= \begin{bmatrix} r\cos(\theta + \phi) \\ r\sin(\theta + \phi) \end{bmatrix} \\ &= \begin{bmatrix} r\cos\theta\cos\phi - r\sin\theta\sin\phi \\ r\sin\theta\cos\phi + r\cos\theta\sin\phi \end{bmatrix} \\ &= \begin{bmatrix} x\cos\theta - y\sin\theta \\ x\sin\theta + y\cos\theta \end{bmatrix} \\ &= \begin{bmatrix} \cos\theta & -\sin\theta \\ \sin\theta & \cos\theta \end{bmatrix} \begin{bmatrix} x \\ y \end{bmatrix} \\ &= A\mathbf{v} = T(\mathbf{v}) \end{aligned}$$

The linear transformation in this example is called the ***rotation of R^2 through the angle θ***.

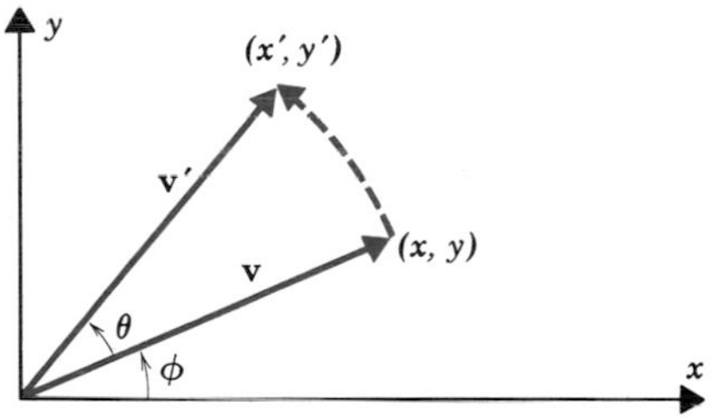

Figure 5.1

Example 3

Let V and W be any two vector spaces. The mapping $T: V \longrightarrow W$ such that $T(\mathbf{v}) = \mathbf{0}$ for every $\mathbf{v}$ in V is a linear transformation called the ***zero transformation***. To see that T is linear, observe that

$$T(\mathbf{u} + \mathbf{v}) = \mathbf{0},\ T(\mathbf{u}) = \mathbf{0},\ T(\mathbf{v}) = \mathbf{0} \qquad \text{and} \qquad T(k\mathbf{u}) = \mathbf{0}$$

Therefore

$$T(\mathbf{u} + \mathbf{v}) = T(\mathbf{u}) + T(\mathbf{v}) \qquad \text{and} \qquad T(k\mathbf{u}) = kT(\mathbf{u})$$

Example 4

Let V be any vector space. The mapping $T: V \to V$ defined by $T(\mathbf{v}) = \mathbf{v}$ is called the ***identity transformation*** on V. The verification that T is linear is left as an exercise.

If, as in Examples 2 and 4, $T: V \to V$ is a linear transformation from a vector space V into itself, then T is called a ***linear operator*** on V.

Example 5

Let V be any vector space and k any fixed scalar. We leave it as an exercise to check that the function $T: V \to V$ defined by

$$T(\mathbf{v}) = k\mathbf{v}$$

is a linear operator on V. If $k > 1$, T is called a ***dilation*** of V and if $0 < k < 1$, then T is called a ***contraction*** of V. Geometrically, a dilation "stretches" each vector in V by a factor of k, and a contraction of V "compresses" each vector by a factor of k (Figure 5.2).

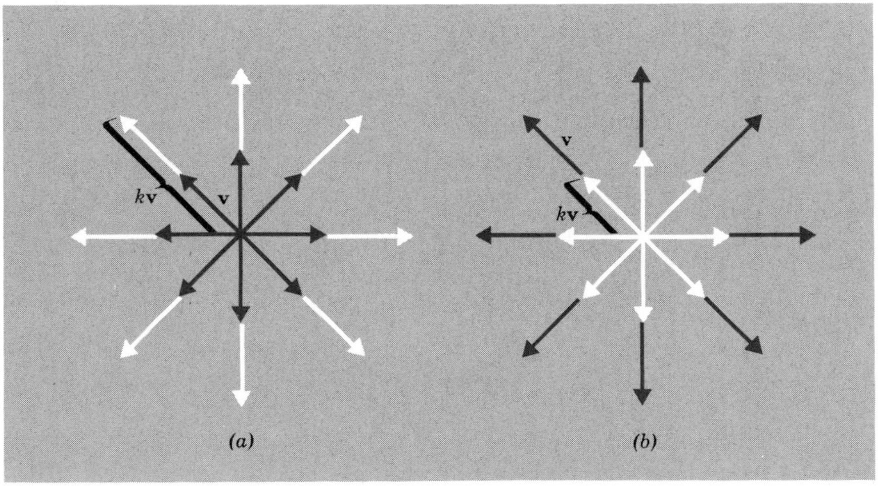

Figure 5.2 (*a*) Dilation of V. (*b*) Contraction of V.

Example 6

Let V be an inner product space, and suppose W is a finite dimensional subspace of V having

$$S = \{\mathbf{w}_1, \mathbf{w}_2, \ldots, \mathbf{w}_r\}$$

as an orthonormal basis. Let $T: V \to W$ be the function that maps a vector $\mathbf{v}$ in

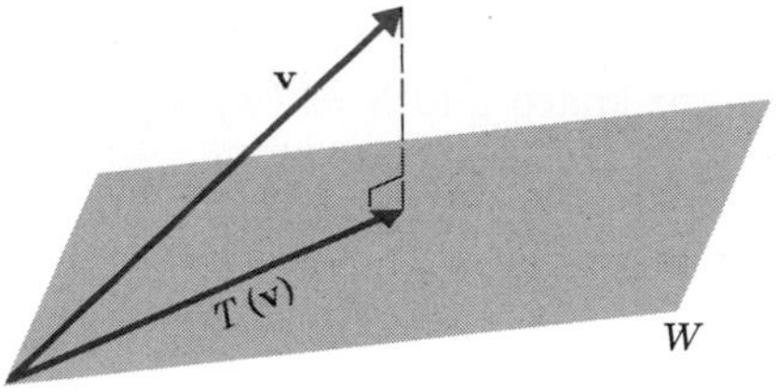

Figure 5.3

V into its orthogonal projection on W (Section 4.9); that is

$$T(\mathbf{v}) = \langle \mathbf{v}, \mathbf{w}_1 \rangle \mathbf{w}_1 + \langle \mathbf{v}, \mathbf{w}_2 \rangle \mathbf{w}_2 + \cdots + \langle \mathbf{v}, \mathbf{w}_r \rangle \mathbf{w}_r$$

(See Figure 5.3.)

The mapping T is called the ***orthogonal projection of V onto W***; its linearity follows from the basic properties of the inner product. For example

$$\begin{aligned} T(\mathbf{u} + \mathbf{v}) &= \langle \mathbf{u} + \mathbf{v}, \mathbf{w}_1 \rangle \mathbf{w}_1 + \langle \mathbf{u} + \mathbf{v}, \mathbf{w}_2 \rangle \mathbf{w}_2 + \cdots + \langle \mathbf{u} + \mathbf{v}, \mathbf{w}_r \rangle \mathbf{w}_r \\ &= \langle \mathbf{u}, \mathbf{w}_1 \rangle \mathbf{w}_1 + \langle \mathbf{u}, \mathbf{w}_2 \rangle \mathbf{w}_2 + \cdots + \langle \mathbf{u}, \mathbf{w}_r \rangle \mathbf{w}_r \\ &\quad + \langle \mathbf{v}, \mathbf{w}_1 \rangle \mathbf{w}_1 + \langle \mathbf{v}, \mathbf{w}_2 \rangle \mathbf{w}_2 + \cdots + \langle \mathbf{v}, \mathbf{w}_r \rangle \mathbf{w}_r \\ &= T(\mathbf{u}) + T(\mathbf{v}) \end{aligned}$$

Similarly, $T(k\mathbf{u}) = kT(\mathbf{u})$.

Example 7

As a special case of the previous example, let $V = R^3$ have the Euclidean inner product. The vectors $\mathbf{w}_1 = (1, 0, 0)$ and $\mathbf{w}_2 = (0, 1, 0)$ form an orthonormal basis for the xy-plane. Thus, if $\mathbf{v} = (x, y, z)$ is any vector in R^3, the orthogonal projection of R^3 onto the xy-plane is given by

$$\begin{aligned} T(\mathbf{v}) &= \langle \mathbf{v}, \mathbf{w}_1 \rangle \mathbf{w}_1 + \langle \mathbf{v}, \mathbf{w}_2 \rangle \mathbf{w}_2 \\ &= x(1, 0, 0) + y(0, 1, 0) \\ &= (x, y, 0) \end{aligned}$$

(See Figure 5.4.)

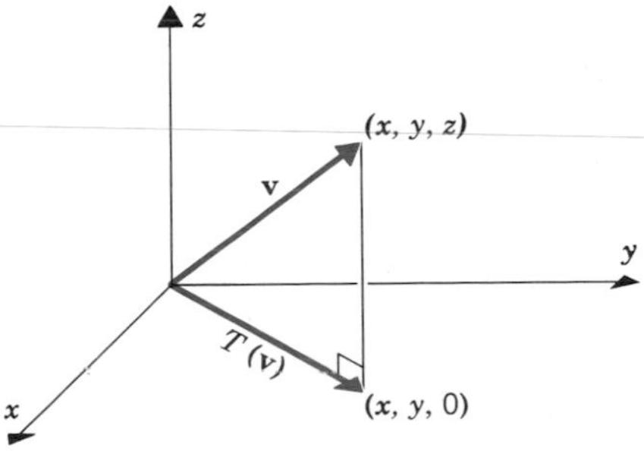

Figure 5.4

Example 8

Let V be an n-dimensional vector space and $S = \{\mathbf{w}_1, \mathbf{w}_2, \ldots, \mathbf{w}_n\}$ a fixed basis for V. By Theorem 24 of Section 4.10 any two vectors $\mathbf{u}$ and $\mathbf{v}$ in V can be written uniquely in the form

$$\mathbf{u} = c_1\mathbf{w}_1 + c_2\mathbf{w}_2 + \cdots + c_n\mathbf{w}_n \qquad \text{and} \qquad \mathbf{v} = d_1\mathbf{w}_1 + d_2\mathbf{w}_2 + \cdots + d_n\mathbf{w}_n$$

Thus

$$(\mathbf{u})_S = (c_1, c_2, \ldots, c_n)$$

$$(\mathbf{v})_S = (d_1, d_2, \ldots, d_n)$$

But

$$\mathbf{u} + \mathbf{v} = (c_1 + d_1)\mathbf{w}_1 + (c_2 + d_2)\mathbf{w}_2 + \cdots + (c_n + d_n)\mathbf{w}_n$$

$$k\mathbf{u} = (kc_1)\mathbf{w}_1 + (kc_2)\mathbf{w}_2 + \cdots + (kc_n)\mathbf{w}_n$$

so that

$$(\mathbf{u} + \mathbf{v})_S = (c_1 + d_1, c_2 + d_2, \ldots, c_n + d_n)$$

$$(k\mathbf{u})_S = (kc_1, kc_2, \ldots, kc_n)$$

Therefore

$$(\mathbf{u} + \mathbf{v})_S = (\mathbf{u})_S + (\mathbf{v})_S \qquad \text{and} \qquad (k\mathbf{u})_S = k(\mathbf{u})_S \tag{5.3}$$

Similarly, for coordinate matrices we have

$$[\mathbf{u} + \mathbf{v}]_S = [\mathbf{u}]_S + [\mathbf{v}]_S \qquad \text{and} \qquad [k\mathbf{u}]_S = k[\mathbf{u}]_S$$

Suppose we let $T: V \to R^n$ be the function that maps a vector $\mathbf{v}$ in V into its coordinate vector with respect to S; that is

$$T(\mathbf{v}) = (\mathbf{v})_S$$

Then in terms of T, (5.3) states

$$T(\mathbf{u} + \mathbf{v}) = T(\mathbf{u}) + T(\mathbf{v})$$

and

$$T(k\mathbf{u}) = kT(\mathbf{u})$$

Thus T is a linear transformation from V into R^n.

Example 9

Let V be an inner product space and let $\mathbf{v}_0$ be any fixed vector in V. Let $T: V \to R$ be the transformation that maps a vector $\mathbf{v}$ into its inner product with $\mathbf{v}_0$; that is

$$T(\mathbf{v}) = \langle \mathbf{v}, \mathbf{v}_0 \rangle$$

From the properties of an inner product

$$T(\mathbf{u} + \mathbf{v}) = \langle \mathbf{u} + \mathbf{v}, \mathbf{v}_0 \rangle = \langle \mathbf{u}, \mathbf{v}_0 \rangle + \langle \mathbf{v}, \mathbf{v}_0 \rangle = T(\mathbf{u}) + T(\mathbf{v})$$

and

$$T(k\mathbf{u}) = \langle k\mathbf{u}, \mathbf{v}_0 \rangle = k\langle \mathbf{u}, \mathbf{v}_0 \rangle = kT(\mathbf{u})$$

so that T is a linear transformation.

Example 10

(For readers who have studied calculus.)
Let $V = C[0, 1]$ be the vector space of all real-valued functions continuous on the interval $0 \leq x \leq 1$, and let W be the subspace of $C[0, 1]$ consisting of all functions with continuous first derivatives on the interval $0 \leq x \leq 1$.

Let $D: W \to V$ be the transformation that maps $\mathbf{f}$ into its derivative; that is

$$D(\mathbf{f}) = \mathbf{f}'$$

From the properties of differentiation, we have

$$D(\mathbf{f} + \mathbf{g}) = D(\mathbf{f}) + D(\mathbf{g})$$

and

$$D(k\mathbf{f}) = kD(\mathbf{f})$$

Thus D is a linear transformation.

Example 11

(For readers who have studied calculus.)
Let $V = C[0, 1]$ be as in the previous example, and let $J: V \to R$ be defined by

$$J(\mathbf{f}) = \int_0^1 f(x)\, dx$$

For example, if $f(x) = x^2$, then

$$J(\mathbf{f}) = \int_0^1 x^2\, dx = \frac{1}{3}$$

Since

$$\int_0^1 (f(x) + g(x))\, dx = \int_0^1 f(x)\, dx + \int_0^1 g(x)\, dx$$

and

$$\int_0^1 kf(x)\, dx = k \int_0^1 f(x)\, dx$$

for any constant k, it follows that

$$J(\mathbf{f} + \mathbf{g}) = J(\mathbf{f}) + J(\mathbf{g})$$
$$J(k\mathbf{f}) = kJ(\mathbf{f})$$

Thus J is a linear transformation.

EXERCISE SET 5.1

In Exercises 1–8 a formula is given for a function $F: R^2 \to R^2$. In each exercise determine if F is linear.

1. $F(x, y) = (2x, y)$

2. $F(x, y) = (x^2, y)$

3. $F(x, y) = (y, x)$

4. $F(x, y) = (0, y)$

5. $F(x, y) = (x, y + 1)$ 6. $F(x, y) = (2x + y, x - y)$

7. $F(x, y) = (y, y)$ 8. $F(x, y) = (\sqrt[3]{x}, \sqrt[3]{y})$

In Exercises 9–12, a formula is given for a function $F:R^3 \to R^2$. In each exercise determine if F is linear.

9. $F(x, y, z) = (x, x + y + z)$ 10. $F(x, y, z) = (0, 0)$

11. $F(x, y, z) = (1, 1)$ 12. $F(x, y, z) = (2x + y, 3y - 4z)$

In Exercises 13–16, a formula is given for a function $F:M_{22} \to R$. In each exercise determine if F is linear.

13. $F\left(\begin{bmatrix} a & b \\ c & d \end{bmatrix}\right) = a + d$ 14. $F\left(\begin{bmatrix} a & b \\ c & d \end{bmatrix}\right) = \det \begin{bmatrix} a & b \\ c & d \end{bmatrix}$

15. $F\left(\begin{bmatrix} a & b \\ c & d \end{bmatrix}\right) = 2a + 3b + c - d$ 16. $F\left(\begin{bmatrix} a & b \\ c & d \end{bmatrix}\right) = a^2 + b^2$

In Exercises 17–20, a formula is given for a function $F:P_2 \to P_2$. In each exercise determine if F is linear.

17. $F(a_0 + a_1x + a_2x^2) = a_0 + (a_1 + a_2)x + (2a_0 - 3a_1)x^2$

18. $F(a_0 + a_1x + a_2x^2) = a_0 + a_1(x + 1) + a_2(x + 1)^2$

19. $F(a_0 + a_1x + a_2x^2) = 0$

20. $F(a_0 + a_1x + a_2x^2) = (a_0 + 1) + a_1x + a_2x^2$

21. Let $F:R^2 \to R^2$ be the function that maps each point in the plane into its reflection about the y-axis. Find a formula for F, and show that F is a linear operator on R^2.

22. Let B be a fixed 2×3 matrix. Show that the function $T:M_{22} \to M_{23}$ defined by $T(A) = AB$ is a linear transformation.

23. Let $T:R^3 \to R^2$ be a matrix transformation, and suppose

$$T\left(\begin{bmatrix} 1 \\ 0 \\ 0 \end{bmatrix}\right) = \begin{bmatrix} 1 \\ 1 \end{bmatrix}, \quad T\left(\begin{bmatrix} 0 \\ 1 \\ 0 \end{bmatrix}\right) = \begin{bmatrix} 3 \\ 0 \end{bmatrix}, \text{ and } T\left(\begin{bmatrix} 0 \\ 0 \\ 1 \end{bmatrix}\right) = \begin{bmatrix} 4 \\ -7 \end{bmatrix}.$$

(a) Find the matrix.

(b) Find $T\left(\begin{bmatrix} 1 \\ 3 \\ 8 \end{bmatrix}\right)$

(c) Find $T\left(\begin{bmatrix} x \\ y \\ z \end{bmatrix}\right)$

24. Let $T:R^3 \to W$ be the orthogonal projection of R^3 onto the xz-plane W.
(a) Find $T(2, 7, -1)$.
(b) Find a formula for T.

25. Let $T: R^3 \to W$ be the orthogonal projection of R^3 onto the plane W having the equation $x + y + z = 0$.
(a) Find $T(3, 8, 4)$.
(b) Find a formula for T.

26. In each part let $T: R^2 \to R^2$ be the linear operator that rotates each vector in the plane through the angle θ. Find $T(-1, 2)$ and $T(x, y)$ when

(a) $\theta = \dfrac{\pi}{4}$ (b) $\theta = \pi$ (c) $\theta = \dfrac{\pi}{6}$ (d) $\theta = -\dfrac{\pi}{3}$

27. Prove that if $T: V \to W$ is a linear transformation, then $T(\mathbf{u} - \mathbf{v}) = T(\mathbf{u}) - T(\mathbf{v})$ for all vectors $\mathbf{u}$ and $\mathbf{v}$ in V.

28. Let $\{\mathbf{v}_1, \mathbf{v}_2, \ldots, \mathbf{v}_n\}$ be a basis for a vector space V and let $T: V \to W$ be a linear transformation. Show that if $T(\mathbf{v}_1) = T(\mathbf{v}_2) = \cdots = T(\mathbf{v}_n) = \mathbf{0}$, then T is the zero transformation.

29. Let $\{\mathbf{v}_1, \mathbf{v}_2, \ldots, \mathbf{v}_n\}$ be a basis for a vector space V and let $T: V \to V$ be a linear operator. Show that if $T(\mathbf{v}_1) = \mathbf{v}_1, T(\mathbf{v}_2) = \mathbf{v}_2, \ldots, T(\mathbf{v}_n) = \mathbf{v}_n$, then T is the identity transformation on V.

30. Let S be a basis for an n-dimensional vector space V. Show that if $\mathbf{v}_1, \mathbf{v}_2, \ldots, \mathbf{v}_r$ form a linearly independent set of vectors in V, then the coordinate vectors $(\mathbf{v}_1)_S, (\mathbf{v}_2)_S, \ldots, (\mathbf{v}_r)_S$ form a linearly independent set in R^n, and conversely.

31. Using the notation from Exercise 30, show that if $\mathbf{v}_1, \mathbf{v}_2, \ldots, \mathbf{v}_r$ span V, then the coordinate vectors $(\mathbf{v}_1)_S, (\mathbf{v}_2)_S, \ldots, (\mathbf{v}_r)_S$ span R^n, and conversely.

32. Find a basis for the subspace of P_2 spanned by the given vectors.
(a) $-1 + x - 2x^2, \quad 3 + 3x + 6x^2, \quad 9$
(b) $1 + x, \quad x^2, \quad -2 + 2x^2, \quad -3x$
(c) $1 + x - 3x^2, \quad 2 + 2x - 6x^2, \quad 3 + 3x - 9x^2$
(***Hint.*** Let S be the standard basis for P_2 and work with the coordinate vectors relative to S; note Exercises 30, 31.)

5.2 PROPERTIES OF LINEAR TRANSFORMATIONS; KERNEL AND RANGE

In this section we develop some basic properties of linear transformations. In particular, we show that once the images of the basis vectors under a linear transformation are known, it is possible to find the images of the remaining vectors in the space.

Theorem 1. *If $T: V \to W$ is a linear transformation, then:*

(a) $T(\mathbf{0}) = \mathbf{0}$
(b) $T(-\mathbf{v}) = -T(\mathbf{v})$ *for all* $\mathbf{v}$ *in* V
(c) $T(\mathbf{v} - \mathbf{w}) = T(\mathbf{v}) - T(\mathbf{w})$ *for all* $\mathbf{v}$ *and* $\mathbf{w}$ *in* V

Proof. Let $\mathbf{v}$ be any vector in V. Since $0\mathbf{v} = \mathbf{0}$ we have

$$T(\mathbf{0}) = T(0\mathbf{v}) = 0T(\mathbf{v}) = \mathbf{0}$$

which proves (*a*).

Also, $T(-\mathbf{v}) = T((-1)\mathbf{v}) = (-1)T(\mathbf{v}) = -T(\mathbf{v})$, which proves (*b*).

Finally, $\mathbf{v} - \mathbf{w} = \mathbf{v} + (-1)\mathbf{w}$; thus

$$\begin{aligned} T(\mathbf{v} - \mathbf{w}) &= T(\mathbf{v} + (-1)\mathbf{w}) \\ &= T(\mathbf{v}) + (-1)T(\mathbf{w}) \\ &= T(\mathbf{v}) - T(\mathbf{w}) \quad \blacksquare \end{aligned}$$

Definition. If $T: V \to W$ is a linear transformation, then the set of vectors in V that T maps into $\mathbf{0}$ is called the ***kernel*** (or ***nullspace***) of T; it is denoted by $\ker(T)$. The set of all vectors in W that are images under T of at least one vector in V is called the ***range*** of T; it is denoted by $R(T)$.

Example 12

Let $T: V \to W$ be the zero transformation. Since T maps every vector into $\mathbf{0}$, $\ker(T) = V$. Since $\mathbf{0}$ is the only possible image under T, $R(T)$ consists only of the zero vector.

Example 13

Let $T: R^n \to R^m$ be multiplication by

$$A = \begin{bmatrix} a_{11} & a_{12} & \cdots & a_{1n} \\ a_{21} & a_{22} & \cdots & a_{2n} \\ \vdots & \vdots & & \vdots \\ a_{m1} & a_{m2} & \cdots & a_{mn} \end{bmatrix}$$

The kernel of T consists of all

$$\mathbf{x} = \begin{bmatrix} x_1 \\ x_2 \\ \vdots \\ x_n \end{bmatrix}$$

that are solution vectors of the homogeneous system

$$A \begin{bmatrix} x_1 \\ x_2 \\ \vdots \\ x_n \end{bmatrix} = \begin{bmatrix} 0 \\ 0 \\ \vdots \\ 0 \end{bmatrix}$$

The range of T consists of vectors

$$\mathbf{b} = \begin{bmatrix} b_1 \\ b_2 \\ \vdots \\ b_m \end{bmatrix}$$

such that the system

$$A \begin{bmatrix} x_1 \\ x_2 \\ \vdots \\ x_n \end{bmatrix} = \begin{bmatrix} b_1 \\ b_2 \\ \vdots \\ b_m \end{bmatrix}$$

is consistent

Theorem 2. *If $T: V \rightarrow W$ is a linear transformation then:*

(a) The kernel of T is a subspace of V.
(b) The range of T is a subspace of W.

Proof.

(*a*) To show that ker(T) is a subspace, we must show it is closed under addition and scalar multiplication. Let $\mathbf{v}_1$ and $\mathbf{v}_2$ be vectors in ker(T), and let k be any scalar. Then

$$\begin{aligned} T(\mathbf{v}_1 + \mathbf{v}_2) &= T(\mathbf{v}_1) + T(\mathbf{v}_2) \\ &= \mathbf{0} + \mathbf{0} = \mathbf{0} \end{aligned}$$

so that $\mathbf{v}_1 + \mathbf{v}_2$ is in ker(T). Also

$$T(k\mathbf{v}_1) = kT(\mathbf{v}_1) = k\mathbf{0} = \mathbf{0}$$

so that $k\mathbf{v}_1$ is in ker(T).

(*b*) Let $\mathbf{w}_1$ and $\mathbf{w}_2$ be vectors in the range of T. To prove this part we must show that $\mathbf{w}_1 + \mathbf{w}_2$ and $k\mathbf{w}_1$ are in the range of T for any scalar k; that is, we must find vectors $\mathbf{a}$ and $\mathbf{b}$ in V such that $T(\mathbf{a}) = \mathbf{w}_1 + \mathbf{w}_2$ and $T(\mathbf{b}) = k\mathbf{w}_1$.

Since $\mathbf{w}_1$ and $\mathbf{w}_2$ are in the range of T, there are vectors $\mathbf{a}_1$ and $\mathbf{a}_2$ in V such that $T(\mathbf{a}_1) = \mathbf{w}_1$ and $T(\mathbf{a}_2) = \mathbf{w}_2$. Let $\mathbf{a} = \mathbf{a}_1 + \mathbf{a}_2$ and $\mathbf{b} = k\mathbf{a}_1$. Then

$$T(\mathbf{a}) = T(\mathbf{a}_1 + \mathbf{a}_2) = T(\mathbf{a}_1) + T(\mathbf{a}_2) = \mathbf{w}_1 + \mathbf{w}_2$$

and

$$T(\mathbf{b}) = T(k\mathbf{a}_1) = kT(\mathbf{a}_1) = k\mathbf{w}_1$$

which completes the proof. ∎

Example 14

Let $T: R^n \rightarrow R^m$ be multiplication by an $m \times n$ matrix A. From Example 13 the kernel of T consists of all solutions of $A\mathbf{x} = \mathbf{0}$; thus the kernel is the *solution space* of this system. Also from Example 13, the range of T consists of all vectors $\mathbf{b}$ such that $A\mathbf{x} = \mathbf{b}$ is consistent. Thus, by Theorem 14 of Section 4.6, the range of T is the *column space* of the matrix A.

Suppose $\{\mathbf{v}_1, \mathbf{v}_2, \ldots, \mathbf{v}_n\}$ is a basis for a vector space V and $T: V \to W$ is a linear transformation. If we happen to know the images of the basis vectors, that is,

$$T(\mathbf{v}_1), T(\mathbf{v}_2), \ldots, T(\mathbf{v}_n)$$

then we can obtain the image $T(\mathbf{v})$ of any vector $\mathbf{v}$ by first expressing $\mathbf{v}$ in terms of the basis, say

$$\mathbf{v} = k_1\mathbf{v}_1 + k_2\mathbf{v}_2 + \cdots + k_n\mathbf{v}_n$$

and then using relation (5.2) of Section 5.1 to write

$$T(\mathbf{v}) = k_1T(\mathbf{v}_1) + k_2T(\mathbf{v}_2) + \cdots + k_nT(\mathbf{v}_n)$$

In short, *a linear transformation is completely determined by its "values" at a basis.*

Example 15

Consider the basis $S = \{\mathbf{v}_1, \mathbf{v}_2, \mathbf{v}_3\}$ for R^3, where $\mathbf{v}_1 = (1, 1, 1)$, $\mathbf{v}_2 = (1, 1, 0)$, $\mathbf{v}_3 = (1, 0, 0)$, and let $T: R^3 \to R^2$ be a linear transformation such that

$$T(\mathbf{v}_1) = (1, 0) \qquad T(\mathbf{v}_2) = (2, -1) \qquad T(\mathbf{v}_3) = (4, 3)$$

Find $T(2, -3, 5)$.

Solution. We first express $\mathbf{v} = (2, -3, 5)$ as a linear combination of $\mathbf{v}_1 = (1, 1, 1)$, $\mathbf{v}_2 = (1, 1, 0)$, and $\mathbf{v}_3 = (1, 0, 0)$. Thus

$$(2, -3, 5) = k_1(1, 1, 1) + k_2(1, 1, 0) + k_3(1, 0, 0)$$

or on equating corresponding components

$$\begin{aligned} k_1 + k_2 + k_3 &= 2 \\ k_1 + k_2 &= -3 \\ k_1 &= 5 \end{aligned}$$

which yields $k_1 = 5$, $k_2 = -8$, $k_3 = 5$ so that

$$(2, -3, 5) = 5\mathbf{v}_1 - 8\mathbf{v}_2 + 5\mathbf{v}_3$$

Thus

$$\begin{aligned} T(2, -3, 5) &= 5T(\mathbf{v}_1) - 8T(\mathbf{v}_2) + 5T(\mathbf{v}_3) \\ &= 5(1, 0) - 8(2, -1) + 5(4, 3) \\ &= (9, 23) \end{aligned}$$

Definition. If $T: V \to W$ is a linear transformation, then the dimension of the range of T is called the ***rank of T*** and the dimension of the kernel is called the ***nullity of T***.

Example 16

Let $T: R^2 \to R^2$ be the rotation of R^2 through the angle $\pi/4$. It is geometrically obvious that the range of T is all of R^2 and the kernel of T is $\{\mathbf{0}\}$. Therefore, T has rank $= 2$ and nullity $= 0$.

Example 17

Let $T: R^n \to R^m$ be multiplication by an $m \times n$ matrix A. In Example 14 we observed that the range of T is the column space of A. Thus the rank of T is the dimension of the column space of A, which is just the rank of A. In short,

$$\text{rank}(T) = \text{rank}(A)$$

Also in Example 14, we saw that the kernel of T is the solution space of $A\mathbf{x} = \mathbf{0}$. Thus the nullity of T is the dimension of this solution space.

Our next theorem establishes a relationship between the rank and nullity of a linear transformation defined on a finite dimensional vector space. We shall defer the proof to the end of the section.

Theorem 3. (Dimension Theorem). *If $T: V \to W$ is a linear transformation from an n-dimensional vector space V to a vector space W, then*

$$(\textit{rank of } T) + (\textit{nullity of } T) = n$$

In the special case where $V = R^n$, $W = R^m$, and $T: R^n \to R^m$ is multiplication by an $m \times n$ matrix A, the dimension theorem yields the following result:

$$\begin{aligned} \text{nullity of } T &= n - (\text{rank of } T) \\ &= (\text{number of columns of } A) - (\text{rank of } T) \end{aligned} \tag{5.4}$$

However, we noted in Example 17 that the nullity of T is the dimension of the solution space of $A\mathbf{x} = \mathbf{0}$, and the rank of T is the rank of the matrix A. Thus (5.4) yields the following theorem.

Theorem 4. *If A is an $m \times n$ matrix then the dimension of the solution space of* $A\mathbf{x} = \mathbf{0}$ is

$$n - \textit{rank}(A)$$

Example 18

In Example 35 of Section 4.5 we showed that the homogeneous system

$$\begin{aligned} 2x_1 + 2x_2 - x_3 \phantom{{}- 3x_4} + x_5 &= 0 \\ -x_1 - x_2 + 2x_3 - 3x_4 + x_5 &= 0 \\ x_1 + x_2 - 2x_3 \phantom{{}- 3x_4} - x_5 &= 0 \\ x_3 + x_4 + x_5 &= 0 \end{aligned}$$

has a two dimensional solution space, by solving the system and finding a basis.

Since the coefficient matrix

$$A = \begin{bmatrix} 2 & 2 & -1 & 0 & 1 \\ -1 & -1 & 2 & -3 & 1 \\ 1 & 1 & -2 & 0 & -1 \\ 0 & 0 & 1 & 1 & 1 \end{bmatrix}$$

has five columns, it follows from Theorem 4 that the rank of A must satisfy

$$2 = 5 - \text{rank}(A)$$

so that $\text{rank}(A) = 3$. The reader can check this result by reducing A to row-echelon form and showing that the resulting matrix has three nonzero rows.

OPTIONAL

Proof of Theorem 3. We must show that

$$\dim(R(T)) + \dim(\ker(T)) = n$$

We shall give the proof for the case where $1 \leq \dim(\ker(T)) < n$. The cases $\dim(\ker(T)) = 0$ and $\dim(\ker(T)) = n$ are left as exercises. Assume $\dim(\ker(T)) = r$, and let $\mathbf{v}_1, \ldots, \mathbf{v}_r$ be a basis for the kernel. Since $\{\mathbf{v}_1, \ldots, \mathbf{v}_r\}$ is linearly independent, part (*c*) of Theorem 9 in Chapter 4 states that there are $n - r$ vectors, $\mathbf{v}_{r+1}, \ldots, \mathbf{v}_n$, such that $\{\mathbf{v}_1, \ldots, \mathbf{v}_r, \mathbf{v}_{r+1}, \ldots, \mathbf{v}_n\}$ is a basis for V. To complete the proof, we shall show that the $n - r$ vectors in the set $S = \{T(\mathbf{v}_{r+1}), \ldots, T(\mathbf{v}_n)\}$ form a basis for the range of T. It will then follow that

$$\dim(R(T)) + \dim(\ker(T)) = (n - r) + r = n$$

First we show that S spans the range of T. If $\mathbf{b}$ is any vector in the range of T, then $\mathbf{b} = T(\mathbf{v})$ for some vector $\mathbf{v}$ in V. Since $\{\mathbf{v}_1, \ldots, \mathbf{v}_r, \mathbf{v}_{r+1}, \ldots, \mathbf{v}_n\}$ is a basis for V, $\mathbf{v}$ can be written in the form

$$\mathbf{v} = c_1\mathbf{v}_1 + \cdots + c_r\mathbf{v}_r + c_{r+1}\mathbf{v}_{r+1} + \cdots + c_n\mathbf{v}_n$$

Since $\mathbf{v}_1, \ldots, \mathbf{v}_r$ lie in the kernel of T, $T(\mathbf{v}_1) = \cdots = T(\mathbf{v}_r) = \mathbf{0}$, so that

$$\mathbf{b} = T(\mathbf{v}) = c_{r+1}T(\mathbf{v}_{r+1}) + \cdots + c_nT(\mathbf{v}_n)$$

Thus S spans the range of T.

Finally, we show that S is a linearly independent set and consequently forms a basis for the range of T. Suppose some linear combination of the vectors in S is zero, that is

$$k_{r+1}T(\mathbf{v}_{r+1}) + \cdots + k_nT(\mathbf{v}_n) = \mathbf{0} \tag{5.5}$$

We must show $k_{r+1} = \cdots = k_n = 0$. Since T is linear, (5.5) can be rewritten as

$$T(k_{r+1}\mathbf{v}_{r+1} + \cdots + k_n\mathbf{v}_n) = \mathbf{0}$$

which says that $k_{r+1}\mathbf{v}_{r+1} + \cdots + k_n\mathbf{v}_n$ is in the kernel of T. This vector can therefore be written as a linear combination of the basis vectors $\{\mathbf{v}_1, \ldots, \mathbf{v}_r\}$ say,

$$k_{r+1}\mathbf{v}_{r+1} + \cdots + k_n\mathbf{v}_n = k_1\mathbf{v}_1 + \cdots + k_r\mathbf{v}_r.$$

Thus

$$k_1\mathbf{v}_1 + \cdots + k_r\mathbf{v}_r - k_{r+1}\mathbf{v}_{r+1} - \cdots - k_n\mathbf{v}_n = \mathbf{0}$$

Since $\{\mathbf{v}_1, \ldots, \mathbf{v}_n\}$ is linearly independent, all the k's are zero; in particular, $k_{r+1} = \cdots = k_n = 0$, which completes the proof. ▪

EXERCISE SET 5.2

1. Let $T:R^2 \to R^2$ be multiplication by

$$\begin{bmatrix} 2 & -1 \\ -8 & 4 \end{bmatrix}$$

Which of the following are in $R(T)$?

(a) $\begin{bmatrix} 1 \\ -4 \end{bmatrix}$ (b) $\begin{bmatrix} 5 \\ 0 \end{bmatrix}$ (c) $\begin{bmatrix} -3 \\ 12 \end{bmatrix}$

2. Let $T:R^2 \to R^2$ be the linear transformation in Exercise 1. Which of the following are in $\ker(T)$?

(a) $\begin{bmatrix} 5 \\ 10 \end{bmatrix}$ (b) $\begin{bmatrix} 3 \\ 2 \end{bmatrix}$ (c) $\begin{bmatrix} 1 \\ 1 \end{bmatrix}$

3. Let $T:R^4 \to R^3$ be multiplication by

$$\begin{bmatrix} 4 & 1 & -2 & -3 \\ 2 & 1 & 1 & -4 \\ 6 & 0 & -9 & 9 \end{bmatrix}$$

Which of the following are in $R(T)$?

(a) $\begin{bmatrix} 0 \\ 0 \\ 6 \end{bmatrix}$ (b) $\begin{bmatrix} 1 \\ 3 \\ 0 \end{bmatrix}$ (c) $\begin{bmatrix} 2 \\ 4 \\ 1 \end{bmatrix}$

4. Let $T:R^4 \to R^3$ be the linear transformation in Exercise 3. Which of the following are in $\ker(T)$?

(a) $\begin{bmatrix} 3 \\ 8 \\ 2 \\ 0 \end{bmatrix}$ (b) $\begin{bmatrix} 0 \\ 0 \\ 0 \\ 1 \end{bmatrix}$ (c) $\begin{bmatrix} 0 \\ -4 \\ 1 \\ 0 \end{bmatrix}$

5. Let $T:P_2 \to P_3$ be the linear transformation defined by $T(p(x)) = xp(x)$. Which of the following are in $\ker(T)$?
(a) x^2 (b) 0 (c) $1 + x$

6. Let $T:P_2 \to P_3$ be the linear transformation in Exercise 5. Which of the following are in $R(T)$?
(a) $x + x^2$ (b) $1 + x$ (c) $3 - x^2$

7. Let V be any vector space, and let $T: V \to V$ be defined by $T(\mathbf{v}) = 3\mathbf{v}$.
(a) What is the kernel of T?
(b) What is the range of T?

8. Find the rank and nullity of the linear transformation in Exercise 1.

9. Find the rank and nullity of the linear transformation in Exercise 5.

10. Let V be an n-dimensional vector space. Find the rank and nullity of the linear transformation $T:V \to V$ defined by
(a) $T(\mathbf{x}) = \mathbf{x}$ (b) $T(\mathbf{x}) = \mathbf{0}$ (c) $T(\mathbf{x}) = 3\mathbf{x}$

11. Consider the basis $S = \{\mathbf{v}_1, \mathbf{v}_2, \mathbf{v}_3\}$ for R^3, where $\mathbf{v}_1 = (1, 2, 3)$, $\mathbf{v}_2 = (2, 5, 3)$, and $\mathbf{v}_3 = (1, 0, 10)$. Find a formula for the linear transformation $T:R^3 \to R^2$ for which $T(\mathbf{v}_1) = (1, 0)$, $T(\mathbf{v}_2) = (1, 0)$, and $T(\mathbf{v}_3) = (0, 1)$. Compute $T(1, 1, 1)$.

12. Find the linear transformation $T:P_2 \to P_2$ for which $T(1) = 1 + x$, $T(x) = 3 - x^2$, and $T(x^2) = 4 + 2x - 3x^2$. Compute $T(2 - 2x + 3x^2)$.

13. In each part use the given information to find the nullity of T.
(a) $T:R^5 \to R^7$ has rank 3.
(b) $T:P_4 \to P_3$ has rank 1.
(c) The range of $T:R^6 \to R^3$ is R^3.
(d) $T:M_{22} \to M_{22}$ has rank 3.

14. Let A be a 7×6 matrix such that $A\mathbf{x} = \mathbf{0}$ has only the trivial solution, and let $T:R^6 \to R^7$ be multiplication by A. Find the rank and nullity of T.

15. Let A be a 5×7 matrix with rank 4.
(a) What is the dimension of the solution space of $A\mathbf{x} = \mathbf{0}$?
(b) Is $A\mathbf{x} = \mathbf{b}$ consistent for all vectors $\mathbf{b}$ in R^5? Explain.

In Exercises 16–19, let T be multiplication by the given matrix. Find:
(a) A basis for the range of T.
(b) A basis for the kernel of T.
(c) The rank and nullity of T.

16. $\begin{bmatrix} 1 & -1 & 3 \\ 5 & 6 & -4 \\ 7 & 4 & 2 \end{bmatrix}$ **17.** $\begin{bmatrix} 2 & 0 & -1 \\ 4 & 0 & -2 \\ 0 & 0 & 0 \end{bmatrix}$

18. $\begin{bmatrix} 4 & 1 & 5 & 2 \\ 1 & 2 & 3 & 0 \end{bmatrix}$ **19.** $\begin{bmatrix} 1 & 4 & 5 & 0 & 9 \\ 3 & -2 & 1 & 0 & -1 \\ -1 & 0 & -1 & 0 & -1 \\ 2 & 3 & 5 & 1 & 8 \end{bmatrix}$

20. Let $T:R^3 \to V$ be a linear transformation from R^3 to any vector space. Show that the kernel of T is a line through the origin, a plane through the origin, the origin only, or all of R^3.

21. Let $T:V \to R^3$ be a linear transformation from any vector space to R^3. Show that the range of T is a line through the origin, a plane through the origin, the origin only, or all of R^3.

22. Let $T:R^3 \to R^3$ be multiplication by

$$\begin{bmatrix} 1 & 3 & 4 \\ 3 & 4 & 7 \\ -2 & 2 & 0 \end{bmatrix}$$

(a) Show that the range of T is a plane through the origin and find an equation for it.
(b) Show that the kernel of T is a line through the origin and find parametric equations for it.

23. Prove: If $\{\mathbf{v}_1, \mathbf{v}_2, \ldots, \mathbf{v}_n\}$ is a basis for V and $\mathbf{w}_1, \mathbf{w}_2, \ldots, \mathbf{w}_n$ are vectors in W, not necessarily distinct, then there exists a linear transformation $T:V \to W$ such that $T(\mathbf{v}_1) = \mathbf{w}_1$, $T(\mathbf{v}_2) = \mathbf{w}_2, \ldots, T(\mathbf{v}_n) = \mathbf{w}_n$.

24. Prove the dimension theorem in the cases:
(a) $\dim(\ker(T)) = 0$
(b) $\dim(\ker(T)) = n$

25. Let $T:V \to V$ be a linear operator on a finite dimensional vector space V. Prove that $R(T) = V$ if and only if $\ker(T) = \{\mathbf{0}\}$.

26. **(For readers who have studied calculus.)** Let $D:P_3 \to P_2$ be the differentiation transformation $D(\mathbf{p}) = \mathbf{p}'$. Describe the kernel of D.

27. **(For readers who have studied calculus.)** Let $J:P_1 \to R$ be the integration transformation $J(\mathbf{p}) = \int_{-1}^{1} p(x)\,dx$. Describe the kernel of J.

5.3 MATRICES OF LINEAR TRANSFORMATIONS

In this section we show that every linear transformation on a finite dimensional vector space can be regarded as a matrix transformation. Our work here will enable us to exploit our knowledge of matrix transformations to study more general linear transformations.

We shall show first that every linear transformation from R^n to R^m is a matrix transformation. More precisely, we shall show that if $T:R^n \to R^m$ is any linear transformation, then we can find an $m \times n$ matrix A such that T is multiplication by A. To see this, let

$$\mathbf{e}_1, \mathbf{e}_2, \ldots, \mathbf{e}_n$$

be the standard basis for R^n, and let A be the $m \times n$ matrix having

$$T(\mathbf{e}_1), T(\mathbf{e}_2), \ldots, T(\mathbf{e}_n)$$

as its column vectors. (We shall assume in this section that all vectors are expressed in matrix notation.) For example, if $T: R^2 \to R^2$ is given by

$$T\left(\begin{bmatrix} x_1 \\ x_2 \end{bmatrix}\right) = \begin{bmatrix} x_1 + 2x_2 \\ x_1 - x_2 \end{bmatrix}$$

then

$$T(\mathbf{e}_1) = T\left(\begin{bmatrix} 1 \\ 0 \end{bmatrix}\right) = \begin{bmatrix} 1 \\ 1 \end{bmatrix} \quad \text{and} \quad T(\mathbf{e}_2) = T\left(\begin{bmatrix} 0 \\ 1 \end{bmatrix}\right) = \begin{bmatrix} 2 \\ -1 \end{bmatrix}$$

$$A = \begin{bmatrix} 1 & 2 \\ 1 & -1 \end{bmatrix}$$
$$\begin{matrix} \uparrow & \uparrow \\ T(\mathbf{e}_1) & T(\mathbf{e}_2) \end{matrix}$$

More generally, if

$$T(\mathbf{e}_1) = \begin{bmatrix} a_{11} \\ a_{21} \\ \vdots \\ a_{m1} \end{bmatrix}, T(\mathbf{e}_2) = \begin{bmatrix} a_{12} \\ a_{22} \\ \vdots \\ a_{m2} \end{bmatrix}, \ldots, T(\mathbf{e}_n) = \begin{bmatrix} a_{1n} \\ a_{2n} \\ \vdots \\ a_{mn} \end{bmatrix}$$

then

$$A = \begin{bmatrix} a_{11} & a_{12} & \cdots & a_{1n} \\ a_{21} & a_{22} & \cdots & a_{2n} \\ \vdots & \vdots & & \vdots \\ a_{m1} & a_{m2} & \cdots & a_{mn} \end{bmatrix} \tag{5.6}$$
$$\begin{matrix} \uparrow & \uparrow & & \uparrow \\ T(\mathbf{e}_1) & T(\mathbf{e}_2) & \cdots & T(\mathbf{e}_n) \end{matrix}$$

We shall show that the linear transformation $T: R^n \to R^m$ is multiplication by A. To see this, observe first that

$$\mathbf{x} = \begin{bmatrix} x_1 \\ x_2 \\ \vdots \\ x_n \end{bmatrix} = x_1\mathbf{e}_1 + x_2\mathbf{e}_2 + \cdots + x_n\mathbf{e}_n$$

Therefore, by the linearity of T,

$$T(\mathbf{x}) = x_1 T(\mathbf{e}_1) + x_2 T(\mathbf{e}_2) + \cdots + x_n T(\mathbf{e}_n) \tag{5.7}$$

On the other hand

$$A\mathbf{x} = \begin{bmatrix} a_{11} & a_{12} & \cdots & a_{1n} \\ a_{21} & a_{22} & \cdots & a_{2n} \\ \vdots & \vdots & & \vdots \\ a_{m1} & a_{m2} & \cdots & a_{mn} \end{bmatrix} \begin{bmatrix} x_1 \\ x_2 \\ \vdots \\ x_n \end{bmatrix} = \begin{bmatrix} a_{11}x_1 + a_{12}x_2 + \cdots + a_{1n}x_n \\ a_{21}x_1 + a_{22}x_2 + \cdots + a_{2n}x_n \\ \vdots \\ a_{m1}x_1 + a_{m2}x_2 + \cdots + a_{mn}x_n \end{bmatrix}$$

$$= x_1 \begin{bmatrix} a_{11} \\ a_{21} \\ \vdots \\ a_{m1} \end{bmatrix} + x_2 \begin{bmatrix} a_{12} \\ a_{22} \\ \vdots \\ a_{m2} \end{bmatrix} + \cdots + x_n \begin{bmatrix} a_{1n} \\ a_{2n} \\ \vdots \\ a_{mn} \end{bmatrix}$$

$$= x_1 T(\mathbf{e}_1) + x_2 T(\mathbf{e}_2) + \cdots + x_n T(\mathbf{e}_n) \qquad (5.8)$$

Comparing (5.7) and (5.8) yields $T(\mathbf{x}) = A\mathbf{x}$, that is, T is multiplication by A.

We shall refer to the matrix A in (5.6) as the ***standard matrix for T***.

Example 19

Find the standard matrix for the transformation $T: R^3 \rightarrow R^4$ defined by

$$T\left(\begin{bmatrix} x_1 \\ x_2 \\ x_3 \end{bmatrix}\right) = \begin{bmatrix} x_1 + x_2 \\ x_1 \quad\; x_2 \\ x_3 \\ x_1 \end{bmatrix}$$

Solution.

$$T(\mathbf{e}_1) = T\left(\begin{bmatrix} 1 \\ 0 \\ 0 \end{bmatrix}\right) = \begin{bmatrix} 1 \\ 1 \\ 0 \\ 1 \end{bmatrix} \qquad T(\mathbf{e}_2) = T\left(\begin{bmatrix} 0 \\ 1 \\ 0 \end{bmatrix}\right) = \begin{bmatrix} 1 \\ -1 \\ 0 \\ 0 \end{bmatrix} \qquad T(\mathbf{e}_3) = T\left(\begin{bmatrix} 0 \\ 0 \\ 1 \end{bmatrix}\right) = \begin{bmatrix} 0 \\ 0 \\ 1 \\ 0 \end{bmatrix}$$

Using $T(\mathbf{e}_1)$, $T(\mathbf{e}_2)$, and $T(\mathbf{e}_3)$ as column vectors, we obtain

$$A = \begin{bmatrix} 1 & 1 & 0 \\ 1 & -1 & 0 \\ 0 & 0 & 1 \\ 1 & 0 & 0 \end{bmatrix}$$

As a check, observe that

$$A \begin{bmatrix} x_1 \\ x_2 \\ x_3 \end{bmatrix} = \begin{bmatrix} x_1 + x_2 \\ x_1 - x_2 \\ x_3 \\ x_1 \end{bmatrix}$$

which agrees with the given formula for T.

Example 20

Let $T: R^2 \to R^2$ be the linear transformation that maps each vector into its symmetric image about the y-axis (Figure 5.5). Find the standard matrix for T.

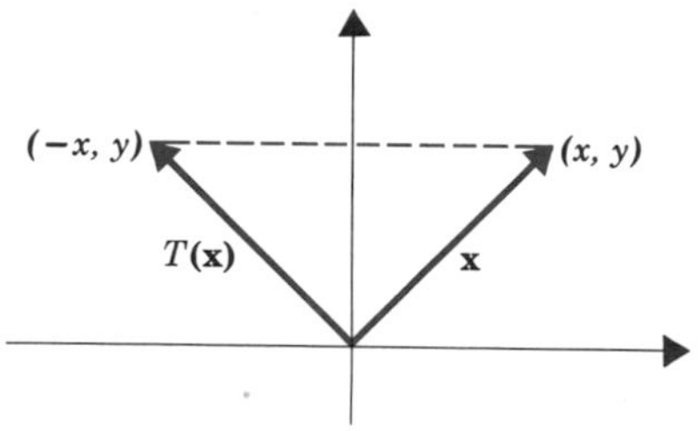

Figure 5.5

Solution.

$$T(\mathbf{e}_1) = T\left(\begin{bmatrix} 1 \\ 0 \end{bmatrix}\right) = \begin{bmatrix} -1 \\ 0 \end{bmatrix} \qquad T(\mathbf{e}_2) = T\left(\begin{bmatrix} 0 \\ 1 \end{bmatrix}\right) = \begin{bmatrix} 0 \\ 1 \end{bmatrix}$$

Using $T(\mathbf{e}_1)$ and $T(\mathbf{e}_2)$ as column vectors we obtain the standard matrix

$$A = \begin{bmatrix} -1 & 0 \\ 0 & 1 \end{bmatrix}$$

As a check, let

$$\mathbf{x} = \begin{bmatrix} x \\ y \end{bmatrix}$$

be any vector in R^2. Then

$$A\mathbf{x} = \begin{bmatrix} -1 & 0 \\ 0 & 1 \end{bmatrix}\begin{bmatrix} x \\ y \end{bmatrix} = \begin{bmatrix} -x \\ y \end{bmatrix}$$

so that $A\mathbf{x}$ is the symmetric image of $\mathbf{x}$ about the y-axis.

We show next that if V and W are any finite dimensional vector spaces (not necessarily R^n and R^m), then with a little ingenuity, any linear transformation $T: V \to W$ can be regarded as a matrix transformation. The basic idea is to choose bases for V and W and to work with the coordinate matrices relative to these bases rather than with the vectors themselves. To be specific, suppose V is n-dimensional and W is m-dimensional. If we choose bases B and B' for V and W, respectively, then for each $\mathbf{x}$ in V, the coordinate matrix $[\mathbf{x}]_B$ will be a vector in R^n and the coordinate matrix $[T(\mathbf{x})]_{B'}$ will be some vector in R^m. Thus in the process of mapping $\mathbf{x}$ into $T(\mathbf{x})$, the linear transformation T "generates" a mapping from R^n into R^m by sending $[\mathbf{x}]_B$ into $[T(\mathbf{x})]_{B'}$. It can be shown that this generated mapping is always a linear transformation. As such it can be carried

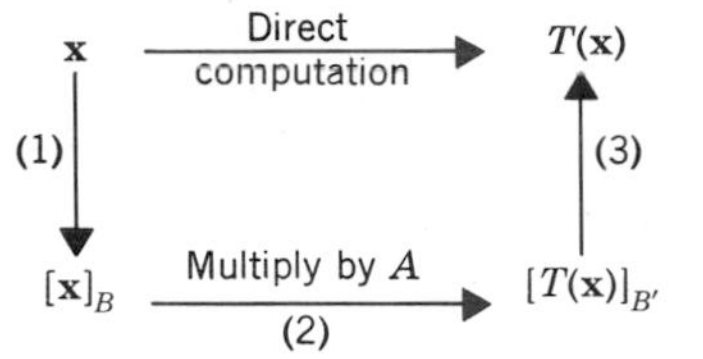

Figure 5.6

out by using the standard matrix A for this transformation; that is

$$A[\mathbf{x}]_B = [T(\mathbf{x})]_{B'} \tag{5.9}$$

If, somehow, we can find the matrix A, then as shown in Figure 5.6, $T(\mathbf{x})$ can be computed in three steps by the following indirect procedure.

(1) Compute the coordinate matrix $[\mathbf{x}]_B$
(2) Multiply $[\mathbf{x}]_B$ on the left by A to produce $[T(\mathbf{x})]_{B'}$
(3) Reconstruct $T(\mathbf{x})$ from its coordinate matrix $[T(\mathbf{x})]_{B'}$

There are two major reasons why this indirect procedure is important. First, it provides an efficient way of carrying out linear transformations on a digital computer. The second reason is theoretical, but with important practical consequences. The matrix A depends on the bases B and B'. Ordinarily one would choose B and B' to make the computation of coordinate matrices as easy as possible. However, one could instead try to choose the bases B and B' to make the matrix A as simple as possible, say with lots of zero entries. When this is done in the right way the matrix A can provide important information about the linear transformation T. We will pursue this idea in later sections.

We turn now to the problem of finding a matrix A satisfying (5.9). Suppose V is an n-dimensional space with basis $B = \{\mathbf{u}_1, \mathbf{u}_2, \ldots, \mathbf{u}_n\}$ and W is an m-dimensional space with basis $B' = \{\mathbf{v}_1, \mathbf{v}_2, \ldots, \mathbf{v}_m\}$. We are looking for an $m \times n$ matrix

$$A = \begin{bmatrix} a_{11} & a_{12} & \cdots & a_{1n} \\ a_{21} & a_{22} & \cdots & a_{2n} \\ \vdots & \vdots & & \vdots \\ a_{m1} & a_{m2} & \cdots & a_{mn} \end{bmatrix}$$

such that (5.9) holds for all vectors $\mathbf{x}$ in V. In particular, when $\mathbf{x}$ is the basis vector $\mathbf{u}_1$ we want

$$A[\mathbf{u}_1]_B = [T(\mathbf{u}_1)]_{B'} \tag{5.10}$$

But

$$[\mathbf{u}_1]_B = \begin{bmatrix} 1 \\ 0 \\ 0 \\ \vdots \\ 0 \end{bmatrix}$$

so that

$$A[\mathbf{u}_1]_B = \begin{bmatrix} a_{11} & a_{12} & \cdots & a_{1n} \\ a_{21} & a_{22} & \cdots & a_{2n} \\ \vdots & \vdots & & \vdots \\ a_{m1} & a_{m2} & \cdots & a_{mn} \end{bmatrix} \begin{bmatrix} 1 \\ 0 \\ 0 \\ \vdots \\ 0 \end{bmatrix} = \begin{bmatrix} a_{11} \\ a_{21} \\ \vdots \\ a_{m1} \end{bmatrix}$$

Thus (5.10) implies that

$$\begin{bmatrix} a_{11} \\ a_{21} \\ \vdots \\ a_{m1} \end{bmatrix} = [T(\mathbf{u}_1)]_{B'}$$

That is, the first column of A is the coordinate matrix for the vector $T(\mathbf{u}_1)$ with respect to the basis B'. Similarly, if we let $\mathbf{x} = \mathbf{u}_2$ in (5.9) we obtain

$$A[\mathbf{u}_2]_B = [T(\mathbf{u}_2)]_{B'}$$

But

$$[\mathbf{u}_2]_B = \begin{bmatrix} 0 \\ 1 \\ 0 \\ \vdots \\ 0 \end{bmatrix}$$

so that

$$A[\mathbf{u}_2]_B = \begin{bmatrix} a_{11} & a_{12} & \cdots & a_{1n} \\ a_{21} & a_{22} & \cdots & a_{2n} \\ \vdots & \vdots & & \vdots \\ a_{m1} & a_{m2} & \cdots & a_{mn} \end{bmatrix} \begin{bmatrix} 0 \\ 1 \\ 0 \\ \vdots \\ 0 \end{bmatrix} = \begin{bmatrix} a_{12} \\ a_{22} \\ \vdots \\ a_{m2} \end{bmatrix}$$

Thus

$$\begin{bmatrix} a_{12} \\ a_{22} \\ \vdots \\ a_{m2} \end{bmatrix} = [T(\mathbf{u}_2)]_{B'}$$

That is, the second column of A is the coordinate matrix for the vector $T(\mathbf{u}_2)$ with respect to the basis B'. Continuing in this way, we find that *the jth column of A is the coordinate matrix for the vector $T(\mathbf{u}_j)$ with respect to B'.* The unique matrix A obtained in this way is called the ***matrix of T with respect to the bases B and B′***. Symbolically, we can denote this matrix by

$$A = \begin{array}{c}\text{matrix of } T \text{ with} \\ \text{respect to the} \\ \text{bases } B \text{ and } B'\end{array} = \Big[[T(\mathbf{u}_1)]_{B'} \;\Big|\; [T(\mathbf{u}_2)]_{B'} \;\Big|\; \cdots \;\Big|\; [T(\mathbf{u}_n)]_{B'} \Big]$$

Example 21

Let $T: P_1 \to P_2$ be the linear transformation defined by

$$T(p(x)) = xp(x)$$

Find the matrix for T with respect to the bases

$$B = \{\mathbf{u}_1, \mathbf{u}_2\} \qquad \text{and} \qquad B' = \{\mathbf{u}'_1, \mathbf{u}'_2, \mathbf{u}'_3\}$$

where

$$\mathbf{u}_1 = 1, \quad \mathbf{u}_2 = x; \quad \mathbf{u}'_1 = 1, \quad \mathbf{u}'_2 = x, \quad \mathbf{u}'_3 = x^2$$

Solution. From the formula for T we obtain

$$T(\mathbf{u}_1) = T(1) = (x)(1) = x$$
$$T(\mathbf{u}_2) = T(x) = (x)(x) = x^2$$

By inspection, we can determine the coordinate matrices for $T(\mathbf{u}_1)$ and $T(\mathbf{u}_2)$ relative to B'; they are

$$[T(\mathbf{u}_1)]_{B'} = \begin{bmatrix} 0 \\ 1 \\ 0 \end{bmatrix}, \ [T(\mathbf{u}_2)]_{B'} = \begin{bmatrix} 0 \\ 0 \\ 1 \end{bmatrix}$$

Thus the matrix for T with respect to B and B' is

$$A = \Big[[T(\mathbf{u}_1)]_{B'} \mid [T(\mathbf{u}_2)]_{B'}\Big] = \begin{bmatrix} 0 & 0 \\ 1 & 0 \\ 0 & 1 \end{bmatrix}$$

Example 22

Let $T: P_1 \to P_2$, B, and B' be as in Example 21, and let

$$\mathbf{x} = 1 - 2x$$

Use the matrix obtained in Example 21 to compute $T(\mathbf{x})$ by the indirect procedure in Figure 5.6.

Solution. By inspection, the coordinate matrix of $\mathbf{x}$ with respect to B is

$$[\mathbf{x}]_B = \begin{bmatrix} 1 \\ -2 \end{bmatrix}$$

Therefore

$$[T(\mathbf{x})]_{B'} = A[\mathbf{x}]_B = \begin{bmatrix} 0 & 0 \\ 1 & 0 \\ 0 & 1 \end{bmatrix} \begin{bmatrix} 1 \\ -2 \end{bmatrix} = \begin{bmatrix} 0 \\ 1 \\ -2 \end{bmatrix}$$

Thus

$$\begin{aligned} T(\mathbf{x}) &= 0\mathbf{u}'_1 + 1\mathbf{u}'_2 - 2\mathbf{u}'_2 = 0(1) + 1(x) - 2(x^2) \\ &= x - 2x^2 \end{aligned}$$

As a check, note that the direct computation of $T(\mathbf{x})$ is

$$T(\mathbf{x}) = T(1 - 2x) = x(1 - 2x) = x - 2x^2$$

which agrees with result obtained by the indirect procedure.

Example 23

If $T:R^n \to R^m$ is a linear transformation and if B and B' are the standard bases for R^n and R^m respectively, then the matrix for T with respect to B and B' is just the standard matrix for T discussed in the beginning of this section. (We leave the verification of this as an exercise.)

In the special case where $V = W$ (so that $T:V \to V$ is a linear operator) it is usual to take $B = B'$ when constructing a matrix of T. The resulting matrix is called the ***matrix of T with respect to the basis B***.

Example 24

If $B = \{\mathbf{u}_1, \mathbf{u}_2, \ldots, \mathbf{u}_n\}$ is any basis for a finite dimensional vector space V and $I:V \to V$ is the identity operator on V, then $I(\mathbf{u}_1) = \mathbf{u}_1, I(\mathbf{u}_2) = \mathbf{u}_2, \ldots, I(\mathbf{u}_n) = \mathbf{u}_n$. Therefore

$$[I(\mathbf{u}_1)]_B = \begin{bmatrix} 1 \\ 0 \\ 0 \\ \vdots \\ 0 \end{bmatrix}, [I(\mathbf{u}_2)]_B = \begin{bmatrix} 0 \\ 1 \\ 0 \\ \vdots \\ 0 \end{bmatrix}, \ldots, [I(\mathbf{u}_n)]_B = \begin{bmatrix} 0 \\ 0 \\ 0 \\ \vdots \\ 1 \end{bmatrix}$$

thus

$$[I]_B = \begin{bmatrix} 1 & 0 & \cdots & 0 \\ 0 & 1 & \cdots & 0 \\ 0 & 0 & \cdots & 0 \\ \vdots & \vdots & & \vdots \\ 0 & 0 & \cdots & 1 \end{bmatrix}$$

Consequently, the matrix of the identity operator with respect to any basis is the $n \times n$ identity matrix.

Example 25

Let $T:R^2 \to R^2$ be the linear operator defined by

$$T\left(\begin{bmatrix} x_1 \\ x_2 \end{bmatrix}\right) = \begin{bmatrix} x_1 + x_2 \\ -2x_1 + 4x_2 \end{bmatrix}$$

Find the matrix of T with respect to the basis $B = \{\mathbf{u}_1, \mathbf{u}_2\}$ where

$$\mathbf{u}_1 = \begin{bmatrix} 1 \\ 1 \end{bmatrix} \quad \text{and} \quad \mathbf{u}_2 = \begin{bmatrix} 1 \\ 2 \end{bmatrix}$$

Solution. From the definition of T

$$T(\mathbf{u}_1) = \begin{bmatrix} 2 \\ 2 \end{bmatrix} = 2\mathbf{u}_1 \quad \text{and} \quad T(\mathbf{u}_2) = \begin{bmatrix} 3 \\ 6 \end{bmatrix} = 3\mathbf{u}_2$$

Therefore

$$[T(\mathbf{u}_1)]_B = \begin{bmatrix} 2 \\ 0 \end{bmatrix} \quad \text{and} \quad [T(\mathbf{u}_2)]_B = \begin{bmatrix} 0 \\ 3 \end{bmatrix}$$

Consequently, the matrix of T with respect to B is

$$A = \begin{bmatrix} 2 & 0 \\ 0 & 3 \end{bmatrix}$$

EXERCISE SET 5.3

1. Find the standard matrix of each of the following linear operators.

(a) $T\left(\begin{bmatrix} x_1 \\ x_2 \end{bmatrix}\right) = \begin{bmatrix} 2x_1 - x_2 \\ x_1 + x_2 \end{bmatrix}$ (b) $T\left(\begin{bmatrix} x_1 \\ x_2 \end{bmatrix}\right) = \begin{bmatrix} x_1 \\ x_2 \end{bmatrix}$

(c) $T\left(\begin{bmatrix} x_1 \\ x_2 \\ x_3 \end{bmatrix}\right) = \begin{bmatrix} x_1 + 2x_2 + x_3 \\ x_1 + 5x_2 \\ x_3 \end{bmatrix}$ (d) $T\left(\begin{bmatrix} x_1 \\ x_2 \\ x_3 \end{bmatrix}\right) = \left(\begin{bmatrix} 4x_1 \\ 7x_2 \\ -8x_3 \end{bmatrix}\right)$

2. Find the standard matrix of each of the following linear transformations.

(a) $T\left(\begin{bmatrix} x_1 \\ x_2 \end{bmatrix}\right) = \begin{bmatrix} x_2 \\ -x_1 \\ x_1 + 3x_2 \\ x_1 - x_2 \end{bmatrix}$ (b) $T\left(\begin{bmatrix} x_1 \\ x_2 \\ x_3 \\ x_4 \end{bmatrix}\right) = \begin{bmatrix} 7x_1 + 2x_2 - x_3 + x_4 \\ x_2 + x_3 \\ -x_1 \end{bmatrix}$

(c) $T\left(\begin{bmatrix} x_1 \\ x_2 \\ x_3 \end{bmatrix}\right) = \begin{bmatrix} 0 \\ 0 \\ 0 \\ 0 \\ 0 \end{bmatrix}$ (d) $T\left(\begin{bmatrix} x_1 \\ x_2 \\ x_3 \\ x_4 \end{bmatrix}\right) = \begin{bmatrix} x_4 \\ x_1 \\ x_3 \\ x_2 \\ x_1 - x_3 \end{bmatrix}$

3. Find the standard matrix for the linear operator $T: R^2 \to R^2$ which maps a vector $\mathbf{v} = (x, y)$ into:

(a) its reflection through the x-axis;

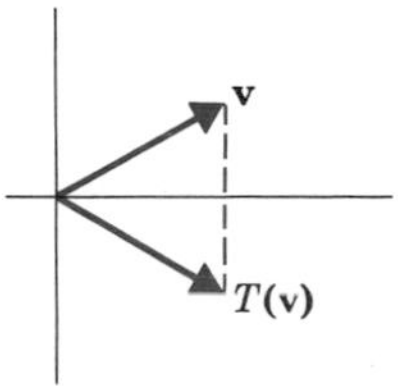

(b) its reflection through the line $y = x$;

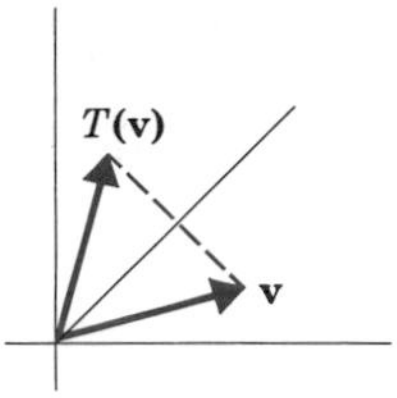

(c) its reflection through the origin;

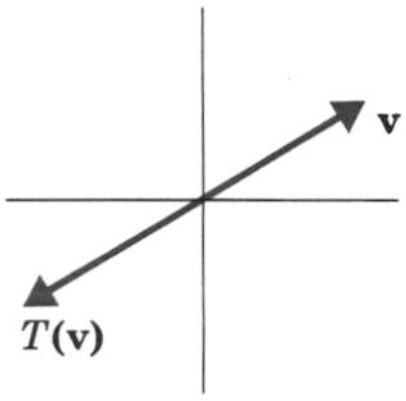

(d) its orthogonal projection on the x-axis.

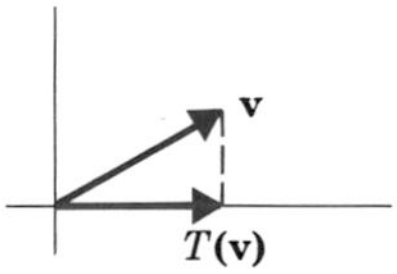

4. For each part of Exercise 3, use the matrix you have obtained to compute $T(2, 1)$. Check your answers geometrically by sketching the vectors $(2, 1)$ and $T(2, 1)$.

5. Find the standard matrix for the linear operator $T: R^3 \to R^3$ which maps a vector $\mathbf{v} = (x, y, z)$ into:
(a) its reflection through the xy-plane;
(b) its reflection through the xz-plane;
(c) its reflection through the yz-plane.

6. For each part of Exercise 5, use the matrix you have obtained to compute $T(1, 1, 1)$. Check your answers geometrically by sketching the vectors $(1, 1, 1)$ and $T(1, 1, 1)$.

7. Find the standard matrix for the linear operator $T: R^3 \to R^3$ which
(a) rotates each vector 90° counterclockwise about the z-axis (looking down the positive z-axis toward the origin);
(b) rotates each vector 90° counterclockwise about the x-axis (looking along the positive x-axis toward the origin);
(c) rotates each vector 90° counterclockwise about the y-axis (looking along the positive y-axis toward the origin).

8. Let $T: P_2 \to P_1$ be the linear transformation defined by

$$T(a_0 + a_1 x + a_2 x^2) = (a_0 + a_1) - (2a_1 + 3a_2)x$$

Find the matrix of T with respect to the standard bases for P_2 and P_1.

9. Let $T:R^2 \to R^3$ be defined by

$$T\left(\begin{bmatrix} x_1 \\ x_2 \end{bmatrix}\right) = \begin{bmatrix} x_1 + 2x_2 \\ -x_1 \\ 0 \end{bmatrix}$$

(a) Find the matrix of T with respect to the bases $B = \{\mathbf{u}_1, \mathbf{u}_2\}$ and $B' = \{\mathbf{v}_1, \mathbf{v}_2, \mathbf{v}_3\}$, where

$$\mathbf{u}_1 = \begin{bmatrix} 1 \\ 3 \end{bmatrix} \quad \mathbf{u}_2 = \begin{bmatrix} -2 \\ 4 \end{bmatrix} \quad \mathbf{v}_1 = \begin{bmatrix} 1 \\ 1 \\ 1 \end{bmatrix} \quad \mathbf{v}_2 = \begin{bmatrix} 2 \\ 2 \\ 0 \end{bmatrix} \quad \mathbf{v}_3 = \begin{bmatrix} 3 \\ 0 \\ 0 \end{bmatrix}$$

(b) Use the matrix obtained in (a) to compute

$$T\left(\begin{bmatrix} 8 \\ 3 \end{bmatrix}\right)$$

10. Let $T:R^3 \to R^3$ be defined by

$$T\left(\begin{bmatrix} x_1 \\ x_2 \\ x_3 \end{bmatrix}\right) = \begin{bmatrix} x_1 - x_2 \\ x_2 - x_1 \\ x_1 - x_3 \end{bmatrix}$$

(a) Find the matrix of T with respect to the basis $B = \{\mathbf{v}_1, \mathbf{v}_2, \mathbf{v}_3\}$, where

$$\mathbf{v}_1 = \begin{bmatrix} 1 \\ 0 \\ 1 \end{bmatrix} \quad \mathbf{v}_2 = \begin{bmatrix} 0 \\ 1 \\ 1 \end{bmatrix} \quad \mathbf{v}_3 = \begin{bmatrix} 1 \\ 1 \\ 0 \end{bmatrix}$$

(b) Use the matrix obtained in (a) to compute

$$T\left(\begin{bmatrix} 2 \\ 0 \\ 0 \end{bmatrix}\right)$$

11. Let $T:P_2 \to P_4$ be the linear transformation defined by $T(p(x)) = x^2p(x)$.
(a) Find the matrix of T with respect to the bases $B = \{\mathbf{p}_1, \mathbf{p}_2, \mathbf{p}_3\}$ and B', where $\mathbf{p}_1 = 1 + x^2$, $\mathbf{p}_2 = 1 + 2x + 3x^2$, $\mathbf{p}_3 = 4 + 5x + x^2$, and B' is the standard basis for P_4.
(b) Use the matrix obtained in (a) to compute $T(-3 + 5x - 2x^2)$.

12. Let $\mathbf{v}_1 = \begin{bmatrix} 1 \\ 3 \end{bmatrix}$ and $\mathbf{v}_2 = \begin{bmatrix} -1 \\ 4 \end{bmatrix}$, and let

$$A = \begin{bmatrix} 1 & 3 \\ -2 & 5 \end{bmatrix}$$

be the matrix of $T:R^2 \to R^2$ with respect to the basis $B = \{\mathbf{v}_1, \mathbf{v}_2\}$.
(a) Find $[T(\mathbf{v}_1)]_B$ and $[T(\mathbf{v}_2)]_B$.
(b) Find $T(\mathbf{v}_1)$ and $T(\mathbf{v}_2)$.
(c) Find $T\left(\begin{bmatrix} 1 \\ 1 \end{bmatrix}\right)$.

13. Let $A = \begin{bmatrix} 3 & -2 & 1 & 0 \\ 1 & 6 & 2 & 1 \\ -3 & 0 & 7 & 1 \end{bmatrix}$ be the matrix of $T: R^4 \to R^3$ with respect to the bases $B = \{\mathbf{v}_1, \mathbf{v}_2, \mathbf{v}_3, \mathbf{v}_4\}$ and $B' = \{\mathbf{w}_1, \mathbf{w}_2, \mathbf{w}_3\}$, where

$$\mathbf{v}_1 = \begin{bmatrix} 0 \\ 1 \\ 1 \\ 1 \end{bmatrix} \quad \mathbf{v}_2 = \begin{bmatrix} 2 \\ 1 \\ -1 \\ -1 \end{bmatrix} \quad \mathbf{v}_3 = \begin{bmatrix} 1 \\ 4 \\ -1 \\ 2 \end{bmatrix} \quad \mathbf{v}_4 = \begin{bmatrix} 6 \\ 9 \\ 4 \\ 2 \end{bmatrix}$$

$$\mathbf{w}_1 = \begin{bmatrix} 0 \\ 8 \\ 8 \end{bmatrix} \quad \mathbf{w}_2 = \begin{bmatrix} -7 \\ 8 \\ 1 \end{bmatrix} \quad \mathbf{w}_3 = \begin{bmatrix} -6 \\ 9 \\ 1 \end{bmatrix}$$

(a) Find $[T(\mathbf{v}_1)]_{B'}$, $[T(\mathbf{v}_2)]_{B'}$, $[T(\mathbf{v}_3)]_{B'}$, and $[T(\mathbf{v}_4)]_{B'}$.
(b) Find $T(\mathbf{v}_1)$, $T(\mathbf{v}_2)$, $T(\mathbf{v}_3)$, and $T(\mathbf{v}_4)$.

(c) Find $T\begin{bmatrix} 2 \\ 2 \\ 0 \\ 0 \end{bmatrix}$

14. Let $A = \begin{bmatrix} 1 & 3 & -1 \\ 2 & 0 & 5 \\ 6 & -2 & 4 \end{bmatrix}$ be the matrix of $T: P_2 \to P_2$ with respect to the basis $B = \{\mathbf{v}_1, \mathbf{v}_2, \mathbf{v}_3\}$, where $\mathbf{v}_1 = 3x + 3x^2$, $\mathbf{v}_2 = -1 + 3x + 2x^2$, $\mathbf{v}_3 = 3 + 7x + 2x^2$.
(a) Find $[T(\mathbf{v}_1)]_B$, $[T(\mathbf{v}_2)]_B$, and $[T(\mathbf{v}_3)]_B$.
(b) Find $T(\mathbf{v}_1)$, $T(\mathbf{v}_2)$, and $T(\mathbf{v}_3)$.
(c) Find $T(1 + x^2)$.

15. Show that if $T: V \to W$ is the zero transformation (Example 3), then the matrix of T with respect to any bases for V and W is a zero matrix.

16. Show that if $T: V \to V$ is a contraction or a dilation of V (Example 5), then the matrix of T with respect to any basis for V is a diagonal matrix.

17. Let $B = \{\mathbf{v}_1, \mathbf{v}_2, \mathbf{v}_3, \mathbf{v}_4\}$ be a basis for a vector space V. Find the matrix with respect to B of the linear operator $T: V \to V$ defined by $T(\mathbf{v}_1) = \mathbf{v}_2$, $T(\mathbf{v}_2) = \mathbf{v}_3$, $T(\mathbf{v}_3) = \mathbf{v}_4$, $T(\mathbf{v}_4) = \mathbf{v}_1$.

18. **(For readers who have studied calculus.)** Let $D: P_2 \to P_2$ be the differentiation operator $D(\mathbf{p}) = \mathbf{p}'$. In parts (a) and (b) find the matrix of D with respect to the basis $B = \{\mathbf{p}_1, \mathbf{p}_2, \mathbf{p}_3\}$.
(a) $\mathbf{p}_1 = 1$, $\mathbf{p}_2 = x$, $\mathbf{p}_3 = x^2$
(b) $\mathbf{p}_1 = 2$, $\mathbf{p}_2 = 2 - 3x$, $\mathbf{p}_3 = 2 - 3x + 8x^2$
(c) Use the matrix in part (a) to compute $D(6 - 6x + 24x^2)$.
(d) Repeat the directions of part (c) for the matrix in part (b).

19. (For readers who studied calculus.) In each part, $B = (\mathbf{f}_1, \mathbf{f}_2, \mathbf{f}_3)$ is a basis for a subspace V of the vector space of real-valued functions defined on the real line. Find the matrix with respect to B of the differentiation operator $D: V \to V$.
(a) $\mathbf{f}_1 = 1, \mathbf{f}_2 = \sin x, \mathbf{f}_3 = \cos x$
(b) $\mathbf{f}_1 = 1, \mathbf{f}_2 = e^x, \mathbf{f}_3 = e^{2x}$
(c) $\mathbf{f}_1 = e^{2x}, \mathbf{f}_2 = xe^{2x}, \mathbf{f}_3 = xe^{2x}$

5.4 SIMILARITY

The matrix of a linear operator $T: V \to V$ depends on the basis selected for V. One of the fundamental problems of linear algebra is to choose a basis for V that makes the matrix of T as simple as possible. Often this problem is attacked by first finding a matrix for T relative to some "simple" basis like a standard basis. Usually, this choice does not yield the simplest matrix for T, so that one then looks for a way to change the basis in order to simplify the matrix. In order to attack a problem like this, we must know how a change of basis affects the matrix of a linear operator; we will study this problem in this section.

The following theorem is the key result in this section.

Theorem 5. *Let $T: V \to V$ be a linear operator on a finite dimensional vector space V. If A is the matrix of T with respect to a basis B, and A' is the matrix of T with respect to a basis B', then*

$$A' = P^{-1}AP \tag{5.11}$$

where P is the transition matrix from B' to B.

To establish this theorem, it will be convenient to describe the relationship

$$A\mathbf{u} = \mathbf{v}$$

pictorially by writing

$$\mathbf{u} \xrightarrow{A} \mathbf{v}$$

Since A is the matrix of T with respect to B, and A' is the matrix of T with respect to B', the following relationships hold for all $\mathbf{x}$ in V.

$$A[\mathbf{x}]_B = [T(\mathbf{x})]_B$$

and

$$A'[\mathbf{x}]_{B'} = [T(\mathbf{x})]_{B'}$$

These can be written

$$[\mathbf{x}]_B \xrightarrow{A} [T(\mathbf{x})]_B$$

and (5.12)

$$[\mathbf{x}]_{B'} \xrightarrow{A'} [T(\mathbf{x})]_{B'}$$

To see how the matrices A and A' are related, let P be the transition matrix from the B' basis to the B basis, so that P^{-1} is the transition matrix from B to B'.

Thus

$$P[\mathbf{x}]_{B'} = [\mathbf{x}]_B$$

and

$$P^{-1}[T(\mathbf{x})]_B = [T(\mathbf{x})]_{B'}$$

which can be written as

$$[\mathbf{x}]_{B'} \xrightarrow{P} [\mathbf{x}]_B \tag{5.13}$$

and

$$[T(\mathbf{x})]_B \xrightarrow{P^{-1}} [T(\mathbf{x})]_{B'}$$

For compactness, relationships (5.12) and (5.13) can be linked together in a single figure as follows:

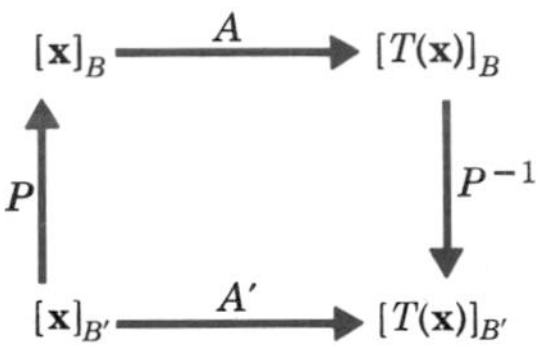

This figure illustrates that there are two ways to obtain the matrix $[T(\mathbf{x})]_{B'}$ from the matrix $[\mathbf{x}]_{B'}$. We can take the bottom path across the figure, that is

$$A'[\mathbf{x}]_{B'} = [T(\mathbf{x})]_{B'} \tag{5.14}$$

or we can go up the left side, across the top, and down the right side, that is

$$P^{-1}AP[\mathbf{x}]_{B'} = [T(\mathbf{x})]_{B'} \tag{5.15}$$

It follows from (5.14) and (5.15) that

$$P^{-1}AP[\mathbf{x}]_{B'} = A'[\mathbf{x}]_{B'} \tag{5.16}$$

for all $\mathbf{x}$ in V. It follows from (5.16) and part (b) of Exercise 11 that

$$P^{-1}AP = A'$$

This proves Theorem 5. ▮

Warning. *When applying Theorem 5, it is easy to forget whether P is the transition matrix from B to B' (incorrect) or from B' to B (correct). It may help to call B the old basis, B' the new basis, A the old matrix, and A' the new matrix. Since P is the transition matrix from B' to B, P^{-1} is the transition matrix from B to B'. Thus (5.11) can be expressed as:*

$$\textit{new matrix} = P^{-1}\,(\textit{old matrix})\,P$$

where P is the transition matrix from the new basis to the old basis.

Example 26

Let $T: R^2 \to R^2$ be defined by

$$T\left(\begin{bmatrix} x_1 \\ x_2 \end{bmatrix}\right) = \begin{bmatrix} x_1 + x_2 \\ -2x_1 + 4x_2 \end{bmatrix}$$

Find the standard matrix for T, that is the matrix for T relative to the basis $B = \{\mathbf{e}_1, \mathbf{e}_2\}$ where

$$\mathbf{e}_1 = \begin{bmatrix} 1 \\ 0 \end{bmatrix} \qquad \mathbf{e}_2 = \begin{bmatrix} 0 \\ 1 \end{bmatrix}$$

and then use Theorem 5 to transform this matrix into the matrix for T relative to the basis $B' = \{\mathbf{u}_1, \mathbf{u}_2\}$ where

$$\mathbf{u}_1 = \begin{bmatrix} 1 \\ 1 \end{bmatrix} \quad \text{and} \quad \mathbf{u}_2 = \begin{bmatrix} 1 \\ 2 \end{bmatrix}$$

Solution. From the formula for T

$$T(\mathbf{e}_1) = T\left(\begin{bmatrix} 1 \\ 0 \end{bmatrix}\right) = \begin{bmatrix} 1 \\ -2 \end{bmatrix}$$

and

$$T(\mathbf{e}_2) = T\left(\begin{bmatrix} 0 \\ 1 \end{bmatrix}\right) = \begin{bmatrix} 1 \\ 4 \end{bmatrix}$$

so that the standard matrix for T is

$$A = \begin{bmatrix} 1 & 1 \\ -2 & 4 \end{bmatrix}$$

Next we need the transition matrix from B' to B. For this transition matrix we must find the coordinate matrices for the B' basis vectors relative to the basis B. By inspection

$$\mathbf{u}_1 = \mathbf{e}_1 + \mathbf{e}_2$$
$$\mathbf{u}_2 = \mathbf{e}_1 + 2\mathbf{e}_2$$

so that

$$[\mathbf{u}_1]_B = \begin{bmatrix} 1 \\ 1 \end{bmatrix} \quad \text{and} \quad [\mathbf{u}_2]_B = \begin{bmatrix} 1 \\ 2 \end{bmatrix}$$

Thus the transition matrix from B' to B is

$$P = \begin{bmatrix} 1 & 1 \\ 1 & 2 \end{bmatrix}$$

The reader can check that

$$P^{-1} = \begin{bmatrix} 2 & -1 \\ -1 & 1 \end{bmatrix}$$

so that by Theorem 5 the matrix of T relative to the basis B' is

$$P^{-1}AP = \begin{bmatrix} 2 & -1 \\ -1 & 1 \end{bmatrix} \begin{bmatrix} 1 & 1 \\ -2 & 4 \end{bmatrix} \begin{bmatrix} 1 & 1 \\ 1 & 2 \end{bmatrix} = \begin{bmatrix} 2 & 0 \\ 0 & 3 \end{bmatrix}.$$

This example illustrates that the standard basis for a vector space does not necessarily produce the simplest matrix for a linear operator; in this example the standard matrix

$$A = \begin{bmatrix} 1 & 1 \\ -2 & 4 \end{bmatrix}$$

was not as simple in structure as the matrix

$$\begin{bmatrix} 2 & 0 \\ 0 & 3 \end{bmatrix} \tag{5.17}$$

relative to the basis B'. Matrix (5.17) is an example of a ***diagonal matrix***; that is, a square matrix all of whose *non*diagonal entries are zeros. Diagonal matrices have many desirable properties. For example, the kth power of a diagonal matrix

$$D = \begin{bmatrix} d_1 & 0 & \cdots & 0 \\ 0 & d_2 & \cdots & 0 \\ \vdots & \vdots & & \vdots \\ 0 & 0 & \cdots & d_n \end{bmatrix}$$

is

$$D^k = \begin{bmatrix} d_1{}^k & 0 & \cdots & 0 \\ 0 & d_2{}^k & \cdots & 0 \\ \vdots & \vdots & & \vdots \\ 0 & 0 & \cdots & d_n{}^k \end{bmatrix}$$

Thus, to raise a diagonal matrix to the kth power, we need only raise each diagonal entry to the kth power. For a nondiagonal matrix there is much more computation involved in obtaining the kth power. Diagonal matrices also have other useful properties.

In the next chapter we discuss the problem of finding bases that produce diagonal matrices for linear operators.

Theorem 5 motivates the following definition.

Definition. If A and B are square matrices, we say that ***B is similar to A*** if there is an invertible matrix P such that $B = P^{-1}AP$.

Note that the equation $B = P^{-1}AP$ can be rewritten as

$$A = PBP^{-1} \qquad \text{or} \qquad A = (P^{-1})^{-1}BP^{-1}$$

Letting $Q = P^{-1}$ yields

$$A = Q^{-1}BQ$$

which says that A is similar to B. Therefore, B is similar to A if and only if A is similar to B; consequently, we shall usually say simply that ***A* and *B* are similar**.

In this terminology, Theorem 5 asserts that *two matrices representing the same linear operator $T: V \to V$ with respect to different bases are similar.*

EXERCISE SET 5.4

In Exercises 1–7 find the matrix of T with respect to B, and use Theorem 5 to compute the matrix of T with respect to B'.

1. $T: R^2 \to R^2$ is defined by

$$T\left(\begin{bmatrix} x_1 \\ x_2 \end{bmatrix}\right) = \begin{bmatrix} x_1 - 2x_2 \\ -x_2 \end{bmatrix}$$

$B = \{\mathbf{u}_1, \mathbf{u}_2\}$ and $B' = \{\mathbf{v}_1, \mathbf{v}_2\}$, where

$$\mathbf{u}_1 = \begin{bmatrix} 1 \\ 0 \end{bmatrix} \quad \mathbf{u}_2 = \begin{bmatrix} 0 \\ 1 \end{bmatrix} \quad \mathbf{v}_1 = \begin{bmatrix} 2 \\ 1 \end{bmatrix} \quad \text{and} \quad \mathbf{v}_2 = \begin{bmatrix} -3 \\ 4 \end{bmatrix}$$

2. $T: R^2 \to R^2$ is defined by

$$T\left(\begin{bmatrix} x_1 \\ x_2 \end{bmatrix}\right) = \begin{bmatrix} x_1 + 7x_2 \\ 3x_1 - 4x_2 \end{bmatrix}$$

$B = \{\mathbf{u}_1, \mathbf{u}_2\}$ and $B' = \{\mathbf{v}_1, \mathbf{v}_2\}$, where

$$\mathbf{u}_1 = \begin{bmatrix} 2 \\ 2 \end{bmatrix} \quad \mathbf{u}_2 = \begin{bmatrix} 4 \\ -1 \end{bmatrix} \quad \mathbf{v}_1 = \begin{bmatrix} 1 \\ 3 \end{bmatrix} \quad \mathbf{v}_2 = \begin{bmatrix} -1 \\ -1 \end{bmatrix}$$

3. $T: R^2 \to R^2$ is the rotation about the origin through $45°$; B and B' are the bases in Exercise 1.

4. $T: R^3 \to R^3$ is defined by

$$T\left(\begin{bmatrix} x_1 \\ x_2 \\ x_3 \end{bmatrix}\right) = \begin{bmatrix} x_1 + 2x_2 - x_3 \\ -x_2 \\ x_1 + 7x_3 \end{bmatrix}$$

B is the standard basis for R^3 and $B' = \{\mathbf{v}_1, \mathbf{v}_2, \mathbf{v}_3\}$, where

$$\mathbf{v}_1 = \begin{bmatrix} 1 \\ 0 \\ 0 \end{bmatrix}, \mathbf{v}_2 = \begin{bmatrix} 1 \\ 1 \\ 0 \end{bmatrix}, \text{and } \mathbf{v}_3 = \begin{bmatrix} 1 \\ 1 \\ 1 \end{bmatrix}.$$

5. $T: R^3 \to R^3$ is the orthogonal projection on the xy-plane; B and B' are as in Exercise 4.

6. $T: R^2 \to R^2$ is defined by $T(\mathbf{x}) = 5\mathbf{x}$; B and B' are the bases in Exercise 2.

7. $T: P_1 \to P_1$ is defined by $T(a_0 + a_1 x) = a_0 + a_1(x + 1)$; $B = \{\mathbf{p}_1, \mathbf{p}_2\}$ and $B' = \{\mathbf{q}_1, \mathbf{q}_2\}$ where $\mathbf{p}_1 = 6 + 3x$, $\mathbf{p}_2 = 10 + 2x$, $\mathbf{q}_1 = 2$, $\mathbf{q}_2 = 3 + 2x$.

8. Prove that if A and B are similar matrices, then $\det(A) = \det(B)$.

9. Prove that similar matrices have the same rank.

10. Prove that if A and B are similar matrices, then A^2 and B^2 are also similar. More generally, prove that A^k and B^k are similar, where k is any positive integer.

11. Let C and D be $m \times n$ matrices. Show that:

(a) If $C\mathbf{x} = D\mathbf{x}$ for all $\mathbf{x}$ in R^n, then $C = D$.

(b) If $B = \{\mathbf{v}_1, \mathbf{v}_2, \ldots, \mathbf{v}_n\}$ is a basis for a vector space V and $C[\mathbf{x}]_B = D[\mathbf{x}]_B$ for all $\mathbf{x}$ in V, then $C = D$.

6 Eigenvalues, Eigenvectors

6.1 EIGENVALUES AND EIGENVECTORS

In many problems in science and mathematics, a linear operator $T: V \rightarrow V$ is given, and it is of importance to determine those scalars λ for which the equation $T\mathbf{x} = \lambda\mathbf{x}$ has nonzero solutions. In this section we discuss this problem, and in later sections we shall investigate some of its applications.

Definition. If A is an $n \times n$ matrix, then a nonzero vector $\mathbf{x}$ in R^n is called an ***eigenvector*** of A if $A\mathbf{x}$ is a scalar multiple of $\mathbf{x}$; that is

$$A\mathbf{x} = \lambda\mathbf{x}$$

for some scalar λ. The scalar λ is called an ***eigenvalue*** of A and $\mathbf{x}$ is said to be an eigenvector ***corresponding*** to λ.

One of the meanings of the word "eigen" in German is "proper"; eigenvalues are also called ***proper values***, ***characteristic values***, or ***latent roots*** by some writers.

Example 1

The vector $\mathbf{x} = \begin{bmatrix} 1 \\ 2 \end{bmatrix}$ is an eigenvector of

$$A = \begin{bmatrix} 3 & 0 \\ 8 & -1 \end{bmatrix}$$

corresponding to the eigenvalue $\lambda = 3$ since

$$A\mathbf{x} = \begin{bmatrix} 3 & 0 \\ 8 & -1 \end{bmatrix}\begin{bmatrix} 1 \\ 2 \end{bmatrix} = \begin{bmatrix} 3 \\ 6 \end{bmatrix} = 3\mathbf{x}$$

Eigenvalues and eigenvectors have a useful geometric interpretation in R^2 and R^3. If λ is an eigenvalue of A corresponding to $\mathbf{x}$ then $A\mathbf{x} = \lambda\mathbf{x}$, so that multiplication by A dilates $\mathbf{x}$, contracts $\mathbf{x}$, or reverses the direction of $\mathbf{x}$, depending on the value of λ (Figure 6.1).

To find the eigenvalues of an $n \times n$ matrix A we rewrite $A\mathbf{x} = \lambda\mathbf{x}$ as

$$A\mathbf{x} = \lambda I\mathbf{x}$$

or equivalently

$$(\lambda I - A)\mathbf{x} = \mathbf{0} \tag{6.1}$$

For λ to be an eigenvalue, there must be a nonzero solution of this equation. However, by Theorem 13 of Section 4.6, Equation 6.1 will have a nonzero solution if and only if

$$\det(\lambda I - A) = 0$$

This is called the ***characteristic equation*** of A; the scalars satisfying this equation are the eigenvalues of A.

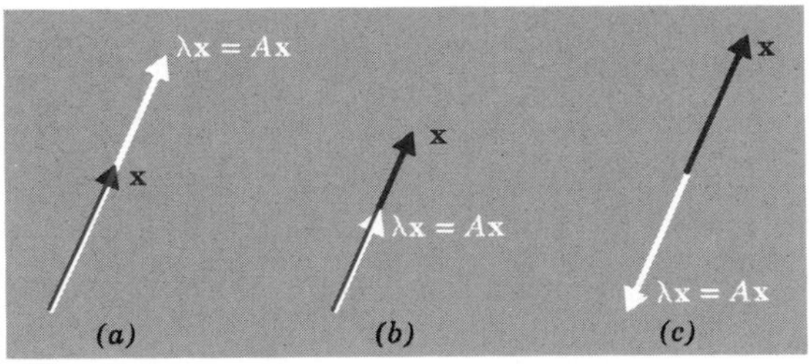

Figure 6.1 (*a*) Dilation $\lambda > 1$. (*b*) Contraction $0 < \lambda < 1$. (*c*) Reversal of direction $\lambda < 0$.

Example 2

Find the eigenvalues of the matrix

$$A = \begin{bmatrix} 3 & 2 \\ -1 & 0 \end{bmatrix}$$

Solution. Since

$$\lambda I - A = \lambda \begin{bmatrix} 1 & 0 \\ 0 & 1 \end{bmatrix} - \begin{bmatrix} 3 & 2 \\ -1 & 0 \end{bmatrix} = \begin{bmatrix} \lambda - 3 & -2 \\ 1 & \lambda \end{bmatrix}$$

and

$$\det(\lambda I - A) = \det \begin{bmatrix} \lambda - 3 & -2 \\ 1 & \lambda \end{bmatrix} = \lambda^2 - 3\lambda + 2$$

the characteristic equation of A is

$$\lambda^2 - 3\lambda + 2 = 0$$

The solutions of this equation are $\lambda = 1$ and $\lambda = 2$; these are the eigenvalues of A.

Example 3

Find the eigenvalues of the matrix

$$A = \begin{bmatrix} -2 & -1 \\ 5 & 2 \end{bmatrix}$$

Solution. Proceeding as in Example 2

$$\det(\lambda I - A) = \det \begin{bmatrix} \lambda + 2 & 1 \\ -5 & \lambda - 2 \end{bmatrix} = \lambda^2 + 1$$

The eigenvalues of A must therefore satisfy the quadratic equation $\lambda^2 + 1 = 0$. Since the only solutions to this equation are the imaginary numbers $\lambda = i$ and $\lambda = -i$, and since we are assuming that all our scalars are real numbers, A has no eigenvalues.*

Example 4

Find the eigenvalues of

$$A = \begin{bmatrix} 2 & 1 & 0 \\ 3 & 2 & 0 \\ 0 & 0 & 4 \end{bmatrix}$$

Solution. As in the preceding examples

$$\det(\lambda I - A) = \det \begin{bmatrix} \lambda - 2 & -1 & 0 \\ -3 & \lambda - 2 & 0 \\ 0 & 0 & \lambda - 4 \end{bmatrix} = \lambda^3 - 8\lambda^2 + 17\lambda - 4$$

The eigenvalues of A must therefore satisfy the cubic equation

$$\lambda^3 - 8\lambda^2 + 17\lambda - 4 = 0 \tag{6.2}$$

To solve this equation, we shall begin by searching for integer solutions. This task can be greatly simplified by exploiting the fact that all integer solutions (if there are any) to a polynomial equation with integer coefficients

$$c_0\lambda^n + c_1\lambda^{n-1} + \cdots + c_n = 0$$

must be divisors of the constant term, c_n. Thus the only possible integer solutions of (6.2) are the divisors of -4, that is, ± 1, ± 2, ± 4. Successively substituting these values in (6.2) shows that $\lambda = 4$ is an integer solution. As a consequence, $\lambda - 4$ must be a factor of the left side of (6.2). Dividing $\lambda - 4$ into $\lambda^3 - 8\lambda^2 + 17\lambda - 4$ shows that (6.2) can be rewritten as

$$(\lambda - 4)(\lambda^2 - 4\lambda + 1) = 0$$

* As we pointed out in Section 4.2, there are some applications that require complex scalars and complex vector spaces. In such cases, matrices are allowed to have complex eigenvalues. In this text, however, we consider only real eigenvalues.

Thus the remaining solutions of (6.2) satisfy the quadratic equation

$$\lambda^2 - 4\lambda + 1 = 0$$

which can be solved by the quadratic formula. The eigenvalues of A are, therefore,

$$\lambda = 4 \qquad \lambda = 2 + \sqrt{3} \qquad \text{and} \qquad \lambda = 2 - \sqrt{3}$$

REMARK. In practical problems, the matrix A is often so large that the determination of the characteristic equation is not practical. As a result, various approximation methods are used to obtain eigenvalues; some of these are discussed in Chapter 8.

The following theorem summarizes our results so far.

Theorem 1. *If A is an $n \times n$ matrix, then the following are equivalent.*

(*a*) *λ is an eigenvalue of A.*
(*b*) *The system of equations $(\lambda I - A)\mathbf{x} = \mathbf{0}$ has nontrivial solutions.*
(*c*) *There is a nonzero vector $\mathbf{x}$ in R^n such that $A\mathbf{x} = \lambda\mathbf{x}$.*
(*d*) *λ is a real solution of the characteristic equation $\det(\lambda I - A) = 0$.*

Now that we know how to find eigenvalues we turn to the problem of finding eigenvectors. The eigenvectors of A corresponding to an eigenvalue λ are the nonzero vectors that satisfy $A\mathbf{x} = \lambda\mathbf{x}$. Equivalently the eigenvectors corresponding to λ are the nonzero vectors in the solution space of $(\lambda I - A)\mathbf{x} = \mathbf{0}$. We call this solution space the ***eigenspace*** of A corresponding to λ.

Example 5

Find bases for the eigenspaces of

$$A = \begin{bmatrix} 3 & -2 & 0 \\ -2 & 3 & 0 \\ 0 & 0 & 5 \end{bmatrix}$$

Solution. The characteristic equation of A is $(\lambda - 1)(\lambda - 5)^2 = 0$ (verify), so that the eigenvalues of A are $\lambda = 1$ and $\lambda = 5$.

By definition

$$\mathbf{x} = \begin{bmatrix} x_1 \\ x_2 \\ x_3 \end{bmatrix}$$

is an eigenvector of A corresponding to λ if and only if $\mathbf{x}$ is a nontrivial solution of $(\lambda I - A)\mathbf{x} = \mathbf{0}$, that is, of

$$\begin{bmatrix} \lambda - 3 & 2 & 0 \\ 2 & \lambda - 3 & 0 \\ 0 & 0 & \lambda - 5 \end{bmatrix} \begin{bmatrix} x_1 \\ x_2 \\ x_3 \end{bmatrix} = \begin{bmatrix} 0 \\ 0 \\ 0 \end{bmatrix} \tag{6.3}$$

If $\lambda = 5$, (6.3) becomes

$$\begin{bmatrix} 2 & 2 & 0 \\ 2 & 2 & 0 \\ 0 & 0 & 0 \end{bmatrix} \begin{bmatrix} x_1 \\ x_2 \\ x_3 \end{bmatrix} = \begin{bmatrix} 0 \\ 0 \\ 0 \end{bmatrix}$$

Solving this system yields (verify)

$$x_1 = -s \qquad x_2 = s \qquad x_3 = t$$

Thus the eigenvectors of A corresponding to $\lambda = 5$ are the nonzero vectors of the form

$$\mathbf{x} = \begin{bmatrix} -s \\ s \\ t \end{bmatrix} = \begin{bmatrix} -s \\ s \\ 0 \end{bmatrix} + \begin{bmatrix} 0 \\ 0 \\ t \end{bmatrix} = s\begin{bmatrix} -1 \\ 1 \\ 0 \end{bmatrix} + t\begin{bmatrix} 0 \\ 0 \\ 1 \end{bmatrix}$$

Since

$$\begin{bmatrix} -1 \\ 1 \\ 0 \end{bmatrix} \quad \text{and} \quad \begin{bmatrix} 0 \\ 0 \\ 1 \end{bmatrix}$$

are linearly independent, they form a basis for the eigenspace corresponding to $\lambda = 5$.

If $\lambda = 1$, then (6.3) becomes

$$\begin{bmatrix} -2 & 2 & 0 \\ 2 & -2 & 0 \\ 0 & 0 & -4 \end{bmatrix} \begin{bmatrix} x_1 \\ x_2 \\ x_3 \end{bmatrix} = \begin{bmatrix} 0 \\ 0 \\ 0 \end{bmatrix}$$

Solving this system yields (verify)

$$x_1 = t \qquad x_2 = t \qquad x_3 = 0$$

Thus the eigenvectors corresponding to $\lambda = 1$ are the nonzero vectors of the form

$$\mathbf{x} = \begin{bmatrix} t \\ t \\ 0 \end{bmatrix} = t\begin{bmatrix} 1 \\ 1 \\ 0 \end{bmatrix}$$

so that

$$\begin{bmatrix} 1 \\ 1 \\ 0 \end{bmatrix}$$

is a basis for the eigenspace corresponding to $\lambda = 1$.

OPTIONAL

Eigenvectors and eigenvalues can be defined for linear operators as well as matrices. A scalar λ is called an ***eigenvalue*** of a linear operator $T: V \rightarrow V$ if there is a nonzero vector $\mathbf{x}$ in V such that $T\mathbf{x} = \lambda\mathbf{x}$. The vector $\mathbf{x}$ is called an ***eigenvector*** of T corresponding to λ. Equivalently, the eigenvectors of T corresponding to λ are the nonzero vectors in the kernel of $\lambda I - T$ (Exercise 19). This kernel is called the ***eigenspace*** of T corresponding to λ.

It can be shown that if V is a finite dimensional space and A is the matrix of T with respect to *any* basis B, then:

1. The eigenvalues of T are the eigenvalues of the matrix A.
2. A vector $\mathbf{x}$ is an eigenvector of T corresponding to λ if and only if its coordinate matrix $[\mathbf{x}]_B$ is an eigenvector of A corresponding to λ.

We leave the proofs for the exercises.

Example 6

Find the eigenvalues and bases for the eigenspaces of the linear operator $T: P_2 \to P_2$ defined by

$$T(a + bx + cx^2) = (3a - 2b) + (-2a + 3b)x + (5c)x^2$$

Solution. The matrix of T with respect to the standard basis $B = \{1, x, x^2\}$ is

$$A = \begin{bmatrix} 3 & -2 & 0 \\ -2 & 3 & 0 \\ 0 & 0 & 5 \end{bmatrix}$$

The eigenvalues of T are the eigenvalues of A; namely $\lambda = 1$ and $\lambda = 5$ (Example 5). Also from Example 5 the eigenspace of A corresponding to $\lambda = 5$ has the basis $\{\mathbf{u}_1, \mathbf{u}_2\}$ and that corresponding to $\lambda = 1$ has the basis $\{\mathbf{u}_3\}$, where

$$\mathbf{u}_1 = \begin{bmatrix} -1 \\ 1 \\ 0 \end{bmatrix} \qquad \mathbf{u}_2 = \begin{bmatrix} 0 \\ 0 \\ 1 \end{bmatrix} \qquad \mathbf{u}_3 = \begin{bmatrix} 1 \\ 1 \\ 0 \end{bmatrix}$$

These matrices are the coordinate matrices with respect to B of

$$\mathbf{p}_1 = -1 + x \qquad \mathbf{p}_2 = x^2 \qquad \mathbf{p}_3 = 1 + x$$

Thus $\{-1 + x, x^2\}$ is a basis for the eigenspace of T corresponding to $\lambda = 5$, and $\{1 + x\}$ is a basis for the eigenspace corresponding to $\lambda = 1$.

EXERCISE SET 6.1

1. Find the characteristic equations of the following matrices:

(a) $\begin{bmatrix} 3 & 0 \\ 8 & -1 \end{bmatrix}$ (b) $\begin{bmatrix} 10 & -9 \\ 4 & -2 \end{bmatrix}$ (c) $\begin{bmatrix} 0 & 3 \\ 4 & 0 \end{bmatrix}$

(d) $\begin{bmatrix} -2 & -7 \\ 1 & 2 \end{bmatrix}$ (e) $\begin{bmatrix} 0 & 0 \\ 0 & 0 \end{bmatrix}$ (f) $\begin{bmatrix} 1 & 0 \\ 0 & 1 \end{bmatrix}$

2. Find the eigenvalues of the matrices in Exercise 1.

3. Find bases for the eigenspaces of the matrices in Exercise 1.

4. In each part of Exercise 1, let $T:R^2 \to R^2$ be multiplication by the given matrix. In each case sketch the lines in R^2 that are mapped into themselves under T.

5. Find the characteristic equations of the following matrices:

(a) $\begin{bmatrix} 4 & 0 & 1 \\ -2 & 1 & 0 \\ -2 & 0 & 1 \end{bmatrix}$ (b) $\begin{bmatrix} 3 & 0 & -5 \\ \frac{1}{5} & -1 & 0 \\ 1 & 1 & -2 \end{bmatrix}$ (c) $\begin{bmatrix} -2 & 0 & 1 \\ -6 & -2 & 0 \\ 19 & 5 & -4 \end{bmatrix}$

(d) $\begin{bmatrix} -1 & 0 & 1 \\ -1 & 3 & 0 \\ -4 & 13 & -1 \end{bmatrix}$ (e) $\begin{bmatrix} 5 & 0 & 1 \\ 1 & 1 & 0 \\ -7 & 1 & 0 \end{bmatrix}$ (f) $\begin{bmatrix} 5 & 6 & 2 \\ 0 & -1 & -8 \\ 1 & 0 & -2 \end{bmatrix}$

6. Find the eigenvalues of the matrices in Exercise 5.

7. Find bases for the eigenspaces of the matrices in Exercise 5.

8. Find the characteristic equations of the following matrices:

(a) $\begin{bmatrix} 0 & 0 & 2 & 0 \\ 1 & 0 & 1 & 0 \\ 0 & 1 & -2 & 0 \\ 0 & 0 & 0 & 1 \end{bmatrix}$ (b) $\begin{bmatrix} 10 & -9 & 0 & 0 \\ 4 & -2 & 0 & 0 \\ 0 & 0 & -2 & -7 \\ 0 & 0 & 1 & 2 \end{bmatrix}$

9. Find the eigenvalues of the matrices in Exercise 8.

10. Find bases for the eigenspaces of the matrices in Exercise 8.

11. **(For readers of the optional material)** Let $T:P_2 \to P_2$ be defined by

$$T(a_0 + a_1x + a_2x^2) = (5a_0 + 6a_1 + 2a_2) - (a_1 + 8a_2)x + (a_0 - 2a_2)x^2.$$

(a) Find the eigenvalues of T.
(b) Find bases for the eigenspaces of T.

12. Let $T:M_{22} \to M_{22}$ be defined by

$$T\left(\begin{bmatrix} a & b \\ c & d \end{bmatrix}\right) = \begin{bmatrix} 2c & a + c \\ b - 2c & d \end{bmatrix}$$

(a) Find the eigenvalues of T.
(b) Find bases for the eigenspaces of T.

13. Prove that $\lambda = 0$ is an eigenvalue of a matrix A if and only if A is not invertible.

14. If A is a square matrix, then $p(\lambda) = \det(\lambda I - A)$ is called the ***characteristic polynomial of*** A. Prove that the constant term in the characteristic polynomial of an $n \times n$ matrix A is $(-1)^n \det(A)$. (***Hint.*** The constant term is the value of the characteristic polynomial when $\lambda = 0$.)

15. The ***trace*** of a square matrix A is the sum of the elements on the main diagonal. Show that the characteristic equation of a 2×2 matrix A is $\lambda^2 - \text{tr}(A)\lambda + \det(A) = 0$, where $\text{tr}(A)$ is the trace of A.

16. Prove that the eigenvalues of a triangular matrix are the entires on the main diagonal.

17. Show that if λ is an eigenvalue of A, then λ^2 is an eigenvalue of A^2; more generally, show that λ^n is an eigenvalue of A^n if n is a positive integer.

18. Use the results of Exercises 16 and 17 to find the eigenvalues of A^9, where

$$A = \begin{bmatrix} 1 & 3 & 7 & 11 \\ 0 & -1 & 3 & 8 \\ 0 & 0 & -2 & 4 \\ 0 & 0 & 0 & 2 \end{bmatrix}$$

19. **(For readers of the optional material)** Let λ be an eigenvalue of a linear operator $T: V \to V$. Prove that the eigenvectors of T corresponding to λ are the nonzero vectors in the kernel of $\lambda I - T$.

6.2 DIAGONALIZATION

In this section and the next, we shall be concerned with the following problems.

Problem 1. Given a linear operator $T: V \to V$ on a finite dimensional vector space, does there exist a basis for V with respect to which the matrix of T is diagonal?

Problem 2. Given a linear operator $T: V \to V$ on a finite dimensional inner product space, does there exist an orthonormal basis for V with respect to which the matrix of T is diagonal?

If A is the matrix of $T: V \to V$ with respect to some basis, then Problem 1 is equivalent to asking if there is a change of basis such that the new matrix for T is diagonal. By Theorem 5 in Section 5.4, the new matrix for T will be $P^{-1}AP$ where P is the appropriate transition matrix. If V is an inner product space and the bases are orthonormal, then by Theorem 27 in Section 4.10, P will be orthogonal. Thus we are led to the following matrix formulations of the above problems.

Problem 1 (matrix form). Given a square matrix A, does there exist an invertible matrix P such that $P^{-1}AP$ is diagonal?

Problem 2 (matrix form). Given a square matrix A, does there exist an orthogonal matrix P such that $P^{-1}AP\ (= P^tAP)$ is diagonal?

These problems suggest the following definitions.

Definition. A square matrix A is called ***diagonalizable*** if there is an invertible matrix P such that $P^{-1}AP$ is diagonal; the matrix P is said to ***diagonalize*** A.

The following theorem is the basic tool in the study of diagonalizability; its proof reveals the technique for diagonalizing a matrix.

Theorem 2. *If A is an $n \times n$ matrix, then the following are equivalent.*

(*a*) *A is diagonalizable.*
(*b*) *A has n linearly independent eigenvectors.*

Proof (*a*) $\Rightarrow$ (*b*). Since A is assumed diagonalizable, there is an invertible matrix

$$P = \begin{bmatrix} p_{11} & p_{12} & \cdots & p_{1n} \\ p_{21} & p_{22} & \cdots & p_{2n} \\ \vdots & \vdots & & \vdots \\ p_{n1} & p_{n2} & \cdots & p_{nn} \end{bmatrix}$$

such that $P^{-1}AP$ is diagonal, say $P^{-1}AP = D$, where

$$D = \begin{bmatrix} \lambda_1 & 0 & \cdots & 0 \\ 0 & \lambda_2 & \cdots & 0 \\ \vdots & \vdots & & \vdots \\ 0 & 0 & \cdots & \lambda_n \end{bmatrix}$$

Therefore, $AP = PD$; that is

$$AP = \begin{bmatrix} p_{11} & p_{12} & \cdots & p_{1n} \\ p_{21} & p_{22} & \cdots & p_{2n} \\ \vdots & \vdots & & \vdots \\ p_{n1} & p_{n2} & \cdots & p_{nn} \end{bmatrix} \begin{bmatrix} \lambda_1 & 0 & \cdots & 0 \\ 0 & \lambda_2 & \cdots & 0 \\ \vdots & \vdots & & \vdots \\ 0 & 0 & \cdots & \lambda_n \end{bmatrix} = \begin{bmatrix} \lambda_1 p_{11} & \lambda_2 p_{12} & \cdots & \lambda_n p_{1n} \\ \lambda_1 p_{21} & \lambda_2 p_{22} & \cdots & \lambda_n p_{2n} \\ \vdots & \vdots & & \vdots \\ \lambda_1 p_{n1} & \lambda_2 p_{n2} & \cdots & \lambda_n p_{nn} \end{bmatrix} \tag{6.4}$$

If we now let $\mathbf{p}_1, \mathbf{p}_2, \ldots, \mathbf{p}_n$ denote the column vectors of P, then from (6.4) the successive columns of AP are $\lambda_1\mathbf{p}_1, \lambda_2\mathbf{p}, \ldots, \lambda_n\mathbf{p}_n$. However, from Example 17 of Section 1.4 the successive columns of AP are $A\mathbf{p}_1, A\mathbf{p}_2, \ldots, A\mathbf{p}_n$. Thus we must have

$$A\mathbf{p}_1 = \lambda_1\mathbf{p}_1, A\mathbf{p}_2 = \lambda_2\mathbf{p}_2, \ldots, A\mathbf{p}_n = \lambda_n\mathbf{p}_n \tag{6.5}$$

Since P is invertible, its column vectors are all nonzero; thus by (6.5), $\lambda_1, \lambda_2, \ldots, \lambda_n$ are eigenvalues of A, and $\mathbf{p}_1, \mathbf{p}_2, \ldots, \mathbf{p}_n$ are corresponding eigenvectors. Since P is invertible, it follows from Theorem 13 in Section 4.6 that $\mathbf{p}_1, \mathbf{p}_2, \ldots, \mathbf{p}_n$ are linearly independent. Thus A has n linearly independent eigenvectors.

(b) ⇒ (a) Assume that A has n linearly independent eigenvectors, $\mathbf{p}_1, \mathbf{p}_2, \ldots, \mathbf{p}_n$, with corresponding eigenvalues $\lambda_1, \lambda_2, \ldots, \lambda_n$, and let

$$P = \begin{bmatrix} p_{11} & p_{12} & \cdots & p_{1n} \\ p_{21} & p_{22} & \cdots & p_{2n} \\ \vdots & \vdots & & \vdots \\ p_{n1} & p_{n2} & \cdots & p_{nn} \end{bmatrix}$$

be the matrix whose column vectors are $\mathbf{p}_1, \mathbf{p}_2, \ldots, \mathbf{p}_n$. By Example 17 in Section 1.4, the columns of the product AP are

$$A\mathbf{p}_1, A\mathbf{p}_2, \ldots, A\mathbf{p}_n$$

But

$$A\mathbf{p}_1 = \lambda_1\mathbf{p}_1, A\mathbf{p}_2 = \lambda_2\mathbf{p}_2, \ldots, A\mathbf{p}_n = \lambda\mathbf{p}_n$$

so that

$$AP = \begin{bmatrix} \lambda_1 p_{11} & \lambda_2 p_{12} & \cdots & \lambda_n p_{1n} \\ \lambda_1 p_{21} & \lambda_2 p_{22} & \cdots & \lambda_n p_{2n} \\ \vdots & \vdots & & \vdots \\ \lambda_1 p_{n1} & \lambda_2 p_{n2} & \cdots & \lambda_n p_{nn} \end{bmatrix} = \begin{bmatrix} p_{11} & p_{12} & \cdots & p_{1n} \\ p_{21} & p_{22} & \cdots & p_{2n} \\ \vdots & \vdots & & \vdots \\ p_{n1} & p_{n2} & \cdots & p_{nn} \end{bmatrix} \begin{bmatrix} \lambda_1 & 0 & \cdots & 0 \\ 0 & \lambda_2 & \cdots & 0 \\ \vdots & \vdots & & \vdots \\ 0 & 0 & \cdots & \lambda_n \end{bmatrix}$$
$$= PD \tag{6.6}$$

where D is the diagonal matrix having the eigenvalues $\lambda_1, \lambda_2, \ldots, \lambda_n$ on the main diagonal. Since the column vectors of P are linearly independent, P is invertible; thus (6.6) can be rewritten as $P^{-1}AP = D$; that is, A is diagonalizable. ∎

From this proof we obtain the following procedure for diagonalizing a diagonalizable $n \times n$ matrix A.

Step 1. Find n linearly independent eigenvectors of A, $\mathbf{p}_1, \mathbf{p}_2, \ldots, \mathbf{p}_n$.
Step 2. Form the matrix P having $\mathbf{p}_1, \mathbf{p}_2, \ldots, \mathbf{p}_n$ as its column vectors.
Step 3. The matrix $P^{-1}AP$ will then be diagonal with $\lambda_1, \lambda_2, \ldots, \lambda_n$ as its successive diagonal entries, where λ_i is the eigenvalue corresponding to $\mathbf{p}_i$, $i = 1, 2, \ldots, n$.

Example 7

Find a matrix P that diagonalizes

$$A = \begin{bmatrix} 3 & -2 & 0 \\ -2 & 3 & 0 \\ 0 & 0 & 5 \end{bmatrix}$$

Solution. From Example 5 the eigenvalues of A are $\lambda = 1$ and $\lambda = 5$. Also from that example the vectors

$$\mathbf{p}_1 = \begin{bmatrix} -1 \\ 1 \\ 0 \end{bmatrix} \quad \text{and} \quad \mathbf{p}_2 = \begin{bmatrix} 0 \\ 0 \\ 1 \end{bmatrix}$$

form a basis for the eigenspace corresponding to $\lambda = 5$ and

$$\mathbf{p}_3 = \begin{bmatrix} 1 \\ 1 \\ 0 \end{bmatrix}$$

is a basis for the eigenspace corresponding to $\lambda = 1$. It is easy to check that $\{\mathbf{p}_1, \mathbf{p}_2, \mathbf{p}_3\}$ is linearly independent, so that

$$P = \begin{bmatrix} -1 & 0 & 1 \\ 1 & 0 & 1 \\ 0 & 1 & 0 \end{bmatrix}$$

diagonalizes A. As a check, the reader should verify that

$$P^{-1}AP = \begin{bmatrix} -\frac{1}{2} & \frac{1}{2} & 0 \\ 0 & 0 & 1 \\ \frac{1}{2} & \frac{1}{2} & 0 \end{bmatrix} \begin{bmatrix} 3 & -2 & 0 \\ -2 & 3 & 0 \\ 0 & 0 & 5 \end{bmatrix} \begin{bmatrix} -1 & 0 & 1 \\ 1 & 0 & 1 \\ 0 & 1 & 0 \end{bmatrix} = \begin{bmatrix} 5 & 0 & 0 \\ 0 & 5 & 0 \\ 0 & 0 & 1 \end{bmatrix}$$

There is no preferred order for the columns of P. Since the ith diagonal entry of $P^{-1}AP$ is an eigenvalue for the ith column vector of P, changing the order of the columns of P just changes the order of the eigenvalues on the diagonal of $P^{-1}AP$. Thus, had we written

$$P = \begin{bmatrix} -1 & 1 & 0 \\ 1 & 1 & 0 \\ 0 & 0 & 1 \end{bmatrix}$$

in the last example, we would have obtained

$$P^{-1}AP = \begin{bmatrix} 5 & 0 & 0 \\ 0 & 1 & 0 \\ 0 & 0 & 5 \end{bmatrix}$$

Example 8

The characteristic equation of

$$A = \begin{bmatrix} -3 & 2 \\ -2 & 1 \end{bmatrix}$$

is

$$\det(\lambda I - A) = \det\begin{bmatrix} \lambda + 3 & -2 \\ 2 & \lambda - 1 \end{bmatrix} = (\lambda + 1)^2 = 0$$

Thus $\lambda = -1$ is the only eigenvalue of A; the eigenvectors corresponding to $\lambda = -1$ are the solutions of $(-I - A)\mathbf{x} = \mathbf{0}$; that is, of

$$\begin{aligned} 2x_1 - 2x_2 &= 0 \\ 2x_1 - 2x_2 &= 0 \end{aligned}$$

The solutions of this system are $x_1 = t$, $x_2 = t$ (verify); hence the eigenspace consists of all vectors of the form

$$\begin{bmatrix} t \\ t \end{bmatrix} = t\begin{bmatrix} 1 \\ 1 \end{bmatrix}$$

Since this space is 1-dimensional, A does not have two linearly independent eigenvectors, and is therefore not diagonalizable.

Example 9

Let $T: R^3 \to R^3$ be the linear operator given by

$$T\left(\begin{bmatrix} x_1 \\ x_2 \\ x_3 \end{bmatrix}\right) = \begin{bmatrix} 3x_1 - 2x_2 \\ -2x_1 + 3x_2 \\ 5x_3 \end{bmatrix}$$

Find a basis for R^3 relative to which the matrix of T is diagonal.

Solution. If $B = \{\mathbf{e}_1, \mathbf{e}_2, \mathbf{e}_3\}$ denotes the standard basis for R^3 then

$$T(\mathbf{e}_1) = T\left(\begin{bmatrix} 1 \\ 0 \\ 0 \end{bmatrix}\right) = \begin{bmatrix} 3 \\ -2 \\ 0 \end{bmatrix}, \quad T(\mathbf{e}_2) = T\left(\begin{bmatrix} 0 \\ 1 \\ 0 \end{bmatrix}\right) = \begin{bmatrix} -2 \\ 3 \\ 0 \end{bmatrix},$$

$$T(\mathbf{e}_3) = T\left(\begin{bmatrix} 0 \\ 0 \\ 1 \end{bmatrix}\right) = \begin{bmatrix} 0 \\ 0 \\ 5 \end{bmatrix}$$

so that the standard matrix for T is

$$A = \begin{bmatrix} 3 & -2 & 0 \\ -2 & 3 & 0 \\ 0 & 0 & 5 \end{bmatrix}$$

We now want to change from the standard basis to a new basis $B' = \{\mathbf{u}_1', \mathbf{u}_2', \mathbf{u}_3'\}$ in order to obtain a diagonal matrix A' for T. If we let P be the transition matrix from the unknown basis B' to the standard basis B, then by Theorem 5 of

Section 5.4, A and A' will be related by

$$A' = P^{-1}AP$$

In other words, the transition matrix P diagonalizes A. We found this matrix in Example 7. From our work in that example

$$P = \begin{bmatrix} -1 & 0 & 1 \\ 1 & 0 & 1 \\ 0 & 1 & 0 \end{bmatrix} \quad \text{and} \quad A' = \begin{bmatrix} 5 & 0 & 0 \\ 0 & 5 & 0 \\ 0 & 0 & 1 \end{bmatrix}$$

Since P represents the transition matrix from the basis $B' = \{\mathbf{u}'_1, \mathbf{u}'_2, \mathbf{u}'_3\}$ to the standard basis $B = \{\mathbf{e}_1, \mathbf{e}_2, \mathbf{e}_3\}$ the columns of P are $[\mathbf{u}'_1]_B$, $[\mathbf{u}'_2]_B$, and $[\mathbf{u}'_3]_B$ so that

$$[\mathbf{u}'_1]_B = \begin{bmatrix} -1 \\ 1 \\ 0 \end{bmatrix}, \quad [\mathbf{u}'_2]_B = \begin{bmatrix} 0 \\ 0 \\ 1 \end{bmatrix}, \quad [\mathbf{u}'_3]_B = \begin{bmatrix} 1 \\ 1 \\ 0 \end{bmatrix}$$

Thus

$$\mathbf{u}'_1 = (-1)\mathbf{e}_1 + (1)\mathbf{e}_2 + (0)\mathbf{e}_3 = \begin{bmatrix} -1 \\ 1 \\ 0 \end{bmatrix}$$

$$\mathbf{u}'_2 = (0)\mathbf{e}_1 + (0)\mathbf{e}_2 + (1)\mathbf{e}_3 = \begin{bmatrix} 0 \\ 0 \\ 1 \end{bmatrix}$$

$$\mathbf{u}'_3 = (1)\mathbf{e}_1 + (1)\mathbf{e}_2 + (0)\mathbf{e}_3 = \begin{bmatrix} 1 \\ 1 \\ 0 \end{bmatrix}$$

are the basis vectors that produce the diagonal matrix A' for T.

Now that we have developed techniques for diagonalizing a diagonalizable matrix, we turn to the question: When is a square matrix A diagonalizable? The following result will help us study this question. We defer its proof to the end of the section.

Theorem 3. *If $\mathbf{v}_1, \mathbf{v}_2, \ldots, \mathbf{v}_k$ are eigenvectors of A corresponding to distinct eigenvalues $\lambda_1, \lambda_2, \ldots, \lambda_k$, then $\{\mathbf{v}_1, \mathbf{v}_2, \ldots, \mathbf{v}_k\}$ is a linearly independent set.*

As a consequence of this theorem, we obtain the following useful result.

Theorem 4. *If an $n \times n$ matrix A has n distinct eigenvalues, then A is diagonalizable.*

Proof. If $\mathbf{v}_1, \mathbf{v}_2, \ldots, \mathbf{v}_n$ are eigenvectors corresponding to the distinct eigenvalues $\lambda_1, \lambda_2, \ldots, \lambda_n$, then by Theorem 3, $\mathbf{v}_1, \mathbf{v}_2, \ldots, \mathbf{v}_n$ are linearly independent. Thus A is diagonalizable, by Theorem 2. ∎

Example 10

We saw in Example 4 that

$$A = \begin{bmatrix} 2 & 1 & 0 \\ 3 & 2 & 0 \\ 0 & 0 & 4 \end{bmatrix}$$

has 3 distinct eigenvalues, $\lambda = 4$, $\lambda = 2 + \sqrt{3}$, $\lambda = 2 - \sqrt{3}$. Therefore, A is diagonalizable. Further

$$P^{-1}AP = \begin{bmatrix} 4 & 0 & 0 \\ 0 & 2+\sqrt{3} & 0 \\ 0 & 0 & 2-\sqrt{3} \end{bmatrix}$$

for some invertible matrix P. If desired, the matrix P can be found using the method shown in Example 7.

Example 11

The converse of Theorem 4 is false; that is, an $n \times n$ matrix A may be diagonalizable even if it does not have n distinct eigenvalues. For example, if

$$A = \begin{bmatrix} 3 & 0 \\ 0 & 3 \end{bmatrix}$$

then the characteristic equation of A is

$$\det(\lambda I - A) = (\lambda - 3)^2 = 0$$

so that $\lambda = 3$ is the only distinct eigenvalue of A. Yet A is obviously diagonalizable since with $P = I$,

$$P^{-1}AP = I^{-1}AI = A = \begin{bmatrix} 3 & 0 \\ 0 & 3 \end{bmatrix}.$$

REMARK. Theorem 3 is a special case of a more general result: Suppose $\lambda_1, \lambda_2, \ldots, \lambda_k$ are distinct eigenvalues and we choose a linearly independent set in each of the corresponding eigenspaces. If we then merge all these vectors into a single set, the result is still a linearly independent set. For example, if we choose three linearly independent vectors from one eigenspace and two linearly independent eigenvectors from another eigenspace, then the five vectors together form a linearly independent set. We omit the proof.

OPTIONAL

We conclude this section with a proof of Theorem 3.

Proof. Let $\mathbf{v}_1, \mathbf{v}_2, \ldots, \mathbf{v}_k$ be eigenvectors of A corresponding to distinct eigenvalues $\lambda_1, \lambda_2, \ldots, \lambda_k$. We shall assume that $\mathbf{v}_1, \mathbf{v}_2, \ldots, \mathbf{v}_k$ are linearly dependent and obtain a contradiction. We can then conclude that $\mathbf{v}_1, \mathbf{v}_2, \ldots, \mathbf{v}_k$ are linearly independent.

Since an eigenvector is, by definition, nonzero, $\{\mathbf{v}_1\}$ is linearly independent. Let r be the largest integer such that $\{\mathbf{v}_1, \mathbf{v}_2, \ldots, \mathbf{v}_r\}$ is linearly independent. Since we are assuming that $\{\mathbf{v}_1, \mathbf{v}_2, \ldots, \mathbf{v}_k\}$ is linearly dependent, r satisfies $1 \leq r < k$. Moreover, by definition of r, $\{\mathbf{v}_1, \mathbf{v}_2, \ldots, \mathbf{v}_{r+1}\}$ is linearly dependent. Thus there are scalars $c_1, c_2, \ldots, c_{r+1}$, not all zero, such that

$$c_1\mathbf{v}_1 + c_2\mathbf{v}_2 + \cdots + c_{r+1}\mathbf{v}_{r+1} = \mathbf{0} \tag{6.7}$$

Multiplying both sides of (6.7) by A and using

$$A\mathbf{v}_1 = \lambda_1\mathbf{v}_1,\ A\mathbf{v}_2 = \lambda_2\mathbf{v}_2, \ldots, A\mathbf{v}_{r+1} = \lambda_{r+1}\mathbf{v}_{r+1}$$

we obtain

$$c_1\lambda_1\mathbf{v}_1 + c_2\lambda_2\mathbf{v}_2 + \cdots + c_{r+1}\lambda_{r+1}\mathbf{v}_{r+1} = \mathbf{0} \tag{6.8}$$

Multiplying both sides of (6.7) by λ_{r+1} and subtracting the resulting equation from (6.8) yields

$$c_1(\lambda_1 - \lambda_{r+1})\mathbf{v}_1 + c_2(\lambda_2 - \lambda_{r+1})\mathbf{v}_2 + \cdots + c_r(\lambda_r - \lambda_{r+1})\mathbf{v}_r = \mathbf{0}$$

Since $\{\mathbf{v}_1, \mathbf{v}_2, \ldots, \mathbf{v}_r\}$ is linearly independent, this equation implies that

$$c_1(\lambda_1 - \lambda_{r+1}) = c_2(\lambda_2 - \lambda_{r+1}) = \cdots = c_r(\lambda_r - \lambda_{r+1}) = 0$$

and since $\lambda_1, \lambda_2, \ldots, \lambda_{r+1}$ are distinct, it follows that

$$c_1 = c_2 = \cdots = c_r = 0 \tag{6.9}$$

Substituting these values in (6.7) yields

$$c_{r+1}\mathbf{v}_{r+1} = \mathbf{0}$$

Since the eigenvector $\mathbf{v}_{r+1}$ is nonzero it follows that

$$c_{r+1} = 0 \tag{6.10}$$

Equations (6.9) and (6.10) contradict the fact that $c_1, c_2, \ldots, c_{r+1}$ are not all zero; this completes the proof. ▮

EXERCISE SET 6.2

Show that the matrices in Exercises 1–4 are not diagonalizable.

1. $\begin{bmatrix} 2 & 0 \\ 1 & 2 \end{bmatrix}$ **2.** $\begin{bmatrix} 2 & -3 \\ 1 & -1 \end{bmatrix}$ **3.** $\begin{bmatrix} 3 & 0 & 0 \\ 0 & 2 & 0 \\ 0 & 1 & 2 \end{bmatrix}$ **4.** $\begin{bmatrix} -1 & 0 & 1 \\ -1 & 3 & 0 \\ -4 & 13 & -1 \end{bmatrix}$

In Exercises 5–8 find a matrix P that diagonalizes A, and determine $P^{-1}AP$.

5. $A = \begin{bmatrix} -14 & 12 \\ -20 & 17 \end{bmatrix}$ **6.** $A = \begin{bmatrix} 1 & 0 \\ 6 & -1 \end{bmatrix}$

7. $A = \begin{bmatrix} 1 & 0 & 0 \\ 0 & 1 & 1 \\ 0 & 1 & 1 \end{bmatrix}$ **8.** $A = \begin{bmatrix} 2 & 0 & -2 \\ 0 & 3 & 0 \\ 0 & 0 & 3 \end{bmatrix}$

In Exercises 9–14 determine if A is diagonalizable. If so, find a matrix P that diagonalizes A, and determine $P^{-1}AP$.

9. $A = \begin{bmatrix} 19 & -9 & -6 \\ 25 & -11 & -9 \\ 17 & -9 & -4 \end{bmatrix}$ **10.** $A = \begin{bmatrix} -1 & 4 & -2 \\ -3 & 4 & 0 \\ -3 & 1 & 3 \end{bmatrix}$

11. $A = \begin{bmatrix} 5 & 0 & 0 \\ 1 & 5 & 0 \\ 0 & 1 & 5 \end{bmatrix}$ **12.** $A = \begin{bmatrix} 0 & 0 & 0 \\ 0 & 0 & 0 \\ 3 & 0 & 1 \end{bmatrix}$

13. $A = \begin{bmatrix} -2 & 0 & 0 & 0 \\ 0 & -2 & 0 & 0 \\ 0 & 0 & 3 & 0 \\ 0 & 0 & 1 & 3 \end{bmatrix}$ **14.** $A = \begin{bmatrix} -2 & 0 & 0 & 0 \\ 0 & -2 & 5 & -5 \\ 0 & 0 & 3 & 0 \\ 0 & 0 & 0 & 3 \end{bmatrix}$

15. Let $T: R^2 \to R^2$ be the linear operator given by

$$T\left(\begin{bmatrix} x_1 \\ x_2 \end{bmatrix}\right) = \begin{bmatrix} 3x_1 + 4x_2 \\ 2x_1 + x_2 \end{bmatrix}$$

Find a basis for R^2 relative to which the matrix of T is diagonal.

16. Let $T: R^3 \to R^3$ be the linear operator given by

$$T\left(\begin{bmatrix} x_1 \\ x_2 \\ x_3 \end{bmatrix}\right) = \begin{bmatrix} 2x_1 - x_2 - x_3 \\ x_1 - x_3 \\ -x_1 + x_2 + 2x_3 \end{bmatrix}$$

Find a basis for R^3 relative to which the matrix of T is diagonal.

17. Let $T: P_1 \to P_1$ be the linear operator defined by

$$T(a_0 + a_1 x) = a_0 + (6a_0 - a_1)x$$

Find a basis for P_1 with respect to which the matrix for T is diagonal.

18. Let A be an $n \times n$ matrix and P an invertible $n \times n$ matrix. Show:
(a) $(P^{-1}AP)^2 = P^{-1}A^2P$.
(b) $(P^{-1}AP)^k = P^{-1}A^kP$ (k a positive integer).

19. Use Exercise 16 to help compute A^{10}, where

$$A = \begin{bmatrix} 1 & 0 \\ -1 & 2 \end{bmatrix}$$

(***Hint.*** Find a matrix P that diagonalizes A and compute $(P^{-1}AP)^{10}$.)

20. Let

$$A = \begin{bmatrix} a & b \\ c & d \end{bmatrix}$$

Show:
(a) A is diagonalizable if $(a - d)^2 + 4bc > 0$.
(b) A is not diagonalizable if $(a - d)^2 + 4bc < 0$.

6.3 ORTHOGONAL DIAGONALIZATION; SYMMETRIC MATRICES

In this section we study the second problem posed at the beginning of Section 6.2. Our study will lead us to the consideration of an important class of matrices called symmetric matrices.

Through this section *orthogonal* means orthogonal with respect to the Euclidean inner product on R^n.

Definition. A square matrix A is called ***orthogonally diagonalizable*** if there is an orthogonal matrix P such that $P^{-1}AP$ $(= P^tAP)$ is diagonal; the matrix P is said to ***orthogonally diagonalize*** A.

We have two questions to consider. First, which matrices are orthogonally diagonalizable; and second, how do we find a matrix P to carry out the orthogonal diagonalization of an orthogonally diagonalizable matrix? The following theorem is concerned with the first question.

Theorem 5. *If A is an $n \times n$ matrix, then the following are equivalent.*

(*a*) *A is orthogonally diagonalizable.*
(*b*) *A has an orthonormal set of n eigenvectors.*

Proof (*a*) $\Rightarrow$ (*b*). Since A is orthogonally diagonalizable, there is an orthogonal matrix P such that $P^{-1}AP$ is diagonal. As shown in the proof of Theorem 2, the n column vectors of P are eigenvectors of A. Since P is orthogonal, these column vectors are orthonormal (see Theorem 28 of Section 4.10) so that A has n orthonormal eigenvectors.

(*b*) $\Rightarrow$ (*a*). Assume that A has an orthonormal set of n eigenvectors $\{\mathbf{p}_1, \mathbf{p}_2, \ldots, \mathbf{p}_n\}$. As shown in the proof of Theorem 2, the matrix P with these eigenvectors as columns diagonalizes A. Since these eigenvectors are orthonormal, P is orthogonal and thus orthogonally diagonalizes A. ∎

The proof of Theorem 5 shows that an orthogonally diagonalizable $n \times n$ matrix A is orthogonally diagonalized by any $n \times n$ matrix P whose columns form an orthonormal set of eigenvectors of A. Let D be the diagonal matrix

$$D = P^{-1}AP$$

Thus

$$A = PDP^{-1}$$

or since P is orthogonal

$$A = PDP^t$$

Therefore

$$A^t = (PDP^t)^t = PD^tP^t = PDP^t = A$$

A matrix with the property

$$A = A^t$$

is said to be ***symmetric***. Thus we have shown that an orthogonally diagonalizable matrix is symmetric. The converse is also true; however, we omit the proof, since it would take us beyond the scope of this text. The following theorem summarizes our discussion.

Theorem 6. *If A is an $n \times n$ matrix, then the following are equivalent*

(*a*) *A is orthogonally diagonalizable.*
(*b*) *A is symmetric.*

Example 12

The matrix

$$A = \begin{bmatrix} 1 & -4 & 5 \\ -4 & 3 & 0 \\ 5 & 0 & 7 \end{bmatrix}$$

is symmetric since $A = A^t$.

We now turn to the problem of finding an orthogonal matrix P to diagonalize a symmetric matrix. The key is the following theorem, whose proof is given at the end of the section.

Theorem 7. *If A is a symmetric matrix, then eigenvectors from different eigenspaces are orthogonal.*

As a consequence of this theorem we obtain the following procedure for orthogonally diagonalizing a symmetric matrix:

Step 1. Find a basis for each eigenspace of A.
Step 2. Apply the Gram-Schmidt process to each of these bases to obtain an orthonormal basis for each eigenspace.
Step 3. Form the matrix P whose columns are the basis vectors constructed in step 2; this matrix orthogonally diagonalizes A.

The justification of this procedure should be clear. Theorem 7 assures that eigenvectors from *distinct* eigenspaces are orthogonal, while the application of the Gram-Schmidt process assures that the eigenvectors obtained within the *same* eigenspace are orthonormal. Thus the *entire* set of eigenvectors obtained by this procedure is orthonormal.

Example 13

Find an orthogonal matrix P that diagonalizes

$$A = \begin{bmatrix} 4 & 2 & 2 \\ 2 & 4 & 2 \\ 2 & 2 & 4 \end{bmatrix}$$

Solution. The characteristic equation of A is

$$\det(\lambda I - A) = \det \begin{bmatrix} \lambda - 4 & -2 & -2 \\ -2 & \lambda - 4 & -2 \\ -2 & -2 & \lambda - 4 \end{bmatrix} = (\lambda - 2)^2(\lambda - 8) = 0$$

Thus the eigenvalues of A are $\lambda = 2$ and $\lambda = 8$. By the method used in Example 5, it can be shown that

$$\mathbf{u}_1 = \begin{bmatrix} -1 \\ 1 \\ 0 \end{bmatrix} \quad \text{and} \quad \mathbf{u}_2 = \begin{bmatrix} -1 \\ 0 \\ 1 \end{bmatrix}$$

form a basis for the eigenspace corresponding to $\lambda = 2$. Applying the Gram-Schmidt process to $\{\mathbf{u}_1, \mathbf{u}_2\}$ yields the orthormal eigenvectors (verify)

$$\mathbf{v}_1 = \begin{bmatrix} -\frac{1}{\sqrt{2}} \\ \frac{1}{\sqrt{2}} \\ 0 \end{bmatrix} \quad \text{and} \quad \mathbf{v}_2 = \begin{bmatrix} -\frac{1}{\sqrt{6}} \\ -\frac{1}{\sqrt{6}} \\ \frac{2}{\sqrt{6}} \end{bmatrix}$$

The eigenspace corresponding to $\lambda = 8$ has

$$\mathbf{u}_3 = \begin{bmatrix} 1 \\ 1 \\ 1 \end{bmatrix}$$

as a basis. Applying the Gram-Schmidt process to $\{\mathbf{u}_3\}$ yields

$$\mathbf{v}_3 = \begin{bmatrix} \frac{1}{\sqrt{3}} \\ \frac{1}{\sqrt{3}} \\ \frac{1}{\sqrt{3}} \end{bmatrix}$$

Finally, using $\mathbf{v}_1$, $\mathbf{v}_2$, and $\mathbf{v}_3$ as column vectors we obtain

$$P = \begin{bmatrix} -\frac{1}{\sqrt{2}} & -\frac{1}{\sqrt{6}} & \frac{1}{\sqrt{3}} \\ \frac{1}{\sqrt{2}} & -\frac{1}{\sqrt{6}} & \frac{1}{\sqrt{3}} \\ 0 & \frac{2}{\sqrt{6}} & \frac{1}{\sqrt{3}} \end{bmatrix}$$

which orthogonally diagonalizes A. (As a check, the reader may wish to verify that P^tAP is a diagonal matrix.)

We conclude this section by stating two important properties of symmetric matrices. We omit the proofs.

Theorem 8.

(*a*) *The characteristic equation of a symmetric matrix A has only real roots.*
(*b*) *If an eigenvalue λ of a symmetric matrix A is repeated k times as a root of the characteristic equation, then the eigenspace corresponding to λ is k-dimensional.*

Example 14

The characteristic equation of the symmetric matrix

$$A = \begin{bmatrix} 3 & 1 & 0 & 0 & 0 \\ 1 & 3 & 0 & 0 & 0 \\ 0 & 0 & 2 & 1 & 1 \\ 0 & 0 & 1 & 2 & 1 \\ 0 & 0 & 1 & 1 & 2 \end{bmatrix}$$

is

$$(\lambda - 4)^2(\lambda - 1)^2(\lambda - 2) = 0$$

so that the eigenvalues are $\lambda = 4$, $\lambda = 1$, and $\lambda = 2$, where $\lambda = 4$ and $\lambda = 1$ are repeated twice and $\lambda = 2$ occurs once. Thus the eigenspaces corresponding to $\lambda = 4$ and $\lambda = 1$ are 2-dimensional and the eigenspace corresponding to $\lambda = 1$ is 1-dimensional.

OPTIONAL

Proof of Theorem 7. Let λ_1 and λ_2 be two different eigenvalues of the $n \times n$ symmetric matrix A, and let

$$\mathbf{v}_1 = \begin{bmatrix} v_1 \\ v_2 \\ \vdots \\ v_n \end{bmatrix} \quad \text{and} \quad \mathbf{v}_2 = \begin{bmatrix} v'_1 \\ v'_2 \\ \vdots \\ v'_n \end{bmatrix}$$

be the corresponding eigenvectors. We want to show that

$$\langle \mathbf{v}_1, \mathbf{v}_2 \rangle = v_1 v'_1 + v_2 v'_2 + \cdots + v_n v'_n = 0$$

Since $\mathbf{v}_1{}^t\mathbf{v}_2$ is a 1×1 matrix having $\langle \mathbf{v}_1, \mathbf{v}_2 \rangle$ as its only entry, we can complete the proof by showing that $\mathbf{v}_1{}^t\mathbf{v}_2 = 0$.

Since $\mathbf{v}_1{}^tA\mathbf{v}_2$ is a 1×1 matrix and every 1×1 matrix is obviously symmetric,

$$\begin{aligned} \mathbf{v}_1{}^tA\mathbf{v}_2 &= (\mathbf{v}_1{}^tA\mathbf{v}_2)^t \\ &= \mathbf{v}_2{}^tA^t\mathbf{v}_1 \qquad \text{(by a property of the transpose; see Section 2.3)} \\ &= \mathbf{v}_2{}^tA\mathbf{v}_1 \qquad \text{(since } A \text{ is symmetric)} \end{aligned}$$

Also

$$\mathbf{v}_1{}^tA\mathbf{v}_2 = \mathbf{v}_1{}^t\lambda_2\mathbf{v}_2 = \lambda_2\mathbf{v}_1{}^t\mathbf{v}_2$$

and

$$\begin{aligned} \mathbf{v}_2{}^tA\mathbf{v}_1 &= \mathbf{v}_2{}^t\lambda_1\mathbf{v}_1 = \lambda_1\mathbf{v}_2{}^t\mathbf{v}_1 \\ &= \lambda_1(\mathbf{v}_2{}^t\mathbf{v}_1)^t = \lambda_1\mathbf{v}_1{}^t\mathbf{v}_2 \end{aligned}$$

Thus

$$\lambda_1\mathbf{v}_1{}^t\mathbf{v}_2 = \lambda_2\mathbf{v}_1{}^t\mathbf{v}_2$$

or

$$(\lambda_1 - \lambda_2)\mathbf{v}_1{}^t\mathbf{v}_2 = 0$$

Since $\lambda_1 \neq \lambda_2$, it follows that $\mathbf{v}_1{}^t\mathbf{v}_2 = 0$. ▮

EXERCISE SET 6.3

1. Use part (*b*) of Theorem 8 to find the dimensions of the eigenspaces of the following symmetric matrices.

(a) $\begin{bmatrix} 1 & 1 \\ 1 & 1 \end{bmatrix}$ (b) $\begin{bmatrix} \frac{7}{25} & 0 & -\frac{24}{25} \\ 0 & -1 & 0 \\ -\frac{24}{25} & 0 & \frac{7}{25} \end{bmatrix}$

(c) $\begin{bmatrix} 1 & 1 & 1 \\ 1 & 1 & 1 \\ 1 & 1 & 1 \end{bmatrix}$ (d) $\begin{bmatrix} 6 & 0 & 0 \\ 0 & 3 & 3 \\ 0 & 3 & 3 \end{bmatrix}$

(e) $\begin{bmatrix} 4 & 4 & 0 & 0 \\ 4 & 4 & 0 & 0 \\ 0 & 0 & 0 & 0 \\ 0 & 0 & 0 & 0 \end{bmatrix}$ (f) $\begin{bmatrix} \frac{10}{3} & -\frac{4}{3} & 0 & -\frac{4}{3} \\ -\frac{4}{3} & -\frac{5}{3} & 0 & \frac{1}{3} \\ 0 & 0 & -2 & 0 \\ -\frac{4}{3} & \frac{1}{3} & 0 & -\frac{5}{3} \end{bmatrix}$

In Exercises 2–9, find a matrix P that orthogonally diagonalizes A, and determine $P^{-1}AP$.

2. $A = \begin{bmatrix} 3 & 1 \\ 1 & 3 \end{bmatrix}$ **3.** $A = \begin{bmatrix} 5 & 3\sqrt{3} \\ 3\sqrt{3} & -1 \end{bmatrix}$

4. $A = \begin{bmatrix} -7 & 24 \\ 24 & 7 \end{bmatrix}$ **5.** $A = \begin{bmatrix} -2 & 0 & -36 \\ 0 & -3 & 0 \\ -36 & 0 & -23 \end{bmatrix}$

6. $A = \begin{bmatrix} 1 & 1 & 0 \\ 1 & 1 & 0 \\ 0 & 0 & 0 \end{bmatrix}$ **7.** $\begin{bmatrix} 2 & -1 & -1 \\ -1 & 2 & -1 \\ -1 & -1 & 2 \end{bmatrix}$

8. $A = \begin{bmatrix} 3 & 1 & 0 & 0 \\ 1 & 3 & 0 & 0 \\ 0 & 0 & 0 & 0 \\ 0 & 0 & 0 & 0 \end{bmatrix} 0$ **9.** $A = \begin{bmatrix} 5 & -2 & 0 & 0 \\ -2 & 2 & 0 & 0 \\ 0 & 0 & 5 & -2 \\ 0 & 0 & -2 & 2 \end{bmatrix}$

10. Find a matrix that orthogonally diagonalizes

$$\begin{bmatrix} a & b \\ b & a \end{bmatrix}$$

where $b \neq 0$.

11. Two $n \times n$ matrices, A and B, are called ***orthogonally similar*** if there is an orthogonal matrix P such that $B = P^{-1}AP$. Show that if A is symmetric and A and B are orthogonally similar, then B is symmetric.

12. Prove Theorem 7 for 2×2 symmetric matrices.

13. Prove Theorem 8*a* for 2×2 symmetric matrices.

7 Applications

7.1 APPLICATION TO DIFFERENTIAL EQUATIONS

Many laws of physics, chemistry, biology, and economics are described in terms of *differential equations*, that is, equations involving functions and their derivatives. The purpose of this section is to illustrate one way in which linear algebra can be applied to solve certain systems of differential equations. The scope of this section is narrow, but it should serve to convince the reader that linear algebra has solid applications.

One of the simplest differential equations is

$$y' = ay \tag{7.1}$$

where $y = f(x)$ is an unknown function to be determined, $y' = dy/dx$ is its derivative, and a is a constant. Like most differential equations, (7.1) has infinitely many solutions; they are the functions of the form

$$y = ce^{ax} \tag{7.2}$$

where c is an arbitrary constant. Each function of this form is a solution of $y' = ay$ since

$$y' = cae^{ax} = ay$$

Conversely, every solution of $y' = ay$ must be a function of the form ce^{ax} (Exercise 7), so that (7.2) describes all solutions of $y' = ay$. We call (7.2) the ***general solution*** of $y' = ay$.

Sometimes the physical problem that generates a differential equation imposes some added condition that enables us to isolate one ***particular solution*** from the general solution. For example, if we require that the solution of $y' = ay$ satisfy the added condition

$$y(0) = 3 \tag{7.3}$$

that is, $y = 3$ when $x = 0$, then on substituting these values in the general solution $y = ce^{ax}$ we obtain a value for c, namely

$$3 = ce^0 = c$$

thus

$$y = 3e^{ax}$$

is the only solution of $y' = ay$ that satisfies the added condition. A condition, such as (7.3), that specifies the value of the solution at a point is called an ***initial condition***, and the problem of solving a differential equation subject to an initial condition is called an ***initial-value problem***.

In this section we will be concerned with solving systems of differential equations having the form

$$\begin{aligned} y_1' &= a_{11}y_1 + a_{12}y_2 + \cdots + a_{1n}y_n \\ y_2' &= a_{21}y_1 + a_{22}y_2 + \cdots + a_{2n}y_n \\ &\vdots \\ y_n' &= a_{n1}y_1 + a_{n2}y_2 + \cdots + a_{nn}y_n \end{aligned} \tag{7.4}$$

where $y_1 = f_1(x)$, $y_2 = f_2(x), \ldots, y_n = f_n(x)$ are functions to be determined, and the a_{ij}'s are constants. In matrix notation (7.4) can be written

$$\begin{bmatrix} y_1' \\ y_2' \\ \vdots \\ y_n' \end{bmatrix} = \begin{bmatrix} a_{11} & a_{12} & \cdots & a_{1n} \\ a_{21} & a_{22} & \cdots & a_{2n} \\ \vdots & \vdots & & \vdots \\ a_{n1} & a_{n2} & \cdots & a_{nn} \end{bmatrix} \begin{bmatrix} y_1 \\ y_2 \\ \vdots \\ y_n \end{bmatrix}$$

or more briefly

$$Y' = AY$$

Example 1

(a) Write the system

$$\begin{aligned} y_1' &= 3y_1 \\ y_2' &= -2y_2 \\ y_3' &= 5y_3 \end{aligned}$$

in matrix form.

(b) Solve the system.

(c) Find a solution of the system which satisfies the initial conditions $y_1(0) = 1$, $y_2(0) = 4$, and $y_3(0) = -2$.

Solution (a).

$$\begin{bmatrix} y_1' \\ y_2' \\ y_3' \end{bmatrix} = \begin{bmatrix} 3 & 0 & 0 \\ 0 & -2 & 0 \\ 0 & 0 & 5 \end{bmatrix} \begin{bmatrix} y_1 \\ y_2 \\ y_3 \end{bmatrix} \tag{7.5}$$

or

$$Y' = \begin{bmatrix} 3 & 0 & 0 \\ 0 & -2 & 0 \\ 0 & 0 & 5 \end{bmatrix} Y$$

(b) Because each equation involves only one unknown function, we can solve the equations individually. From (7.2), we obtain

$$\begin{aligned} y_1 &= c_1 e^{3x} \\ y_2 &= c_2 e^{-2x} \\ y_3 &= c_3 e^{5x} \end{aligned}$$

or in matrix notation

$$Y = \begin{bmatrix} y_1 \\ y_2 \\ y_3 \end{bmatrix} = \begin{bmatrix} c_1 e^{3x} \\ c_2 e^{-2x} \\ c_3 e^{5x} \end{bmatrix}$$

(c) From the given initial conditions, we obtain

$$\begin{aligned} 1 &= y_1(0) = c_1 e^0 = c_1 \\ 4 &= y_2(0) = c_2 e^0 = c_2 \\ -2 &= y_3(0) = c_3 e^0 = c_3 \end{aligned}$$

so that the solution satisfying the initial conditions is

$$y_1 = e^{3x}, \; y_2 = 4e^{-2x}, \; y_3 = -2e^{5x}$$

or in matrix notation

$$Y = \begin{bmatrix} y_1 \\ y_2 \\ y_3 \end{bmatrix} = \begin{bmatrix} e^{3x} \\ 4e^{-2x} \\ -2e^{5x} \end{bmatrix}$$

The system in this example was easy to solve because each equation involved only one unknown function, and this was the case because the matrix of coefficients (7.5) for the system was diagonal. But how do we handle a system

$$Y' = AY$$

in which the matrix A is not diagonal? The idea is simple: Try to make a substitution for Y that will yield a new system with a diagonal coefficient matrix; solve this new simpler system, and then use this solution to determine the solution of the original system.

The kind of substitution we have in mind is

$$\begin{aligned} y_1 &= p_{11}u_1 + p_{12}u_2 + \cdots + p_{1n}u_n \\ y_2 &= p_{21}u_1 + p_{22}u_2 + \cdots + p_{2n}u_n \\ &\vdots \\ y_n &= p_{n1}u_1 + p_{n2}u_2 + \cdots + p_{nn}u_n \end{aligned} \tag{7.6}$$

or in matrix notation

$$\begin{bmatrix} y_1 \\ y_2 \\ \vdots \\ y_n \end{bmatrix} = \begin{bmatrix} p_{11} & p_{12} & \cdots & p_{1n} \\ p_{21} & p_{22} & \cdots & p_{2n} \\ \vdots & \vdots & & \vdots \\ p_{n1} & p_{n2} & \cdots & p_{nn} \end{bmatrix} \begin{bmatrix} u_1 \\ u_2 \\ \vdots \\ u_n \end{bmatrix}$$

or more briefly

$$Y = PU$$

In this substitution the p_{ij}'s are constants to be determined in such a way that the new system involving the unknown functions $u_1, u_2, \ldots, u_n$ has a diagonal coefficient matrix. We leave it as an exercise for the student to differentiate each equation in (7.6) and deduce

$$Y' = PU'$$

If we make the substitutions $Y = PU$ and $Y' = PU'$ in the original system

$$Y' = AY$$

and if we assume P to be invertible, we obtain

$$PU' = A(PU)$$

or

$$U' = (P^{-1}AP)U$$

or

$$U' = DU$$

where $D = P^{-1}AP$. The choice for P is now clear; if we want the new coefficient matrix D to be diagonal, we must choose P to be a matrix that diagonalizes A.

This suggests the following procedure for solving a system

$$Y' = AY$$

with a diagonalizable coefficient matrix A:

Step 1. Find a matrix P that diagonalizes A.

Step 2. Make the substitutions $Y = PU$ and $Y' = PU'$ to obtain a new "diagonal system" $U' = DU$, where $D = P^{-1}AP$.

Step 3. Solve $U' = DU$.

Step 4. Determine Y from the equation $Y = PU$.

Example 2

(a) Solve

$$\begin{aligned} y_1' &= y_1 + y_2 \\ y_2' &= 4y_1 - 2y_2 \end{aligned}$$

(b) Find the solution that satisfies the initial conditions $y_1(0) = 1$, $y_2(0) = 6$.

Solution (a). The coefficient matrix for the system is

$$A = \begin{bmatrix} 1 & 1 \\ 4 & -2 \end{bmatrix}$$

As discussed in Section 6.2, A will be diagonalized by any matrix P whose columns are linearly independent eigenvectors of A. Since

$$\det(\lambda I - A) = \begin{vmatrix} \lambda - 1 & -1 \\ -4 & \lambda + 2 \end{vmatrix} = \lambda^2 + \lambda - 6 = (\lambda + 3)(\lambda - 2)$$

the eigenvalues of A are $\lambda = 2$, $\lambda = -3$. By definition,

$$\mathbf{x} = \begin{bmatrix} x_1 \\ x_2 \end{bmatrix}$$

is an eigenvector of A corresponding to λ if and only if $\mathbf{x}$ is a nontrivial solution of $(\lambda I - A)\mathbf{x} = 0$, that is, of

$$\begin{bmatrix} \lambda - 1 & -1 \\ -4 & \lambda + 2 \end{bmatrix}\begin{bmatrix} x_1 \\ x_2 \end{bmatrix} = \begin{bmatrix} 0 \\ 0 \end{bmatrix}$$

If $\lambda = 2$, this system becomes

$$\begin{bmatrix} 1 & -1 \\ -4 & 4 \end{bmatrix}\begin{bmatrix} x_1 \\ x_2 \end{bmatrix} = \begin{bmatrix} 0 \\ 0 \end{bmatrix}.$$

Solving, yields

$$x_1 = t, \qquad x_2 = t$$

so that

$$\begin{bmatrix} x_1 \\ x_2 \end{bmatrix} = \begin{bmatrix} t \\ t \end{bmatrix} = t\begin{bmatrix} 1 \\ 1 \end{bmatrix}$$

Thus

$$\mathbf{p}_1 = \begin{bmatrix} 1 \\ 1 \end{bmatrix}$$

is a basis for the eigenspace corresponding to $\lambda = 2$. Similarly, the reader can show that

$$\mathbf{p}_2 = \begin{bmatrix} -\frac{1}{4} \\ 1 \end{bmatrix}$$

is a basis for the eigenspace corresponding to $\lambda = -3$. Thus

$$P = \begin{bmatrix} 1 & -\frac{1}{4} \\ 1 & 1 \end{bmatrix}$$

diagonalizes A and

$$D = P^{-1}AP = \begin{bmatrix} 2 & 0 \\ 0 & -3 \end{bmatrix}$$

Therefore the substitution

$$Y = PU \quad \text{and} \quad Y' = PU'$$

yields the new "diagonal system,"

$$U' = DU = \begin{bmatrix} 2 & 0 \\ 0 & -3 \end{bmatrix} U \quad \text{or} \quad \begin{aligned} u_1' &= 2u_1 \\ u_2' &= -3u_1 \end{aligned}$$

From (7.2) the solution of this system is

$$\begin{aligned} u_1 &= c_1e^{2x} \\ u_2 &= c_2e^{-3x} \end{aligned} \quad \text{or} \quad U = \begin{bmatrix} c_1e^{2x} \\ c_2e^{-3x} \end{bmatrix}$$

so that the equation $Y = PU$ yields as the solution for Y

$$Y = \begin{bmatrix} y_1 \\ y_2 \end{bmatrix} = \begin{bmatrix} 1 & -\frac{1}{4} \\ 1 & 1 \end{bmatrix} \begin{bmatrix} c_1e^{2x} \\ c_2e^{-3x} \end{bmatrix} = \begin{bmatrix} c_1e^{2x} - \frac{1}{4}c_2e^{-3x} \\ c_1e^{2x} + c_2e^{-3x} \end{bmatrix}$$

or

$$\begin{aligned} y_1 &= c_1e^{2x} - \tfrac{1}{4}c_2e^{-3x} \\ y_2 &= c_1e^{2x} + c_2e^{-3x} \end{aligned} \tag{7.7}$$

(b) If we substitute the given initial conditions in (7.7) we obtain

$$\begin{aligned} c_1 - \tfrac{1}{4}c_2 &= 1 \\ c_1 + c_2 &= 6 \end{aligned}$$

Solving this system we obtain

$$c_1 = 2, \qquad c_2 = 4$$

so that from (7.7) the solution satisfying the initial conditions is

$$\begin{aligned} y_1 &= 2e^{2x} - e^{-3x} \\ y_2 &= 2e^{2x} + 4e^{-3x} \end{aligned}$$

We have assumed in this section that the coefficient matrix of $Y' = AY$ is diagonalizable. If this is not the case, other methods must be used to solve the system. Such methods are discussed in more advanced texts.

EXERCISE SET 7.1

1. (a) Solve the system

$$\begin{aligned} y_1' &= y_1 + 4y_2 \\ y_2' &= 2y_1 + 3y_2 \end{aligned}$$

(b) Find the solution that satisfies the initial conditions $y_1(0) = 0$, $y_2(0) = 0$.

2. (a) Solve the system

$$\begin{aligned} y_1' &= y_1 + 3y_2 \\ y_2' &= 4y_1 + 5y_2 \end{aligned}$$

(b) Find the solution that satisfies the conditions $y_1(0) = 2$, $y_2'(0) = 1$

3. (a) Solve the system

$$\begin{aligned} y_1' &= 4y_1 + y_3 \\ y_2' &= -2y_1 + y_2 \\ y_3' &= -2y_1 + y_3 \end{aligned}$$

(b) Find the solution that satisfies the initial conditions $y_1(0) = -1$, $y_2(0) = 1$, $y_3(0) = 0$.

4. Solve the system

$$\begin{aligned} y_1' &= 4y_1 + 2y_2 + 2y_3 \\ y_2' &= 2y_1 + 4y_2 + 2y_3 \\ y_3' &= 2y_1 + 2y_2 + 4y_3 \end{aligned}$$

5. Solve the differential equation $y'' - y' - 6y = 0$. ***Hint.*** Let $y_1 = y$, $y_2 = y'$ and then show

$$\begin{aligned} y_1' &= y_2 \\ y_2' &= y'' = y' + 6y = y_1' + 6y_1 = 6y_1 + y_2 \end{aligned}$$

6. Solve the differential equation $y''' - 6y'' + 11y' - 6y = 0$. ***Hint.*** Let $y_1 = y$, $y_2 = y'$, $y_3 = y''$ and then show

$$\begin{aligned} y_1' &= y_2 \\ y_2' &= y_3 \\ y_3' &= 6y_1 - 11y_2 + 6y_3 \end{aligned}$$

7. Prove: Every solution of $y' = ay$ has the form $y = ce^{ax}$. (***Hint.*** Let $y = f(x)$ be a solution and show that $f(x)e^{-ax}$ is constant.)

8. Prove: If A is diagonalizable and

$$Y = \begin{bmatrix} y_1 \\ y_2 \\ \vdots \\ y_n \end{bmatrix}$$

satisfies $Y' = AY$ then each y_i is a linear combination of $e^{\lambda_1 x}, e^{\lambda_2 x}, \ldots, e^{\lambda_n x}$, where $\lambda_1, \lambda_2, \ldots, \lambda_n$ are eigenvalues of A.

7.2 APPLICATION TO APPROXIMATION PROBLEMS; FOURIER SERIES

In many applications one is concerned with finding the best possible approximation over an interval to a function f by another function in some specified class; for example:

(a) Find the best possible approximation to e^x over $[0, 1]$ by a polynomial of the form $a_0 + a_1x + a_2x^2$.

(b) Find the best possible approximation to $\sin \pi x$ over $[-1, 1]$ by a function of the form $a_0 + a_1 e^x + a_2 e^{2x} + a_3 e^{3x}$.
(c) Find the best possible approximation to $|x|$ over $[0, 2\pi]$ by a function of the form $a_0 + a_1 \sin x + a_2 \sin 2x + b_1 \cos x + b_2 \cos 2x$.

Observe that in each of these examples the approximating functions were drawn from some subspace of the vector space $C[a, b]$ (continuous functions on $[a, b]$). In the first example it was the subspace of $C[0, 1]$ spanned by 1, x, and x^2; in the second example it was the subspace of $C[-1, 1]$ spanned by 1, e^x, e^{2x}, and e^{3x}; and in the third example it was the subspace of $C[0, 2\pi]$ spanned by 1, $\sin x$, $\sin 2x$, $\cos x$, and $\cos 2x$. Thus each of these examples states a problem of the following form:

Approximation Problem. Find the best possible approximation over $[a, b]$ to a given function f using only approximations from a specified subspace W of $C[a, b]$.

To solve this problem we must make the term, "best possible approximation over $[a, b]$" more precise. Intuitively, the best possible approximation over $[a, b]$ is one that produces the smallest error. But what do we mean by "error"? If we were concerned only with approximating $f(x)$ at a single point x_0 then the error at x_0 by an approximation $g(x)$ would be simply

$$\text{error} = |f(x_0) - g(x_0)|$$

sometimes called the ***deviation*** between f and g at x_0. (Figure 7.1). However, we are concerned with approximation over an entire interval $[a, b]$, not a single point. Consequently, in one part of the interval an approximation $g_1(x)$ may have smaller deviations from f than an approximation $g_2(x)$, and in another part of the interval it might be the other way around. How does one decide which is the better overall approximation? What we need is some way of measuring the overall error in an approximation $g(x)$. One possible measure of overall error is obtained by integrating the deviation $|f(x) - g(x)|$ over the entire interval;

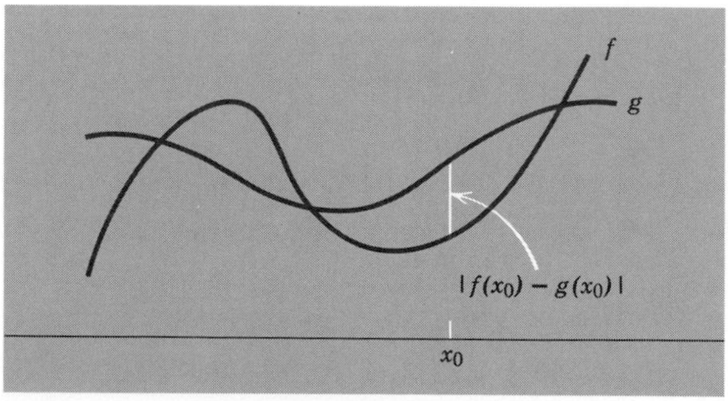

Figure 7.1

that is

$$\text{error} = \int_a^b |f(x) - g(x)|\, dx \tag{7.8}$$

Geometrically, (7.8) is the area between the graphs of $f(x)$ and $g(x)$ over the interval $[a, b]$ (Figure 7.2); the greater the area, the greater the overall error.

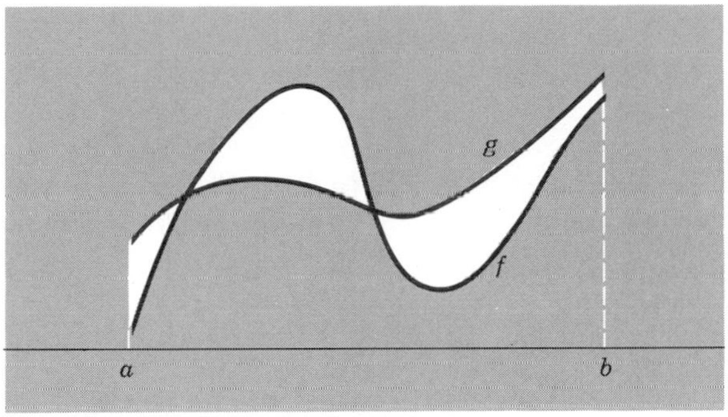

Figure 7.2

While (7.8) is natural and appealing geometrically, the occurrence of an absolute value sign is sufficiently bothersome in computations that most mathematicians and scientists generally favor the following alternative measure of error, called the ***mean square error***

$$\text{mean square error} = \int_a^b [f(x) - g(x)]^2\, dx$$

Mean square error has the added advantage that it allows us to bring to bear the theory of inner product spaces in problems of approximation. To see how, consider the inner product

$$\langle \mathbf{f}, \mathbf{g} \rangle = \int_a^b f(x)g(x)\, dx \tag{7.9}$$

on the vector space $C[a, b]$. With this inner product

$$\|\mathbf{f} - \mathbf{g}\|^2 = \langle \mathbf{f} - \mathbf{g}, \mathbf{f} - \mathbf{g} \rangle = \int_a^b [f(x) - g(x)]^2\, dx$$

which states that the mean square error that results from approximating **f** by **g** over $[a, b]$ is the square of the distance between **f** and **g** when these functions are viewed as vectors in $C[a, b]$ with the inner product (7.9). Thus an approximation g from a subspace W of $C[a, b]$ minimizes the mean square error if and only if it minimizes $\|\mathbf{f} - \mathbf{g}\|^2$, or equivalently if and only if it minimizes $\|\mathbf{f} - \mathbf{g}\|$. In short, the approximation **g** in W which minimizes the mean square error is the vector **g** in W *closest* to **f** using inner product 7.9. But we already know what the vector **g** is; it is the orthogonal projection of **f** on the subspace W (Theorem 23, Section 4.9). See Figure 7.3. In summary, we have the following result.

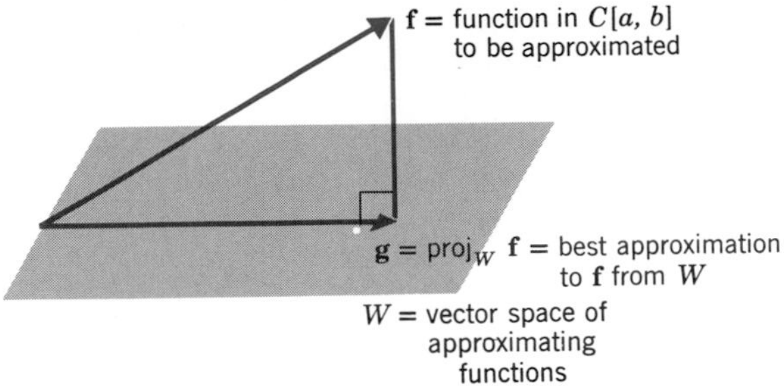

Figure 7.3

Solution of the Least Squares Problem. If $\mathbf{f}$ is a continuous function on $[a, b]$ and W is a finite dimensional subspace of $C[a, b]$, then the function $\mathbf{g}$ in W that minimizes the mean square error

$$\int_a^b [f(x) - g(x)]^2 \, dx$$

is $\mathbf{g} = \text{proj}_W \, \mathbf{f}$, the orthogonal projection of $\mathbf{f}$ on W, relative to inner product (7.9). The function $\mathbf{g} = \text{proj}_W \, \mathbf{f}$ is called the ***least squares approximation*** to $\mathbf{f}$ from W.

Fourier Series

A function of the form

$$\begin{aligned} t(x) = {} & c_0 + c_1 \cos x + c_2 \cos 2x + \cdots + c_n \cos nx \\ & + d_1 \sin x + d_2 \sin 2x + \cdots + d_n \sin nx \end{aligned} \tag{7.10}$$

is called a ***trigonometric polynomial***; if c_n and d_n are not both zero then $t(x)$ is said to have ***order*** n.

Example 3

$$t(x) = 2 + \cos x - 3 \cos 2x + 7 \sin 4x$$

is a trigonometric polynomial with

$$c_0 = 2, c_1 = 1, c_2 = -3, d_1 = 0, d_2 = 0, d_3 = 0, d_4 = 7$$

The order of $t(x)$ is 4.

It is evident from (7.10) that the trigonometric polynomials of order n or less are the various possible linear combinations of

$$1, \quad \cos x, \quad \cos 2x, \ldots, \cos nx, \quad \sin x, \quad \sin 2x, \ldots, \sin nx \tag{7.11}$$

Thus the trigonometric polynomials of order n or less form a subspace W of the vector of continuous functions; namely the subspace spanned by the $2n + 1$ functions listed in (7.11). It can be shown that these functions are linearly independent, and consequently form a basis for W.

Let us consider the problem of approximating a continuous function $f(x)$ over the interval $[0, 2\pi]$ by a trigonometric polynomial of order n or less. As noted earlier, the least squares approximation to **f** from W is the orthogonal projection of **f** on W. To find this orthogonal projection, we must find an orthonormal basis $\mathbf{g}_0, \mathbf{g}_1, \ldots, \mathbf{g}_{2n}$ for W, after which we can compute the orthogonal projection on W from the formula

$$\text{proj}_W \mathbf{f} = \langle \mathbf{f}, \mathbf{g}_0 \rangle \mathbf{g}_0 + \langle \mathbf{f}, \mathbf{g}_1 \rangle \mathbf{g}_1 + \cdots + \langle \mathbf{f}, \mathbf{g}_{2n} \rangle \mathbf{g}_{2n} \tag{7.12}$$

(see Theorem 20 of Section 4.9). An orthonormal basis for W can be obtained by applying the Gram-Schmidt process to the basis (7.11), using the inner product

$$\langle \mathbf{u}, \mathbf{v} \rangle = \int_0^{2\pi} u(x)v(x)\,dx$$

This yields (Exercise 6) the orthonormal basis

$$\mathbf{g}_0 = \frac{1}{\sqrt{2\pi}},\ \mathbf{g}_1 = \frac{1}{\sqrt{\pi}}\cos x, \ldots, \mathbf{g}_n = \frac{1}{\sqrt{\pi}}\cos nx,$$

$$\mathbf{g}_{n+1} = \frac{1}{\sqrt{\pi}}\sin x, \ldots, \mathbf{g}_{2n} = \frac{1}{\sqrt{\pi}}\sin nx \tag{7.13}$$

If we introduce the notation

$$a_0 = \frac{2}{\sqrt{2\pi}}\langle \mathbf{f}, \mathbf{g}_0 \rangle,\ a_1 = \frac{1}{\sqrt{\pi}}\langle \mathbf{f}, \mathbf{g}_1 \rangle, \ldots, a_n = \frac{1}{\sqrt{\pi}}\langle \mathbf{f}, \mathbf{g}_n \rangle$$

$$b_1 = \frac{1}{\sqrt{\pi}}\langle \mathbf{f}, \mathbf{g}_{n+1} \rangle, \ldots, \frac{1}{\sqrt{\pi}}\langle \mathbf{f}, \mathbf{g}_{2n} \rangle$$

then on substituting (7.13) in (7.12) we obtain

$$\text{proj}_W(\mathbf{f}) = \frac{a_0}{2} + [a_1 \cos x + \cdots + a_n \cos nx] + [b_1 \sin x + \cdots + b_n \sin nx]$$

where

$$a_0 = \frac{2}{\sqrt{2\pi}}\langle \mathbf{f}, \mathbf{g}_0 \rangle = \frac{2}{\sqrt{2\pi}}\int_0^{2\pi} f(x)\frac{1}{\sqrt{2\pi}}\,dx = \frac{1}{\pi}\int_0^{2\pi} f(x)\,dx$$

$$a_1 = \frac{1}{\sqrt{\pi}}\langle \mathbf{f}, \mathbf{g}_1 \rangle = \frac{1}{\sqrt{\pi}}\int_0^{2\pi} f(x)\frac{1}{\sqrt{\pi}}\cos x\,dx = \frac{1}{\pi}\int_0^{2\pi} f(x)\cos x\,dx$$

$$\vdots$$

$$a_n = \frac{1}{\sqrt{\pi}}\langle \mathbf{f}, \mathbf{g}_n \rangle = \frac{1}{\sqrt{\pi}}\int_0^{2\pi} f(x)\frac{1}{\sqrt{\pi}}\cos nx\,dx = \frac{1}{\pi}\int_0^{2\pi} f(x)\cos nx\,dx$$

$$b_1 = \frac{1}{\sqrt{\pi}}\langle \mathbf{f}, \mathbf{g}_{n+1} \rangle = \frac{1}{\sqrt{\pi}}\int_0^{2\pi} f(x)\frac{1}{\sqrt{\pi}}\sin x\,dx = \frac{1}{\pi}\int_0^{2\pi} f(x)\sin x\,dx$$

$$\vdots$$

$$b_n = \frac{1}{\sqrt{\pi}}\langle \mathbf{f}, \mathbf{g}_{2n} \rangle = \frac{1}{\sqrt{\pi}}\int_0^{2\pi} f(x)\frac{1}{\sqrt{\pi}}\sin nx\,dx = \frac{1}{\pi}\int_0^{2\pi} f(x)\sin nx\,dx$$

In short

$$a_k = \frac{1}{\pi}\int_0^{2\pi} f(x)\cos kx\,dx, \qquad b_k = \frac{1}{\pi}\int_0^{2\pi} f(x)\sin kx\,dx$$

The numbers $a_0, a_1, \ldots, a_n, b_1, \ldots, b_n$ are called the *Fourier* coefficients* of **f**.

Example 4

Find the least squares approximation of $f(x) = x$ on $[0, 2\pi]$ by
(a) a trigonometric polynomial of order 2 or less
(b) a trigonometric polynomial of order n or less.

Solution.

$$a_0 = \frac{1}{\pi}\int_0^{2\pi} f(x)\,dx = \frac{1}{\pi}\int_0^{2\pi} x\,dx = 2\pi$$

For $k = 1, 2, \ldots$ integration by parts yields (verify):

$$a_k = \frac{1}{\pi}\int_0^{2\pi} f(x)\cos kx\,dx = \frac{1}{\pi}\int_0^{2\pi} x\cos kx\,dx = 0$$

$$b_k = \frac{1}{\pi}\int_0^{2\pi} f(x)\sin kx\,dx = \frac{1}{\pi}\int_0^{2\pi} x\sin kx\,dx = -\frac{2}{k} \qquad (7.14)$$

Thus the least squares approximation to x on $[0, 2\pi]$ by a trigonometric polynomial of order 2 or less is

$$x \simeq \frac{a_0}{2} + a_1\cos x + a_2\cos 2x + b_1\sin x + b_2\sin 2x$$

or from (7.14)

$$x \simeq \pi - 2\sin x - \sin 2x$$

(b) The least squares approximation to x on $[0, 2\pi]$ by a trigonometric polynomial of order n or less is

$$x \simeq \frac{a_0}{2} + [a_1\cos x + \cdots + a_n\cos nx] + [b_1\sin x + \cdots + b_n\sin nx]$$

* Jean Baptiste Joseph Fourier (1768–1830)—French mathematician and physicist. Fourier discovered Fourier series and related ideas while working on problems of heat diffusion. This discovery is one of the most influential in the history of mathematics; it is the cornerstone of many fields of mathematical research and a basic tool in many branches of engineering.

Fourier, a political activist during the French revolution, spent time in jail for his defense of many victims during the Terror. He later became a favorite of Napoleon and was named both a Baron and a Count.

or from (7.14)

$$x \simeq \pi - 2\left(\sin x + \frac{\sin 2x}{2} + \frac{\sin 3x}{3} + \cdots + \frac{\sin nx}{n}\right)$$

It is natural to expect that the mean square error will diminish as the number of terms in the least squares approximation

$$f(x) \simeq \frac{a_0}{2} + \sum_{k=1}^{n} (a_k \cos kx + b_k \sin kx)$$

increases. It can be proved that the mean square error approaches zero as $n \to +\infty$; this is denoted by writing

$$f(x) = \frac{a_0}{2} + \sum_{k=1}^{\infty} (a_k \cos kx + b_k \sin kx)$$

The right side of this equation is called the ***Fourier series*** for f. Such series are of major importance in engineering, science, and mathematics.

EXERCISE SET 7.2

1. Find the least squares approximation of $f(x) = 1 + x$ over the interval $[0, 2\pi]$ by
(a) a trigonometric polynomial of order 2 or less.
(b) a trigonometric polynomial of order n or less.

2. Find the least squares approximation of $f(x) = x^2$ over the interval $[0, 2\pi]$ by
(a) a trigonometric polynomial of order 3 or less.
(b) a trigonometric polynomial of order n or less.

3. (a) Find the least squares approximation of x over the interval $[0, 1]$ by a function of the form $a + be^x$.
(b) Find the mean square error of the approximation.

4. (a) Find the least squares approximation of e^x over the interval $[0, 1]$ by a polynomial of the form $a_0 + a_1x$.
(b) Sketch the graphs of e^x and your approximation.

5. (a) Find the least squares approximation of $\sin \pi x$ over the interval $[-1, 1]$ by a polynomial of the form $a_0 + a_1x + a_2x^2$.
(b) Sketch the graph of $\sin \pi x$ and your approximation.

6. Use the Gram-Schmidt process to obtain the orthonormal basis (7.13) from the basis (7.11).

7. Carry out the integrations in (7.14).

8. Find the Fourier series of $f(x) = \pi - x$.

7.3 QUADRATIC FORMS; APPLICATION TO CONIC SECTIONS

In this section we apply our results on orthogonal coordinate transformations to the study of quadratic equations, quadratic forms, and conic sections. Quadratic forms arise in a variety of important problems concerned with such diverse areas as vibrations, relativity, geometry, and statistics.

An equation of the form

$$ax^2 + 2bxy + cy^2 + dx + ey + f = 0 \tag{7.15}$$

where $a, b, \ldots, f$ are real numbers and at least one of the numbers a, b, c is not zero, is called a ***quadratic equation in x and y***; the expression

$$ax^2 + 2bxy + cy^2$$

is called the ***associated quadratic form***.

Example 15

In the quadratic equation

$$3x^2 + 5xy - 7y^2 + 2x + 7 = 0$$

the constants in (7.15) are

$$a = 3 \qquad b = \tfrac{5}{2} \qquad c = -7 \qquad d = 2 \qquad e = 0 \qquad f = 7$$

Example 16

Quadratic equation	**Associated quadratic form**
$3x^2 + 5xy - 7y^2 + 2x + 7 = 0$	$3x^2 + 5xy - 7y^2$
$4x^2 - 5y^2 + 8y + 9 = 0$	$4x^2 - 5y^2$
$xy + y = 0$	xy

Graphs of quadratic equations in x and y are called ***conics*** or ***conic sections***. The most important conics are ellipses, circles, hyperbolas, and parabolas; these are called the ***nondegenerate*** conics. The remaining conics are called ***degenerate*** and include single points and pairs of lines (see Exercise 13).

A nondegenerate conic is said to be in ***standard position*** relative to the coordinate axes if its equation can be expressed in one of the forms given in Figure 7.4.

Graph

$\dfrac{x^2}{k^2} + \dfrac{y^2}{l^2} = 1;\ k, l > 0$

(Ellipse or Circle)

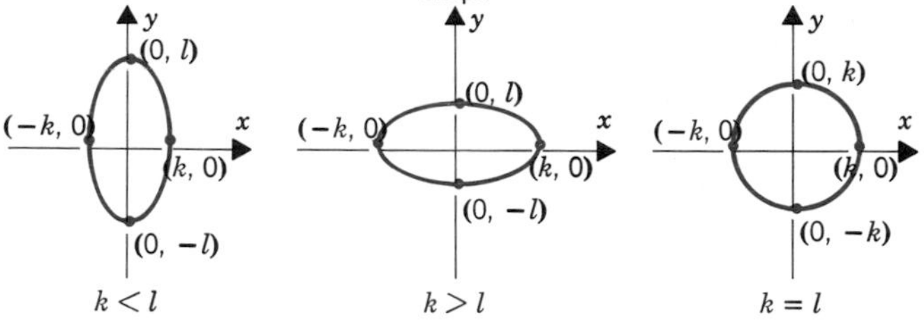

$\dfrac{x^2}{k^2} - \dfrac{y^2}{l^2} = 1;\ k, l > 0$

(Hyperbola)

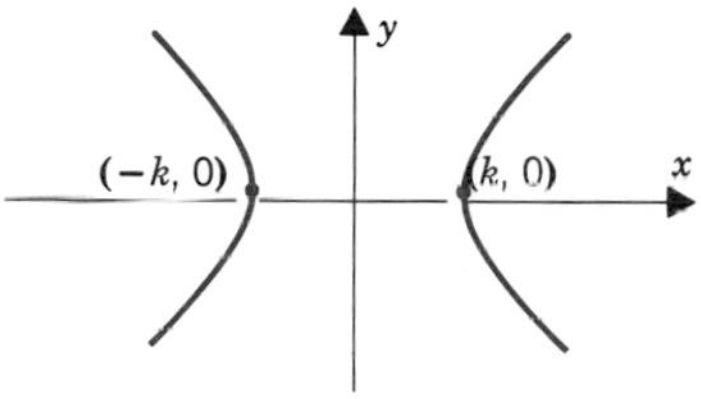

$\dfrac{y^2}{k^2} - \dfrac{x^2}{l^2} = 1;\ k, l > 0$

(Hyperbola)

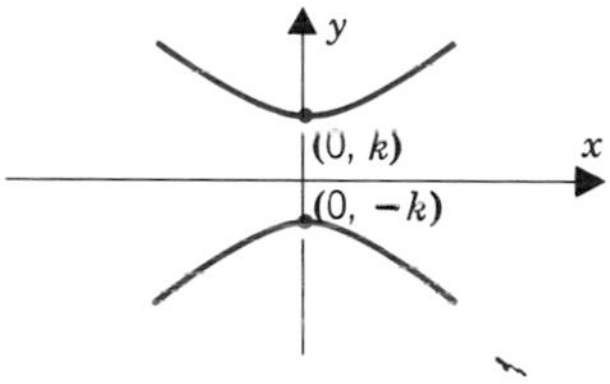

$y^2 = kx$

(Parabola)

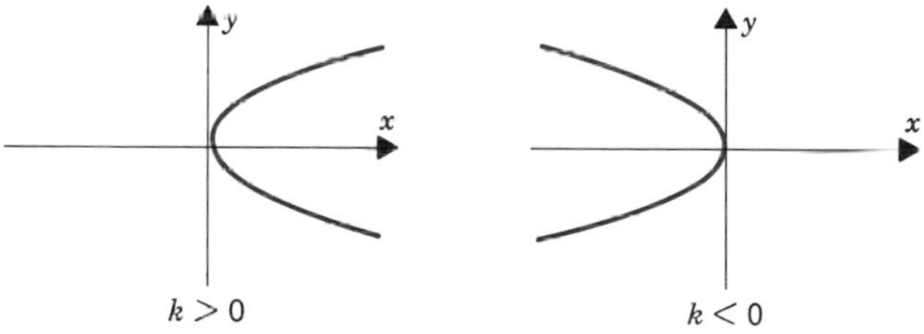

$x^2 = ky$

(Parabola)

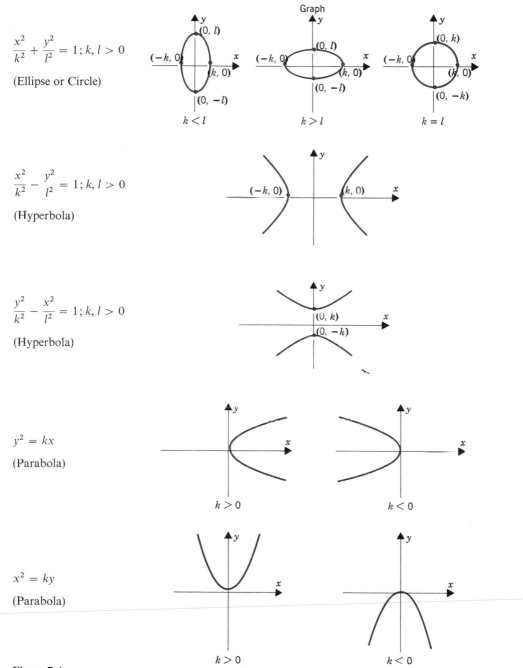

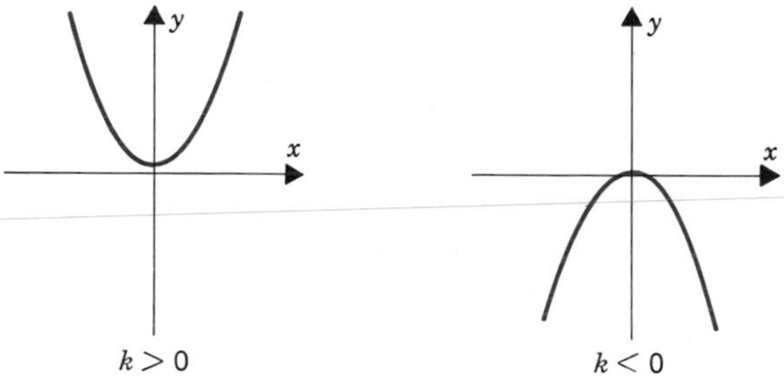

Figure 7.4

Example 17

The equation

$$\frac{x^2}{4} + \frac{y^2}{9} = 1 \text{ is of the form } \frac{x^2}{k^2} + \frac{y^2}{l^2} = 1 \text{ with } k = 2, l = 3$$

Its graph is thus an ellipse in standard position intersecting the x-axis at $(-2, 0)$ and $(2, 0)$ and intersecting the y-axis at $(0, -3)$ and $(0, 3)$.

The equation $x^2 - 8y^2 = -16$ can be rewritten as $y^2/2 - x^2/16 = 1$, which is of the form $y^2/k^2 - x^2/l^2 = 1$ with $k = \sqrt{2}, l = 4$. Its graph is thus a hyperbola in standard position intersecting the y-axis at $(0, -\sqrt{2})$ and $(0, \sqrt{2})$.

The equation $5x^2 + 2y = 0$ can be rewritten as $x^2 = -\frac{2}{5}y$, which is of the form $x^2 = ky$ with $k = -\frac{2}{5}$. Since $k < 0$, its graph is a parabola in standard position opening downward.

Observe that no conic in standard position has an xy-term (called a ***cross-product term***) in its equation; the presence of an xy-term in the equation of a nondegenerate conic indicates that the conic is rotated out of standard position (Fig. 7.5*a*). Also, no conic in standard position has both an x^2 and x term or both a y^2 and y term. The occurrence of either of these pairs in the equation of a nondegenerate conic usually indicates that the conic is translated out of standard position (Fig. 7.5*b*).

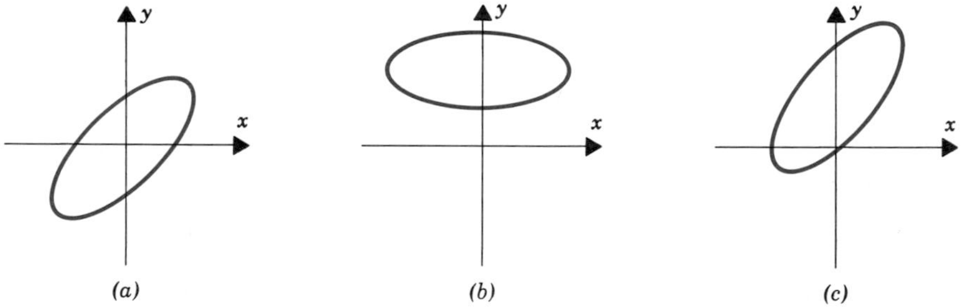

Figure 7.5 (*a*) Rotated. (*b*) Translated. (*c*) Rotated and translated.

One technique for identifying the graph of a nondegenerate conic that is not in standard position consists of rotating and translating the xy-coordinate axes to obtain an $x'y'$-coordinate system relative to which the conic is in standard position. Once this is done, the equation of the conic in the $x'y'$-system will have one of the forms given in Figure 7.4 and can then easily be identified.

Example 18

Since the quadratic equation

$$2x^2 + y^2 - 12x - 4y + 18 = 0$$

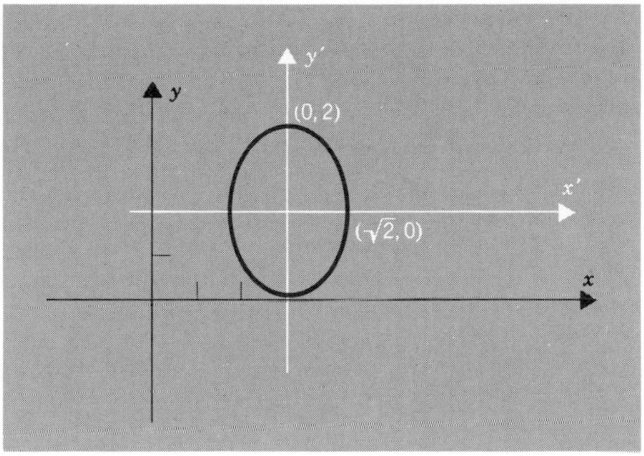

Figure 7.6

contains x^2, x, y^2, and y terms but no cross-product term, its graph is a conic that is translated out of standard position, but not rotated. This conic can be brought into standard position by properly translating the coordinate axes. To do this, first collect x and y terms. This yields

$$(2x^2 - 12x) + (y^2 - 4y) + 18 = 0$$

or

$$2(x^2 - 6x) + (y^2 - 4y) = -18$$

By completing the squares* on the two expressions in parentheses, we obtain

$$2(x^2 - 6x + 9) + (y^2 - 4y + 4) = -18 + 18 + 4$$

or

$$2(x - 3)^2 + (y - 2)^2 = 4 \tag{7.16}$$

If we translate the coordinate axes by means of the translation equations

$$x' = x - 3 \qquad y' = y - 2$$

(Example 3 in Section 3.1), then (7.16) becomes

$$2x'^2 + y'^2 = 4$$

or

$$\frac{x'^2}{2} + \frac{y'^2}{4} = 1$$

which is the equation of an ellipse in standard position in the $x'y'$-system. This ellipse is sketched in Figure 7.6.

* To complete the square on an expression of the form $x^2 + px$, add and subtract the constant $(p/2)^2$ to obtain

$$x^2 + px = x^2 + px + \left(\frac{p}{2}\right)^2 - \left(\frac{p}{2}\right)^2 = \left(x + \frac{p}{2}\right)^2 - \left(\frac{p}{2}\right)^2$$

We now consider the identification of conics that are rotated out of standard position. For the remainder of this text we follow a standard convention of omitting the brackets on all 1 × 1 matrices. Thus the symbol 8 can denote either the number 8 or the 1 × 1 matrix whose entry is the number 8. It will always be possible to tell from the context which is meant. With this convention, (7.15) can be written in the matrix form

$$[x \quad y]\begin{bmatrix} a & b \\ b & c \end{bmatrix}\begin{bmatrix} x \\ y \end{bmatrix} + [d \quad e]\begin{bmatrix} x \\ y \end{bmatrix} + f = 0$$

or

$$\mathbf{x}^t A\mathbf{x} + K\mathbf{x} + f = 0 \tag{7.17}$$

where

$$\mathbf{x} = \begin{bmatrix} x \\ y \end{bmatrix} \qquad A = \begin{bmatrix} a & b \\ b & c \end{bmatrix} \qquad K = [d \quad e]$$

In this notation the quadratic form associated with (7.17) is

$$\mathbf{x}^t A\mathbf{x}$$

The symmetric matrix A is called the ***matrix of the quadratic form*** $\mathbf{x}^t A\mathbf{x}$.

Example 19

The matrices of the quadratic forms

$$3x^2 + 5xy + 7y^2 \qquad \text{and} \qquad 8x^2 - 4y^2$$

are

$$\begin{bmatrix} 3 & \frac{5}{2} \\ \frac{5}{2} & 7 \end{bmatrix} \qquad \text{and} \qquad \begin{bmatrix} 8 & 0 \\ 0 & -4 \end{bmatrix}$$

Consider a conic C with the equation

$$\mathbf{x}^t A\mathbf{x} + K\mathbf{x} + f = 0 \tag{7.18}$$

We now show it is possible to rotate the xy-coordinate axes so that the equation of the conic in the $x'y'$-coordinate system has no cross product term.

Step 1. Find a matrix

$$P = \begin{bmatrix} p_{11} & p_{12} \\ p_{21} & p_{22} \end{bmatrix}$$

which orthogonally diagonalizes A.

Step 2. Interchange the columns of P, if necessary, to make $\det(P) = 1$. This assures that the orthogonal coordinate transformation

$$\mathbf{x} = P\mathbf{x}', \text{ that is, } \begin{bmatrix} x \\ y \end{bmatrix} = P\begin{bmatrix} x' \\ y' \end{bmatrix} \tag{7.19}$$

is a rotation (Section 4.10).

Step 3. To obtain the equation for C in the $x'y'$-system, substitute (7.19) into (7.18). This yields

$$(P\mathbf{x}')^t A(P\mathbf{x}') + K(P\mathbf{x}') + f = 0$$

or

$$\mathbf{x}'^t(P^tAP)\mathbf{x}' + (KP)\mathbf{x}' + f = 0 \qquad (7.20)$$

Since P orthogonally diagonalizes A

$$P^tAP = \begin{bmatrix} \lambda_1 & 0 \\ 0 & \lambda_2 \end{bmatrix}$$

where λ_1 and λ_2 are eigenvalues of A. Thus (7.20) can be rewritten as

$$[x' \quad y']\begin{bmatrix} \lambda_1 & 0 \\ 0 & \lambda_2 \end{bmatrix}\begin{bmatrix} x' \\ y' \end{bmatrix} + [d \quad e]\begin{bmatrix} p_{11} & p_{12} \\ p_{21} & p_{22} \end{bmatrix}\begin{bmatrix} x' \\ y' \end{bmatrix} + f = 0$$

or

$$\lambda_1 x'^2 + \lambda_2 y'^2 + d'x' + e'y' + f = 0$$

(where $d' = dp_{11} + ep_{21}$ and $e' = dp_{12} + ep_{22}$). This equation has no cross-product term.

The following theorem summarizes this discussion.

Theorem 9. (Principal Axes Theorem for $\mathbf{R}^2$). *Let*

$$ax^2 + 2bxy + cy^2 + dx + ey + f = 0$$

be the equation of a conic C, and let

$$\mathbf{x}^t A\mathbf{x} = ax^2 + 2bxy + cy^2$$

be the associated quadratic form. Then the coordinate axes can be rotated so the equation for C in the new $x'y'$-coordinate system has the form

$$\lambda_1 x'^2 + \lambda_2 y'^2 + d'x' + e'y' + f = 0$$

where λ_1 and λ_2 are the eigenvalues of A. The rotation can be accomplished by the substitution

$$\mathbf{x} = P\mathbf{x}'$$

where P orthogonally diagonalizes A and $\det(P) = 1$.

Example 20

Describe the conic C whose equation is $5x^2 - 4xy + 8y^2 - 36 = 0$.

Solution. The matrix form of this equation is

$$\mathbf{x}^t A\mathbf{x} - 36 = 0 \qquad (7.21)$$

where

$$A = \begin{bmatrix} 5 & -2 \\ -2 & 8 \end{bmatrix}$$

The characteristic equation of A is

$$\det(\lambda I - A) = \det\begin{bmatrix} \lambda - 5 & 2 \\ 2 & \lambda - 8 \end{bmatrix} = (\lambda - 9)(\lambda - 4) = 0$$

Thus the eigenvalues of A are $\lambda = 4$ and $\lambda = 9$.

The eigenvectors corresponding to $\lambda = 4$ are the nonzero solutions of

$$\begin{bmatrix} -1 & 2 \\ 2 & -4 \end{bmatrix}\begin{bmatrix} x \\ y \end{bmatrix} = \begin{bmatrix} 0 \\ 0 \end{bmatrix}$$

Solving this system yields

$$\begin{bmatrix} x \\ y \end{bmatrix} = \begin{bmatrix} 2t \\ t \end{bmatrix} = t\begin{bmatrix} 2 \\ 1 \end{bmatrix}$$

Thus

$$\begin{bmatrix} 2 \\ 1 \end{bmatrix}$$

is a basis for the eigenspace corresponding to $\lambda = 4$. Normalizing this vector to obtain an orthonormal basis for this eigenspace, we get

$$\mathbf{v}_1 = \begin{bmatrix} \dfrac{2}{\sqrt{5}} \\ \dfrac{1}{\sqrt{5}} \end{bmatrix}$$

Similarly

$$\mathbf{v}_2 = \begin{bmatrix} -\dfrac{1}{\sqrt{5}} \\ \dfrac{2}{\sqrt{5}} \end{bmatrix}$$

is an orthonormal basis for the eigenspace corresponding to $\lambda = 9$.

Therefore

$$P = \begin{bmatrix} \dfrac{2}{\sqrt{5}} & -\dfrac{1}{\sqrt{5}} \\ \dfrac{1}{\sqrt{5}} & \dfrac{2}{\sqrt{5}} \end{bmatrix}$$

orthogonally diagonalizes A. Moreover, $\det(P) = 1$ so that the orthogonal coordinate transformation

$$\mathbf{x} = P\mathbf{x}' \tag{7.22}$$

is a rotation. Substituting (7.22) into (7.21) yields

$$(P\mathbf{x}')^t A(P\mathbf{x}') - 36 = 0$$

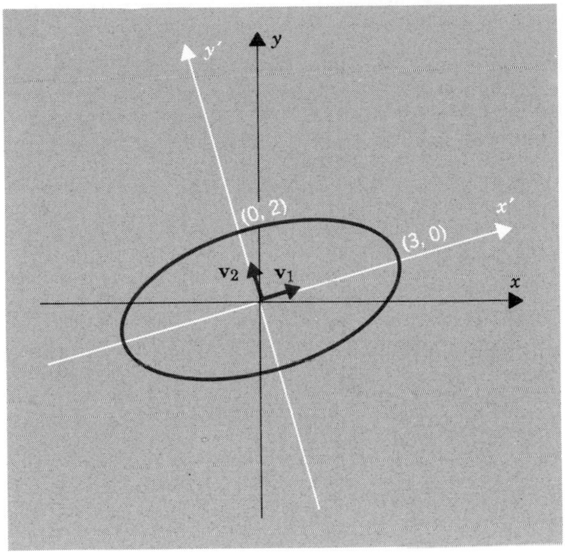

Figure 7.7

or

$$(\mathbf{x}')^t(P^tAP)\mathbf{x}' - 36 = 0$$

Since

$$P^tAP = \begin{bmatrix} 4 & 0 \\ 0 & 9 \end{bmatrix}$$

this equation can be written as

$$\begin{bmatrix} x' & y' \end{bmatrix} \begin{bmatrix} 4 & 0 \\ 0 & 9 \end{bmatrix} \begin{bmatrix} x' \\ y' \end{bmatrix} - 36 = 0$$

or

$$4x'^2 + 9y'^2 - 36 = 0$$

This equation can be rewritten as

$$\frac{x'^2}{9} + \frac{y'^2}{4} = 1$$

which is the equation of the ellipse sketched in Figure 7.7.

Example 21

Describe the conic C whose equation is

$$5x^2 - 4xy + 8y^2 + \frac{20}{\sqrt{5}}x - \frac{80}{\sqrt{5}}y + 4 = 0$$

Solution. The matrix form of this equation is

$$\mathbf{x}^tA\mathbf{x} + K\mathbf{x} + 4 = 0 \tag{7.23}$$

where

$$A = \begin{bmatrix} 5 & -2 \\ -2 & 8 \end{bmatrix} \quad \text{and} \quad K = \begin{bmatrix} \frac{20}{\sqrt{5}} & \frac{-80}{\sqrt{5}} \end{bmatrix}$$

As shown in Example 20

$$P = \begin{bmatrix} \frac{2}{\sqrt{5}} & -\frac{1}{\sqrt{5}} \\ \frac{1}{\sqrt{5}} & \frac{2}{\sqrt{5}} \end{bmatrix}$$

orthogonally diagonalizes A. Substituting $\mathbf{x} = P\mathbf{x}'$ into (7.23) gives

$$(P\mathbf{x}')^t A(P\mathbf{x}') + K(P\mathbf{x}') + 4 = 0$$

or

$$(\mathbf{x}')^t(P^tAP)\mathbf{x}' + (KP)\mathbf{x}' + 4 = 0 \tag{7.24}$$

Since

$$P^tAP = \begin{bmatrix} 4 & 0 \\ 0 & 9 \end{bmatrix} \quad \text{and} \quad KP = \begin{bmatrix} \frac{20}{\sqrt{5}} & \frac{-80}{\sqrt{5}} \end{bmatrix} \begin{bmatrix} \frac{2}{\sqrt{5}} & -\frac{1}{\sqrt{5}} \\ \frac{1}{\sqrt{5}} & \frac{2}{\sqrt{5}} \end{bmatrix} = [-8, -36]$$

(7.24) can be written as

$$4x'^2 + 9y'^2 - 8x' - 36y' + 4 = 0 \tag{7.25}$$

To bring the conic into standard position, the $x'y'$-axes must be translated. Proceeding as in Example 18, we rewrite (7.25) as

$$4(x'^2 - 2x') + 9(y'^2 - 4y') = -4$$

Completing the squares yields

$$4(x'^2 - 2x' + 1) + 9(y'^2 - 4y' + 4) = -4 + 4 + 36$$

or

$$4(x' - 1)^2 + 9(y' - 2)^2 = 36 \tag{7.26}$$

If we translate the coordinate axes by means of the translation equations

$$x'' = x' - 1 \qquad y'' = y' - 2$$

then (7.26) becomes

$$4x''^2 + 9y''^2 = 36$$

or

$$\frac{x''^2}{9} + \frac{y''^2}{4} = 1$$

which is the equation of the ellipse sketched in Figure 7.8.

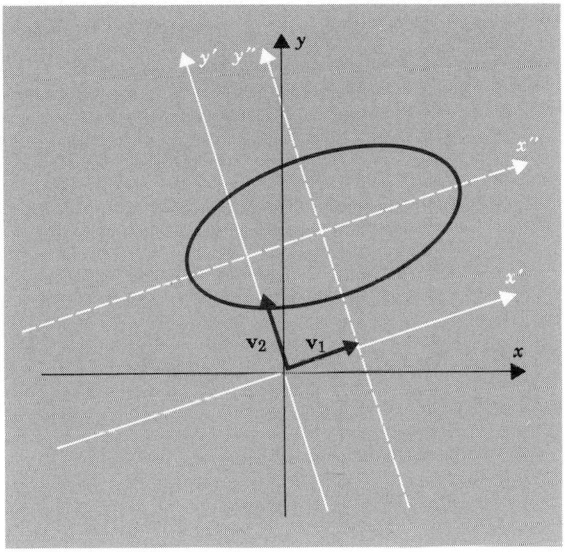

Figure 7.8

EXERCISE SET 7.3

1. Find the quadratic forms associated with the following quadratic equations.
(a) $2x^2 - 3xy + 4y^2 - 7x + 2y + 7 = 0$
(b) $x^2 - xy + 5x + 8y - 3 = 0$
(c) $5xy = 8$
(d) $4x^2 - 2y^2 = 7$
(e) $y^2 + 7x - 8y - 5 = 0$

2. Find the matrices of the quadratic forms in Exercise 1.

3. Express each of the quadratic equations in Exercise 1 in the matrix form
$\mathbf{x}^t A\mathbf{x} + K\mathbf{x} + f = 0$.

4. Name the following conics.
(a) $2x^2 + 5y^2 = 20$
(b) $4x^2 + 9y^2 = 1$
(c) $x^2 - y^2 - 8 = 0$
(d) $4y^2 - 5x^2 = 20$
(e) $x^2 + y^2 - 25 = 0$
(f) $7y^2 - 2x = 0$
(g) $-x^2 = 2y$
(h) $3x - 11y^2 = 0$
(i) $y - x^2 = 0$
(j) $x^2 - 3 = -y^2$

5. In each part a translation will put the conic in standard position. Name the conic and give its equation in the translated coordinate system.
(a) $9x^2 + 4y^2 - 36x - 24y + 36 = 0$
(b) $x^2 - 16y^2 + 8x + 128y = 256$
(c) $y^2 - 8x - 14y + 49 = 0$
(d) $x^2 + y^2 + 6x - 10y + 18 = 0$
(e) $2x^2 - 3y^2 + 6x + 20y = -41$
(f) $x^2 + 10x + 7y = -32$

6. The following nondegenerate conics are rotated out of standard position. In each part rotate the coordinate axes to remove the xy-term. Name the conic and give its equation in the rotated coordinate system.
(a) $2x^2 - 4xy - y^2 + 8 = 0$ (b) $x^2 + 2xy + y^2 + 8x + y = 0$
(c) $5x^2 + 4xy + 5y^2 = 9$ (d) $11x^2 + 24xy + 4y^2 - 15 = 0$

In Exercises 7–12 translate and rotate the coordinate axes, if necessary, to put the conic in standard position. Name the conic and give its equation in the final coordinate system.

7. $9x^2 - 4xy + 6y^2 - 10x - 20y = 5$

8. $3x^2 - 8xy - 12y^2 - 30x - 64y = 0$

9. $2x^2 - 4xy - y^2 - 4x - 8y = -14$

10. $21x^2 + 6xy + 13y^2 - 114x + 34y + 73 = 0$

11. $x^2 - 6xy - 7y^2 + 10x + 2y + 9 = 0$

12. $4x^2 - 20xy + 25y^2 - 15x - 6y = 0$

13. The graph of a quadratic equation in x and y can, in certain cases, be a point, a line, or a pair of lines. These are called ***degenerate*** conics. It is also possible that the equation is not satisfied by any real values of x and y. In such cases the equation has no graph; it is said to represent an ***imaginary conic***. Each of the following represents a degenerate or imaginary conic. Where possible, sketch the graph.
(a) $x^2 - y^2 = 0$ (b) $x^2 + 3y^2 + 7 = 0$
(c) $8x^2 + 7y^2 = 0$ (d) $x^2 - 2xy + y^2 = 0$
(e) $9x^2 + 12xy + 4y^2 - 52 = 0$ (f) $x^2 + y^2 - 2x - 4y = -5$

7.4 APPLICATION TO QUADRIC SURFACES

In this section the techniques of the previous section are extended to quadratic equations in three variables.

An equation of the form

$$ax^2 + by^2 + cz^2 + 2dxy + 2exz + 2fyz + gx + hy + iz + j = 0 \qquad (7.27)$$

where $a, b, \ldots, f$ are not all zero, is called a ***quadratic equation in x, y, and z***; the expression

$$ax^2 + by^2 + cz^2 + 2dxy + 2exz + 2fyz$$

is called the ***associated quadratic form***.

Equation (7.27) can be written in the matrix form

$$[x \quad y \quad z]\begin{bmatrix} a & d & e \\ d & b & f \\ e & f & c \end{bmatrix}\begin{bmatrix} x \\ y \\ z \end{bmatrix} + [g \quad h \quad i]\begin{bmatrix} x \\ y \\ z \end{bmatrix} + j = 0$$

or

$$\mathbf{x}^t A\mathbf{x} + K\mathbf{x} + j = 0$$

where

$$\mathbf{x} = \begin{bmatrix} x \\ y \\ z \end{bmatrix} \qquad A = \begin{bmatrix} a & d & e \\ d & b & f \\ e & f & c \end{bmatrix} \qquad K = [g \quad h \quad i]$$

The symmetric matrix A is called the ***matrix of the quadratic form***

$$\mathbf{x}^t A\mathbf{x} = ax^2 + by^2 + cz^2 + 2dxy + 2exz + 2fyz$$

Example 22

The quadratic form associated with the quadratic equation

$$3x^2 + 2y^2 - z^2 + 4xy + 3xz - 8yz + 7x + 2y + 3z - 7 = 0$$

is

$$3x^2 + 2y^2 - z^2 + 4xy + 3xz - 8yz$$

The matrix of this quadratic form is

$$\begin{bmatrix} 3 & 2 & \frac{3}{2} \\ 2 & 2 & -4 \\ \frac{3}{2} & -4 & -1 \end{bmatrix}$$

Graphs of quadratic equations in x, y, and z are called ***quadrics*** or ***quadric surfaces***. We now give some examples of quadrics and their equations.

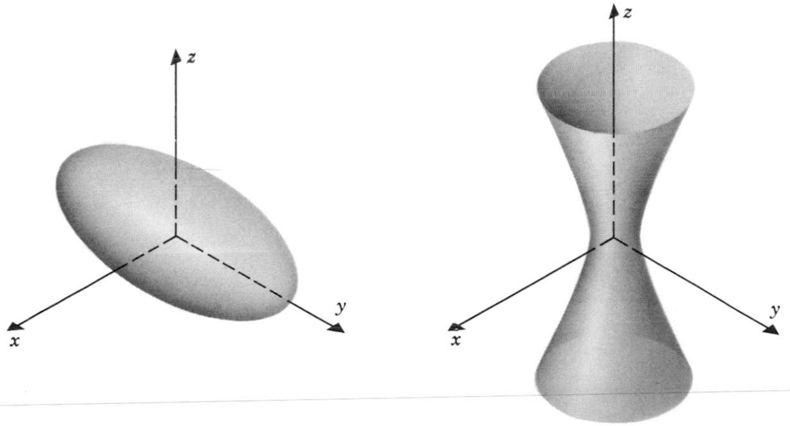

$$\frac{x^2}{l^2} + \frac{y^2}{m^2} + \frac{z^2}{n^2} = 1$$

Ellipsoid

$$\frac{x^2}{l^2} + \frac{y^2}{m^2} - \frac{z^2}{n^2} = 1$$

Hyperboloid of One Sheet

Figure 7.9

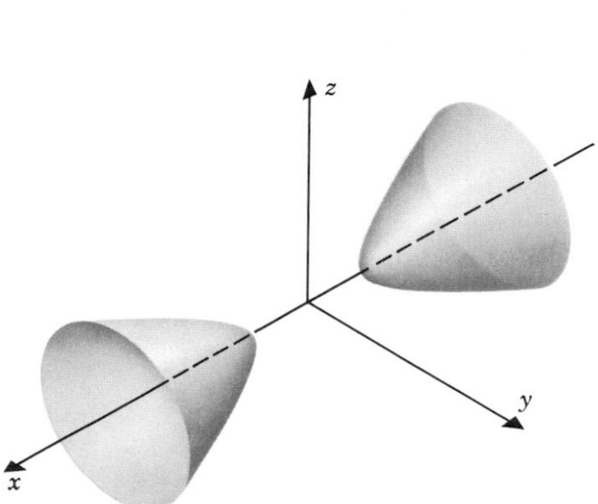

$$\frac{x^2}{l^2}\frac{y^2}{m^2}\frac{z^2}{n^2} = 1$$

Hyperboloid of Two Sheets

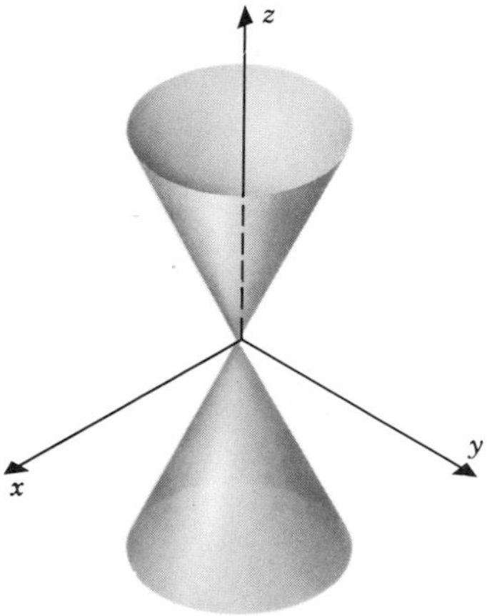

$$\frac{x^2}{l^2} + \frac{y^2}{m^2} - \frac{z^2}{n^2} = 0$$

Elliptic Cone

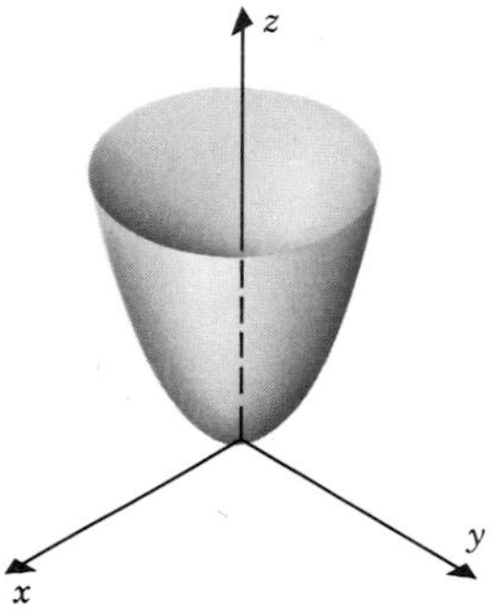

$$\frac{x^2}{l^2} + \frac{y^2}{m^2} - z = 0$$

Elliptic Paraboloid

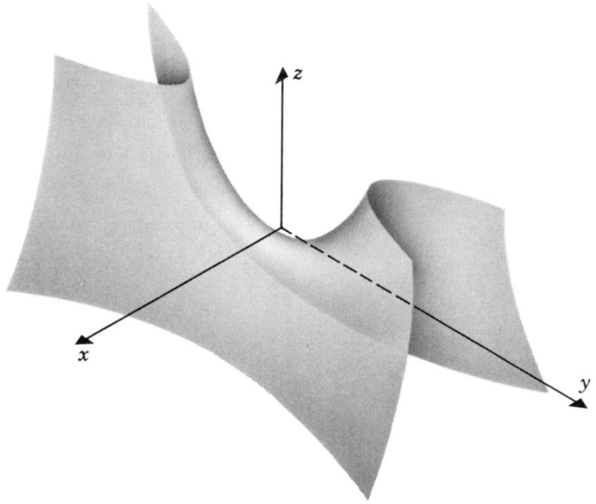

$$\frac{x^2}{l^2} - \frac{y^2}{m^2} + z = 0$$

Hyperbolic Paraboid

Figure 7.9

A quadric whose equation has one of the forms that are given in Figure 7.9 is said to be in ***standard position*** relative to the coordinate axes. The presence of one or more of the cross-product terms xy, xz, and yz in the equation of a nondegenerate quadric indicates that the quadric is rotated out of standard position; the presence of both x^2 and x terms, y^2 and y terms, or z^2 and z terms usually indicates the quadric is translated out of standard position.

Example 23

Describe the quadric surface whose equation is

$$4x^2 + 36y^2 - 9z^2 - 16x - 216y + 304 = 0$$

Solution. Rearranging terms gives

$$4(x^2 - 4x) + 36(y^2 - 6y) - 9z^2 = -304$$

Completing the squares yields

$$4(x^2 - 4x + 4) + 36(y^2 - 6y + 9) - 9z^2 = -304 + 16 + 324$$

or

$$4(x - 2)^2 + 36(y - 3)^2 - 9z^2 = 36$$

or

$$\frac{(x-2)^2}{9} + (y - 3)^2 - \frac{z^2}{4} = 1$$

Translating the axes by means of the translation equations

$$x' = x - 2 \qquad y' = y - 3 \qquad z' = z$$

yields

$$\frac{x'^2}{9} + y'^2 - \frac{z'^2}{4} = 1$$

which is the equation of a hyperboloid of one sheet.

The next result shows that it is always possible to eliminate the cross-product terms from the equation of a quadric by rotating the coordinate axes.

Theorem 10. (Principal Axes Theorem for $\mathbf{R}^3$). *Let*

$$ax^2 + by^2 + cz^2 + 2dxy + 2exz + 2fyz + gx + hy + iz + j = 0 \tag{7.28}$$

be the equation of a quadric Q, and let

$$\mathbf{x}^t A\mathbf{x} = ax^2 + by^2 + cz^2 + 2dxy + 2exz + 2fyz$$

be the associated quadratic form. Then the coordinate axes can be rotated so that the equation of Q in the $x'y'z'$-coordinate system has the form

$$\lambda_1 x'^2 + \lambda_2 y'^2 + \lambda_3 z'^2 + g'x' + h'y' + i'z' + j = 0 \tag{7.29}$$

where λ_1, λ_2, and λ_3 are the eigenvalues of A. The rotation can be accomplished by the substitution

$$\mathbf{x} = P\mathbf{x}'$$

where P orthogonally diagonalizes A and $\det(P) = 1$.

This theorem suggests the following procedure for removing the cross-product terms from a quadratic equation in x, y, and z.

Step 1. Find a matrix P that orthogonally diagonalizes A.

Step 2. Interchange two columns of P, if necessary, to make $\det(P) = 1$. This assures that the orthogonal coordinate transformation

$$\begin{bmatrix} x \\ y \\ z \end{bmatrix} = P \begin{bmatrix} x' \\ y' \\ z' \end{bmatrix} \tag{7.30}$$

is a rotation.

Step 3. Substitute (7.30) into (7.29).

The proof that the new equation has the form given in (7.29) is similar to the proof given in the previous section; it is left as an exercise.

Example 24

Describe the quadric surface whose equation is

$$4x^2 + 4y^2 + 4z^2 + 4xy + 4xz + 4yz - 3 = 0$$

Solution. The matrix form of the above quadratic equation is

$$\mathbf{x}^t A \mathbf{x} - 3 = 0 \tag{7.31}$$

where

$$A = \begin{bmatrix} 4 & 2 & 2 \\ 2 & 4 & 2 \\ 2 & 2 & 4 \end{bmatrix}$$

As shown in Example 13 of Section 6.3, the eigenvalues of A are $\lambda = 2$ and $\lambda = 8$, and A is orthogonally diagonalized by the matrix

$$P = \begin{bmatrix} -\frac{1}{\sqrt{2}} & -\frac{1}{\sqrt{6}} & \frac{1}{\sqrt{3}} \\ \frac{1}{\sqrt{2}} & -\frac{1}{\sqrt{6}} & \frac{1}{\sqrt{3}} \\ 0 & \frac{2}{\sqrt{6}} & \frac{1}{\sqrt{3}} \end{bmatrix}$$

where the first two column vectors in P are eigenvectors corresponding to $\lambda = 2$ and the third column vector is an eigenvector corresponding to $\lambda = 8$.

Since $\det(P) = 1$ (verify), the orthogonal coordinate transformation

$$\mathbf{x} = P\mathbf{x}', \text{ that is, } \begin{bmatrix} x \\ y \\ z \end{bmatrix} = P \begin{bmatrix} x' \\ y' \\ z' \end{bmatrix} \tag{7.32}$$

is a rotation.

Substituting (7.32) into (7.31) yields

$$(P\mathbf{x}')^t A(P\mathbf{x}') - 3 = 0$$

or equivalently

$$(\mathbf{x}')^t(P^t AP)\mathbf{x}' - 3 = 0 \tag{7.33}$$

Since

$$P^t AP = \begin{bmatrix} 2 & 0 & 0 \\ 0 & 2 & 0 \\ 0 & 0 & 8 \end{bmatrix}$$

(7.33) becomes

$$\begin{bmatrix} x' & y' & z' \end{bmatrix} \begin{bmatrix} 2 & 0 & 0 \\ 0 & 2 & 0 \\ 0 & 0 & 8 \end{bmatrix} \begin{bmatrix} x' \\ y' \\ z' \end{bmatrix} - 3 = 0$$

or

$$2x'^2 + 2y'^2 + 8z'^2 = 3$$

This can be rewritten as

$$\frac{x'^2}{3/2} + \frac{y'^2}{3/2} + \frac{z'^2}{3/8} = 1$$

which is the equation of an ellipsoid.

EXERCISE SET 7.4

1. Find the quadratic forms associated with the following quadratic equations.
(a) $x^2 + 2y^2 - z^2 + 4xy - 5yz + 7x + 2z = 3$
(b) $3x^2 + 7z^2 + 2xy - 3xz + 4yz - 3x = 4$
(c) $xy + xz + yz = 1$
(d) $x^2 + y^2 - z^2 = 7$
(e) $3z^2 + 3xz - 14y + 9 = 0$
(f) $2z^2 + 2xz + y^2 + 2x - y + 3z = 0$

2. Find the matrices of the quadratic forms in Exercise 1.

3. Express each of the quadratic equations in Exercise 1 in the matrix form $\mathbf{x}^t A\mathbf{x} + K\mathbf{x} + f = 0$.

4. Name the following quadrics.
(a) $36x^2 + 9y^2 + 4z^2 - 36 = 0$
(b) $2x^2 + 6y^2 - 3z^2 = 18$
(c) $6x^2 - 3y^2 - 2z^2 - 6 = 0$
(d) $9x^2 + 4y^2 - z^2 = 0$
(e) $16x^2 + y^2 = 16z$
(f) $7x^2 - 3y^2 + z = 0$
(g) $x^2 + y^2 + z^2 = 25$

5. In each part determine the translation equations that will put the quadric in standard position. Name the quadric.
(a) $9x^2 + 36y^2 + 4z^2 - 18x - 144y - 24z + 153 = 0$
(b) $6x^2 + 3y^2 - 2z^2 + 12x - 18y - 8z = -7$
(c) $3x^2 - 3y^2 - z^2 + 42x + 144 = 0$
(d) $4x^2 + 9y^2 - z^2 - 54y - 50z = 544$
(e) $x^2 + 16y^2 + 2x - 32y - 16z - 15 = 0$
(f) $7x^2 - 3y^2 + 126x + 72y + z + 135 = 0$
(g) $x^2 + y^2 + z^2 - 2x + 4y - 6z = 11$

6. In each part find a rotation $\mathbf{x} = P\mathbf{x}'$ that removes the cross-product terms. Name the quadric and give its equation in the $x'y'z'$-system.
(a) $2x^2 + 3y^2 + 23z^2 + 7xz + 150 = 0$
(b) $4x^2 + 4y^2 + 4z^2 + 4xy + 4xz + 4yz - 5 = 0$
(c) $144x^2 + 100y^2 + 81z^2 - 216xz - 540x - 720z = 0$
(d) $2xy + z = 0$

In Exercises 7–10 translate and rotate the coordinate axes to put the quadric in standard position. Name the quadric and give its equation in the final coordinate system.

7. $2xy + 2xz + 2yz - 6x - 6y - 4z = -9$

8. $7x^2 + 7y^2 + 10z^2 - 2xy - 4xz + 4yz - 12x + 12y + 60z = 24$

9. $2xy - 6x + 10y + z - 31 = 0$

10. $2x^2 + 2y^2 + 5z^2 - 4xy - 2xz + 2yz + 10x - 26y - 2z = 0$

11. Prove Theorem 10.

8 Introduction to Numerical Methods of Linear Algebra

8.1 GAUSSIAN ELIMINATION WITH PIVOTAL CONDENSATION

In this section we discuss some practical aspects of solving systems of n linear equations in n unknowns. In practice, systems of linear equations are often solved on digital computers. Since computers are limited in the number of decimal places they can carry, they round off or truncate most numerical quantities. For example, a computer designed to store eight decimal places might record 2/3 either as .66666667 (rounded off) or .66666666 (truncated). In either case, an error is introduced that we shall call ***rounding error***.

The main practical considerations in solving systems of linear equations on digital computers are:

1. Minimizing inaccuracies due to rounding errors.
2. Minimizing the computer time (and thus cost) needed to obtain the solution.

Except when the coefficient matrix has a specialized structure (for example, a large number of zeros), Gaussian elimination is usually the best method for solving the system. In this section we introduce a variation of Gaussian elimination designed to minimize the effect of rounding error.

Most computer arithmetic is performed using ***normalized floating point numbers***. This means the numbers are expressed in the form*

$$\pm M \times 10^k \tag{8.1}$$

* Most computers convert decimal numbers (base 10) to binary numbers (base 2). For simplicity, however, we shall think in terms of decimals.

where k is an integer and M is a fraction that satisfies

$$.1 \leq M < 1$$

The fraction M is called the ***mantissa***.

Example 1

The following numbers are expressed in normalized floating point form.

$$73 = .73 \times 10^2$$
$$-.000152 = -.152 \times 10^{-3}$$
$$1{,}579 = .1579 \times 10^4$$
$$-1/4 = -.25 \times 10^0$$

The number of decimal places in the mantissa and the allowable size of the exponent k in (8.1) depend on the computer being used. For example, the IBM 360 stores the equivalent of seven decimal digits in the mantissa and allows 10^k to range from 10^{-75} to 10^{75}. A computer that uses n decimal places in the mantissa is said to round off numbers to ***n* significant digits**.

Example 2

The following numbers are rounded off to three significant digits.

Number	*Normalized Floating Point Form*	*Rounded Value*
7/3	$.233 \times 10^1$	2.33
1,758	$.176 \times 10^4$	1,760
.0000092143	$.921 \times 10^{-5}$	.00000921
$-.12$	$-.120 \times 10^0$	$-.12$
13.850	$.138 \times 10^2$	13.8
$-.08495$	$-.850 \times 10^{-1}$	$-.085$

(If, as in the last two cases, the portion of decimal to be discarded in the rounding process is exactly half a unit, we shall adopt the convention of rounding so that the last retained digit is even. In practice, the treatment of this situation varies from computer to computer.)

We shall now introduce a variation of Gaussian elimination called ***pivotal condensation*** or ***Gaussian elimination with pivoting***; it is designed to minimize the cumulative effect of rounding error in solving n linear equations in n unknowns. We assume that the system has a unique solution. As we describe each step, we shall illustrate the idea using the augmented matrix for the system

$$\begin{aligned} 3x_1 + 2x_2 - x_3 &= 1 \\ 6x_1 + 6x_2 + 2x_3 &= 12 \\ 3x_1 - 2x_2 + x_3 &= 11 \end{aligned}$$

Step 1. In the lefmost column find an entry that has the largest absolute value. This is called the ***pivot entry***.

$$\begin{bmatrix} 3 & 2 & -1 & 1 \\ \textcircled{6} & 6 & 2 & 12 \\ 3 & -2 & 1 & 11 \end{bmatrix}$$

Pivot entry

Leftmost column

Step 2. Perform a row interchange, if necessary, to bring the pivot entry to the top of the column

$$\begin{bmatrix} 6 & 6 & 2 & 12 \\ 3 & 2 & -1 & 1 \\ 3 & -2 & 1 & 11 \end{bmatrix}$$

The first and second rows of the previous matrix were interchanged

Step 3. If the pivot entry is a, multiply the top row by $1/a$.

$$\begin{bmatrix} 1 & 1 & \frac{1}{3} & 2 \\ 3 & 2 & -1 & 1 \\ 3 & -2 & 1 & 11 \end{bmatrix}$$

The first row of the previous matrix was multiplied by 1/6

Step 4. Add suitable multiples of the top row to the rows below so that in the column located in Step 1, all the entries below the top become zeros.

$$\begin{bmatrix} 1 & 1 & \frac{1}{3} & 2 \\ 0 & -1 & -2 & -5 \\ 0 & -5 & 0 & 5 \end{bmatrix}$$

-3 times the first row of the previous matrix was added to the second and third rows

Step 5. Cover the top row in the matrix and begin again with Step 1, applied to the submatrix that remains. Continue in this way until the *entire* matrix is in row-echelon form.

$$\begin{bmatrix} 1 & 1 & \frac{1}{3} & 2 \\ 0 & -1 & -2 & -5 \\ 0 & \textcircled{-5} & 0 & 5 \end{bmatrix}$$

Pivot entry

Leftmost nonzero column in the *sub*matrix

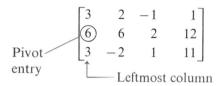

$$\begin{bmatrix} 1 & 1 & \frac{1}{3} & 2 \\ 0 & -5 & 0 & 5 \\ 0 & -1 & -2 & -5 \end{bmatrix}$$

The first and second rows of the *sub*matrix were interchanged

$$\begin{bmatrix} 1 & 1 & \frac{1}{3} & 2 \\ 0 & 1 & 0 & -1 \\ 0 & -1 & -2 & -5 \end{bmatrix}$$

The first row of the *sub*matrix was multiplied by $-1/5$

$$\begin{bmatrix} 1 & 1 & \frac{1}{3} & 2 \\ 0 & 1 & 0 & -1 \\ 0 & 0 & -2 & -6 \end{bmatrix}$$

The first row of the *sub*matrix was added to the second row

$$\begin{bmatrix} 1 & 1 & \frac{1}{3} & 2 \\ 0 & 1 & 0 & -1 \\ 0 & 0 & \textcircled{-2} & -6 \end{bmatrix}$$

The first row of the *sub*matrix was covered and we returned again to Step 1

Pivot entry (the circled -2)

Leftmost column in the new submatrix (the third column)

$$\begin{bmatrix} 1 & 1 & \frac{1}{3} & 2 \\ 0 & 1 & 0 & -1 \\ 0 & 0 & 1 & 3 \end{bmatrix}$$

The first row of the *new* submatrix was multiplied by $-1/2$

The entire matrix is now in row-echelon form.

Step 6. Solve the corresponding system of equations by back substitution.

The corresponding system of equations is

$$\begin{aligned} x_1 + x_2 + \tfrac{1}{3}x_3 &= 2 \\ x_2 \qquad &= -1 \\ x_3 &= 3 \end{aligned}$$

Solving by back substitution yields

$$x_3 = 3 \qquad x_2 = -1 \qquad x_1 = 2$$

Since the above computations are exact, this example does not illustrate the effectiveness of pivotal condensation in the reduction of rounding error; the next example does.

Example 3

Solve the following system by Gaussian elimination with pivoting. After each calculation round off the result to three significant digits.

$$\begin{aligned} .00044x_1 + .0003x_2 - .0001x_3 &= .00046 \\ 4x_1 + x_2 + x_3 &= 1.5 \\ 3x_1 - 9.2x_2 - .5x_3 &= -8.2 \end{aligned} \tag{8.2}$$

Solution (with pivoting). The augmented matrix is

$$\begin{bmatrix} .00044 & .0003 & -.0001 & .00046 \\ 4 & 1 & 1 & 1.5 \\ 3 & -9.2 & -.5 & -8.2 \end{bmatrix}$$

To bring the pivot entry to the top of the first column, we interchange the first and second rows; this yields

$$\begin{bmatrix} 4 & 1 & 1 & 1.5 \\ .00044 & .0003 & -.0001 & .00046 \\ 3 & -9.2 & -.5 & -8.2 \end{bmatrix}$$

Dividing each entry in the first row by 4 yields

$$\begin{bmatrix} 1 & .25 & .25 & .375 \\ .00044 & .0003 & -.0001 & .00046 \\ 3 & -9.2 & -.5 & -8.2 \end{bmatrix}$$

Adding $-.00044$ times the first row to the second and -3 times the first row to the third yields (after rounding to three significant digits),

$$\begin{bmatrix} 1 & .25 & .25 & .375 \\ 0 & .000190 & -.00021 & .000295 \\ 0 & -9.95 & -1.25 & -9.32 \end{bmatrix}$$

Interchanging the second and third rows yields

$$\begin{bmatrix} 1 & .25 & .25 & .375 \\ 0 & -9.95 & -1.25 & -9.32 \\ 0 & .000190 & -.00021 & .000295 \end{bmatrix}$$

Dividing each entry in the second row by -9.95 yields

$$\begin{bmatrix} 1 & .25 & .25 & .375 \\ 0 & 1 & .126 & .937 \\ 0 & -.000190 & -.00021 & .000295 \end{bmatrix}$$

Adding .000190 times the second row to the third yields

$$\begin{bmatrix} 1 & .25 & .25 & .375 \\ 0 & 1 & .126 & .937 \\ 0 & 0 & -.000234 & .000117 \end{bmatrix}$$

Dividing each entry in the third row by $-.000234$ yields the row-echelon form

$$\begin{bmatrix} 1 & .25 & .25 & .375 \\ 0 & 1 & .126 & .937 \\ 0 & 0 & 1 & -.5 \end{bmatrix}$$

The corresponding system of equations is

$$\begin{aligned} x_1 + .25x_2 + .25x_3 &= .375 \\ x_2 + .126x_3 &= .937 \\ x_3 &= -.5 \end{aligned}$$

Solving by back substitution yields (to three significant digits)

$$x_1 = .250 \qquad x_2 = 1.00 \qquad x_3 = -.500 \tag{8.3}$$

If (8.2) is solved by Gaussian elimination *without* pivoting and each calculation is rounded to three significant digits, one obtains (details omitted)

$$x_1 = .245 \qquad x_2 = 1.01 \qquad x_3 = -.492 \tag{8.4}$$

Comparing (8.3) and (8.4) to the exact solution

$$x_1 = \tfrac{1}{4} \qquad x_2 = 1 \qquad x_3 = -\tfrac{1}{2}$$

shows that the use of pivoting yields more accurate results.

In spite of the fact that pivotal condensation can reduce the cumulative effect of rounding error, there are certain systems of equations, called ***ill conditioned*** systems, that are so extremely sensitive that even the slightest errors in the coefficients can result in major inaccuracies in the solution. For example, consider the system

$$\begin{aligned} x_1 + x_2 &= -3 \\ x_1 + 1.016x_2 &= 5 \end{aligned} \tag{8.5}$$

If we assume that this system is to be solved on a computer that rounds off to three significant digits, the computer will store this system as

$$\begin{aligned} x_1 x_2 &= -3 \\ x_1 + 1.02x_2 &= 5 \end{aligned} \tag{8.6}$$

The exact solution to (8.5) is $x_1 = -503$, $x_2 = 500$, and the exact solution to (8.6) is $x_1 = -403$, $x_2 = 400$. Thus a rounding error of only .004 in one coefficient of (8.5) results in a gross error in the solution.

There is little that can be done computationally to avoid large errors in the solutions of ill-conditioned linear systems. However, in physical problems, where ill-conditioned systems arise, it is sometimes possible to reformulate the problem giving rise to the system to avoid ill-conditioning. Some of the texts referenced at the end of this chapter explain how to recognize ill-conditioned systems.

EXERCISE SET 8.1

1. Express the following in normalized floating point form.

(a) $\frac{14}{5}$ (b) 3,452 (c) .000003879

(d) $-.135$ (e) 17.921 (f) $-.0863$

2. Round off the numbers in Exercise 1 to three significant digits.

3. Round off the numbers in Exercise 1 to two significant digits.

In Exercises 4–7 use Gaussian elimination with pivoting to solve the system exactly. Check your work by using Gaussian elimination without pivoting to solve the system.

4. $$\begin{aligned} 3x_1 + x_2 &= -2 \\ -5x_1 + x_2 &= 22 \end{aligned}$$

5. $$\begin{aligned} x_1 + x_2 + x_3 &= 6 \\ 2x_1 - x_2 + 4x_3 &= 12 \\ -3x_1 + 2x_2 - x_3 &= -4 \end{aligned}$$

6. $$\begin{aligned} 2x_1 + 3x_2 - x_3 &= 5 \\ 4x_1 + 4x_2 - 3x_3 &= 3 \\ 2x_1 - 3x_2 + x_3 &= -1 \end{aligned}$$

7. $$\begin{aligned} 5x_1 + 6x_2 - x_3 + 2x_4 &= -3 \\ 2x_1 - x_2 + x_3 + x_4 &= 0 \\ -8x_1 + x_2 + 2x_3 - x_4 &= 3 \\ 5x_1 + 2x_2 + 3x_3 - x_4 &= 4 \end{aligned}$$

In Exercises 8–9 solve the system by Gaussian elimination with pivoting. Round off all calculations to three significant digits.

8. $$\begin{aligned} .21x_1 + .33x_2 &= .54 \\ .70x_1 + .24x_2 &= .94 \end{aligned}$$

9. $$\begin{aligned} .11x_1 - .13x_2 + .20x_3 &= -.02 \\ .10x_1 + .36x_2 + .45x_3 &= .25 \\ .50x_1 - .01x_2 + .30x_3 &= -.70 \end{aligned}$$

10. Solve

$$\begin{aligned} .0001x_1 + x_2 &= 1 \\ x_1 + x_2 &= 2 \end{aligned}$$

by Gaussian elimination with and without pivoting. Round off all computations to three significant digits. Compare the results with the exact solution.

8.2 THE GAUSS-SEIDEL AND JACOBI METHODS

Gaussian elimination is usually the best technique for solving a system of linear equations. However, when the number of equations is large, say 100 or more, and when the matrix has many zeros, other methods may be more effective; in this section we study two such methods.

Consider a system of n linear equations in n unknowns

$$\begin{aligned} a_{11}x_1 + a_{12}x_2 + \cdots + a_{1n}x_n &= b_1 \\ a_{21}x_1 + a_{22}x_2 + \cdots + a_{2n}x_n &= b_2 \\ \vdots \qquad \vdots \qquad\qquad \vdots \quad &\quad \vdots \\ a_{n1}x_1 + a_{n2}x_2 + \cdots + a_{nn}x_n &= b_n \end{aligned} \tag{8.7}$$

We shall assume that the diagonal entries $a_{11}, a_{22}, \ldots, a_{nn}$ are nonzero and that the system has exactly one solution.

The first method we shall discuss is called ***Jacobi iteration*** or the ***method of simultaneous displacements***. To begin, rewrite system (8.7) by solving the first equation for x_1 in terms of the remaining unknowns, solving the second equation for x_2 in terms of the remaining unknowns, solving the third equation for x_3 in terms of the remaining unknowns, etc. This yields

$$\begin{aligned}
x_1 &= \frac{1}{a_{11}}(b_1 - a_{12}x_2 - a_{13}x_3 - \cdots - a_{1n}x_n) \\
x_2 &= \frac{1}{a_{22}}(b_2 - a_{21}x_1 - a_{23}x_3 - \cdots - a_{2n}x_n) \\
&\vdots \\
x_n &= \frac{1}{a_{nn}}(b_n - a_{n1}x_1 - a_{n2}x_2 - \cdots - a_{nn-1}x_{n-1})
\end{aligned} \tag{8.8}$$

For example, the system

$$\begin{aligned}
20x_1 + x_2 - x_3 &= 17 \\
x_1 - 10x_2 + x_3 &= 13 \\
-x_1 + x_2 + 10x_3 &= 18
\end{aligned} \tag{8.9}$$

would be rewritten

$$\begin{aligned}
x_1 &= \tfrac{17}{20} - \tfrac{1}{20}x_2 + \tfrac{1}{20}x_3 \\
x_2 &= -\tfrac{13}{10} + \tfrac{1}{10}x_1 + \tfrac{1}{10}x_3 \\
x_3 &= \tfrac{18}{10} + \tfrac{1}{10}x_1 - \tfrac{1}{10}x_2
\end{aligned}$$

or

$$\begin{aligned}
x_1 &= .850 - .05x_2 + .05x_3 \\
x_2 &= -1.3 + .1x_1 + .1x_3 \\
x_3 &= 1.8 + .1x_1 - .1x_2
\end{aligned} \tag{8.10}$$

If an approximation to the solution of (8.7) is known, and these approximate values are substituted into the right side of (8.8), it is often the case that the values of $x_1, x_2, \ldots, x_n$ that result on the left side form an even better approximation to the solution. This observation is the key to the Jacobi method.

To solve system (8.7) by Jacobi iteration, make an initial approximation to the solution. When no better choice is available use $x_1 = 0, x_2 = 0, x_3 = 0, \ldots$.

Substitute this initial approximation into the right side of (8.8) and use the values of $x_1, x_2, \ldots$ that result on the left side as a new approximation to the solution.

For example, to solve (8.9) by the Jacobi method, we would substitute the initial approximation $x_1 = 0, x_2 = 0, x_3 = 0$ into the right side of (8.10), and calculate the new approximation

$$x_1 = .850 \qquad x_2 = -1.3 \qquad x_3 = 1.8 \tag{8.11}$$

To improve the approximation, we would repeat the substitution process. For example, in solving (8, 9), we would substitute the approximation (8.11) into the right side of (8.10) to obtain the next approximation

$$\begin{aligned}
x_1 &= .850 - .05(-1.3) + .05(1.8) = 1.005 \\
x_2 &= -1.3 + .1(.850) + .1(1.8) = -1.035 \\
x_3 &= 1.8 + .1(.850) - .1(-1.3) = 2.0105
\end{aligned}$$

In this way a succession of approximations can be generated which, under certain conditions, get closer and closer to the exact solution of the system. In Figure 8.1 we have summarized the results obtained in solving system (8.9) by Jacobi iteration. All computations were rounded off to five significant digits. At the end of six substitutions (called ***iterations***), the exact solution $x_1 = 1, x_2 = -1, x_3 = 2$ is accurately known to five significant digits.

	Initial approx-imation	First approx-imation	Second approx-imation	Third approx-imation	Fourth approx-imation	Fifth approx-imation	Sixth approx-imation
x_1	0	.850	1.005	1.0025	1.0001	.99997	1.0000
x_2	0	-1.3	-1.035	$-.9980$	$-.99935$	$-.99999$	-1.0000
x_3	0	1.8	2.015	2.004	2.0000	1.9999	2.0000

Figure 8.1

We now discuss a minor modification of the Jacobi method that often reduces the number of iterations needed to obtain a given degree of accuracy. The technique is called ***Gauss-Seidel iteration*** or the method of ***successive displacements***.

In each iteration of the Jacobi method, the new approximation is obtained by substituting the previous approximation into the right side of (8.8) and solving for new values of $x_1, x_2, \ldots$. These new x-values are not all computed simultaneously; first x_1 is obtained from the top equation, then x_2 is obtained from the second equation, then x_3, etc. Since the new x-values are generally closer to the exact solution, this suggests that better accuracy might be obtained by using the new x-values as soon as they are known. To illustrate, consider system (8.9). In the first iteration of the Jacobi method, the initial approximation $x_1 = 0, x_2 = 0, x_3 = 0$ was substituted into each equation on the right side of (8.10) to obtain the new approximation

$$x_1 = .850 \qquad x_2 = -1.3 \qquad x_3 = 1.8 \tag{8.12}$$

In the first iteration of the Gauss-Seidel method, the new approximation would be computed as follows. Substitute the initial approximation $x_1 = 0$, $x_2 = 0, x_3 = 0$ into the right side of the first equation in (8.10). This yields the new estimate $x_1 = .850$.

Use this new value of x_1 immediately by substituting

$$x_1 = .850 \qquad x_2 = 0 \qquad x_3 = 0$$

into the right side of the second equation in (8.10). This yields the new estimate $x_2 = -1.215$.

Use this new value of x_2 immediately by substituting

$$x_1 = .850 \qquad x_2 = -1.215 \qquad x_3 = 0$$

into the right side of the third equation in (8.10). This yields the new estimate $x_3 = 2.0065$.

Thus at the end of the first iteration of the Gauss-Seidel method the new approximation is

$$x_1 = .850 \qquad x_2 = -1.215 \qquad x_3 = 2.0065 \tag{8.13}$$

The computations for the second iteration would be carried out as follows.

Substituting (8.13) into the right side of the first equation in (8.10) and rounding off to five significant digits yields

$$x_1 = .850 - .05(-1.215) + .05(2.0065) = 1.0111$$

Substituting

$$x_1 = 1.0111 \qquad x_2 = -1.215 \qquad x_3 = 2.0065$$

into the right side of the second equation in (8.10) and rounding off to five signiffcant digits yields

$$x_2 = -1.3 + .1(1.0111) + .1(2.0065) = -.99824$$

Substituting

$$x_1 = 1.0111 \qquad x_2 = -.99824 \qquad x_3 = 2.0065$$

into the right side of the third equation in (8.10) and rounding off to five significant digits yields

$$x_3 = 1.8 + .1(1.0111) - .1(-.99824) = 2.0009$$

Thus, at the end of the second iteration of the Gauss-Seidel method, the new approximation is

$$x_1 = 1.0111 \qquad x_2 = -.99824 \qquad x_3 = 2.0009$$

In Figure 8.2 we have summarized the results obtained using four iterations of the Gauss-Seidel method to solve (8.9). All numbers were rounded off to five significant digits.

Comparing the tables in Figures 8.1 and 8.2, we see that the Gauss-Seidel method produces the solution of (8.9) (accurate to five significant digits) in four iterations while six iterations are needed to attain the same accuracy with the

Figure 8.2

	Initial approximation	*First approximation*	*Second approximation*	*Third approximation*	*Fourth approximation*
x_1	0	.850	1.0111	.99995	1.0000
x_2	0	−1.215	−.99824	−.99992	−1.0000
x_3	0	2.0065	2.0009	2.0000	2.0000

Jacobi method. It should not be concluded from this example, however, that the Gauss-Seidel method is always better than the Jacobi. Although it may seem surprising, there are examples where the Jacobi method is better than the Gauss-Seidel method.

The Gauss-Seidel and Jacobi methods do not always work. In some cases, one or both of these methods can fail to produce a good approximation to the solution, regardless of the number of iterations performed. In such cases the approximations are said to ***diverge***. If by performing sufficiently many iterations, the solution can be obtained to any desired degree of accuracy, the approximations are said to ***converge***.

We conclude this section by discussing a condition that will insure that the approximations generated by the two methods converge.

A square matrix

$$A = \begin{bmatrix} a_{11} & a_{12} & \cdots & a_{1n} \\ a_{21} & a_{22} & \cdots & a_{2n} \\ \vdots & \vdots & & \vdots \\ a_{n1} & a_{n2} & \cdots & a_{nn} \end{bmatrix}$$

is called ***strictly diagonally dominant*** if the absolute value of each diagonal entry is greater than the sum of the absolute values of the remaining entries in the same row; that is,

$$\begin{aligned} |a_{11}| &> |a_{12}| + |a_{13}| + \cdots + |a_{1n}| \\ |a_{22}| &> |a_{21}| + |a_{23}| + \cdots + |a_{2n}| \\ &\vdots \\ |a_{nn}| &> |a_{n1}| + |a_{n2}| + \cdots + |a_{nn-1}| \end{aligned}$$

Example 4

$$\begin{bmatrix} 7 & -2 & 3 \\ 4 & 1 & -6 \\ 5 & 12 & -4 \end{bmatrix}$$

is not diagonally dominant since in the second row, $|1|$ is not greater than $|4| + |-6|$, and in the third row $|-4|$ is not greater than $|5| + |12|$.

If the second and third rows are interchanged, however, the resulting matrix

$$\begin{bmatrix} 7 & -2 & 3 \\ 5 & 12 & -4 \\ 4 & 1 & -6 \end{bmatrix}$$

is strictly diagonally dominant since

$$\begin{aligned} |7| &> |-2| + |3| \\ |12| &> |5| + |-4| \\ |-6| &> |4| + |1| \end{aligned}$$

It can be shown that if A is strictly diagonally dominant, then the Gauss-Seidel and Jacobi approximations to the solutions $A\mathbf{x} = \mathbf{b}$ both converge.

EXERCISE SET 8.2

In Exercises 1–4 solve the systems by Jacobi iteration. Start with $x_1 = 0$, $x_2 = 0$. Use four iterations and round off the computations to three significant digits. Compare your results to the exact solutions.

1. $\begin{aligned} 2x_1 + x_2 &= 7 \\ x_1 - 2x_2 &= 1 \end{aligned}$

2. $\begin{aligned} 3x_1 - x_2 &= 5 \\ 2x_1 + 3x_2 &= -4 \end{aligned}$

3. $\begin{aligned} 5x_1 - 2x_2 &= -13 \\ x_1 + 7x_2 &= -10 \end{aligned}$

4. $\begin{aligned} .4x_1 + .1x_2 &= .2 \\ .3x_1 + .7x_2 &= 1.4 \end{aligned}$

In Exercises 5–8 solve the systems by Gauss-Seidel iteration. Start with $x_1 = 0$, $x_2 = 0$. Use three iterations and round off the computations to three significant digits. Compare your results to the exact solutions.

5. Solve the system in Exercise 1.

6. Solve the system in Exercise 2.

7. Solve the system in Exercise 3.

8. Solve the system in Exercise 4.

In Exercises 9–10 solve the systems by Jacobi iteration. Start with $x_1 = 0$, $x_2 = 0$, $x_3 = 0$. Use three iterations and round off the computations to three significant digits. Compare your results to the exact solutions.

9. $\begin{aligned} 10x_1 + x_2 + 2x_3 &= 3 \\ x_1 + 10x_2 - x_3 &= \tfrac{3}{2} \\ 2x_1 + x_2 + 10x_3 &= -9 \end{aligned}$

10. $\begin{aligned} 20x_1 - x_2 + x_3 &= 20 \\ 2x_1 + 10x_2 - x_3 &= 11 \\ x_1 + x_2 - 20x_3 &= -18 \end{aligned}$

In Exercises 11–12 solve the systems by Gauss-Seidel iteration. Start with $x_1 = 0$, $x_2 = 0$, $x_3 = 0$. Use three iterations and round off the computations to three significant digits. Compare your results to the exact solutions.

11. Solve the system in Exercise 9.

12. Solve the system in Exercise 10.

13. Which of the following are strictly diagonally dominant?

(a) $\begin{bmatrix} 2 & 1 \\ -1 & 4 \end{bmatrix}$ (b) $\begin{bmatrix} 3 & -5 \\ 1 & 2 \end{bmatrix}$

(c) $\begin{bmatrix} 6 & 0 & 1 \\ 3 & 5 & 3 \\ 0 & 0 & 1 \end{bmatrix}$ (d) $\begin{bmatrix} 4 & 1 & 2 \\ 0 & 3 & 2 \\ 4 & 1 & -7 \end{bmatrix}$ (e) $\begin{bmatrix} 5 & 1 & 2 & 0 \\ 3 & -7 & 2 & 1 \\ 0 & 2 & 5 & 1 \\ 1 & 1 & 2 & -5 \end{bmatrix}$

14. Consider the system

$$\begin{aligned} x_1 + 3x_2 &= 4 \\ x_1 - x_2 &= 0 \end{aligned}$$

(a) Show that the approximations obtained by Jacobi iteration diverge.
(b) Is the coefficient matrix

$$A = \begin{bmatrix} 1 & 3 \\ 1 & -1 \end{bmatrix}$$

strictly diagonally dominant?

15. Show that if one or more of the diagonal entries $a_{11}, a_{22}, \ldots, a_{nn}$ in (8.7) is zero, then it is possible to interchange equations and relabel the unknowns so the diagonal entries in the resulting system are all nonzero.

8.3 APPROXIMATING EIGENVALUES BY THE POWER METHOD

The eigenvalues of a matrix can be found by solving its characteristic equation. In practical problems this method is inefficient. Moreover, in many physical problems only the eigenvalue with the largest absolute value is needed. In this section we discuss a method for approximating this eigenvalue and a corresponding eigenvector. In the next section we shall discuss the approximation of the remaining eigenvalues and eigenvectors.

Definition. An eigenvalue of a matrix A is called the ***dominant eigenvalue*** of A if its absolute value is larger than the absolute values of the remaining eigenvalues. An eigenvector corresponding to the dominant eigenvalue is called a ***dominant eigenvector*** of A.

Example 5

If a 4×4 matrix A has eigenvalues

$$\lambda_1 = -4 \qquad \lambda_2 = 3 \qquad \lambda_3 = -2 \qquad \lambda_4 = 2$$

then $\lambda_1 = -4$ is the dominant eigenvalue since

$$|-4| > |3| \qquad |-4| > |-2| \qquad \text{and} \qquad |-4| > |2|$$

Example 6

A 3×3 matrix A with eigenvalues

$$\lambda_1 = 7 \qquad \lambda_2 = -7 \qquad \lambda_3 = 2$$

has no dominant eigenvalue.

Let A be a diagonalizable $n \times n$ matrix with a dominant eigenvalue. We shall show at the end of this section that if $\mathbf{x}_0$ is an arbitrary nonzero vector in R^n, then the vector

$$A^p\mathbf{x}_0 \tag{8.14}$$

is usually a good approximation to a dominant eigenvector of A when the exponent p is large. The following example illustrates this idea.

Example 7

As shown in Example 2 of Chapter 6, the matrix

$$A = \begin{bmatrix} 3 & 2 \\ -1 & 0 \end{bmatrix}$$

has eigenvalues $\lambda_1 = 2$ and $\lambda_2 = 1$.

The eigenspace corresponding to the dominant eigenvalue $\lambda_1 = 2$ is the solution space of the system

$$(2I - A)\mathbf{x} = \mathbf{0}$$

that is

$$\begin{bmatrix} -1 & -2 \\ 1 & 2 \end{bmatrix}\begin{bmatrix} x_1 \\ x_2 \end{bmatrix} = \begin{bmatrix} 0 \\ 0 \end{bmatrix}$$

Solving this system yields $x_1 = -2t, x_2 = t$. Thus the eigenvectors corresponding to $\lambda_1 = 2$ are the nonzero vectors of the form

$$\mathbf{x} = \begin{bmatrix} -2t \\ t \end{bmatrix} \tag{8.15}$$

We now illustrate a procedure for using (8.14) to estimate a dominant eigenvector of A. To start, we arbitrarily select

$$\mathbf{x}_0 = \begin{bmatrix} 1 \\ 1 \end{bmatrix}$$

Repeatedly multiplying $\mathbf{x}_0$ by A yields

$$A\mathbf{x}_0 = \begin{bmatrix} 3 & 2 \\ -1 & 0 \end{bmatrix}\begin{bmatrix} 1 \\ 1 \end{bmatrix} = \begin{bmatrix} 5 \\ -1 \end{bmatrix}$$

$$A^2\mathbf{x}_0 = A(A\mathbf{x}_0) = \begin{bmatrix} 3 & 2 \\ -1 & 0 \end{bmatrix}\begin{bmatrix} 5 \\ -1 \end{bmatrix} = \begin{bmatrix} 13 \\ -5 \end{bmatrix} = 5\begin{bmatrix} 2.6 \\ -1 \end{bmatrix}$$

$$A^3\mathbf{x}_0 = A(A^2\mathbf{x}_0) = \begin{bmatrix} 3 & 2 \\ -1 & 0 \end{bmatrix}\begin{bmatrix} 13 \\ -5 \end{bmatrix} = \begin{bmatrix} 29 \\ -13 \end{bmatrix} \approx 13\begin{bmatrix} 2.23 \\ -1 \end{bmatrix}$$

$$A^4\mathbf{x}_0 = A(A^3\mathbf{x}_0) = \begin{bmatrix} 3 & 2 \\ -1 & 0 \end{bmatrix}\begin{bmatrix} 29 \\ -13 \end{bmatrix} = \begin{bmatrix} 61 \\ -29 \end{bmatrix} \approx 29\begin{bmatrix} 2.10 \\ -1 \end{bmatrix}$$

$$A^5\mathbf{x}_0 = A(A^4\mathbf{x}_0) = \begin{bmatrix} 3 & 2 \\ -1 & 0 \end{bmatrix}\begin{bmatrix} 61 \\ -29 \end{bmatrix} = \begin{bmatrix} 125 \\ -61 \end{bmatrix} \approx 61\begin{bmatrix} 2.05 \\ -1 \end{bmatrix}$$

$$A^6\mathbf{x}_0 = A(A^5\mathbf{x}_0) = \begin{bmatrix} 3 & 2 \\ -1 & 0 \end{bmatrix}\begin{bmatrix} 125 \\ -61 \end{bmatrix} = \begin{bmatrix} 253 \\ -125 \end{bmatrix} \approx 125\begin{bmatrix} 2.02 \\ -1 \end{bmatrix}$$

$$A^7\mathbf{x}_0 = A(A^6\mathbf{x}_0) = \begin{bmatrix} 3 & 2 \\ -1 & 0 \end{bmatrix}\begin{bmatrix} 253 \\ -125 \end{bmatrix} = \begin{bmatrix} 509 \\ -253 \end{bmatrix} \approx 253\begin{bmatrix} 2.01 \\ -1 \end{bmatrix}$$

It is evident from these calculations that the products are getting closer and closer to scalar multiples of

$$\begin{bmatrix} 2 \\ -1 \end{bmatrix}$$

which is the dominant eigenvector of A obtained by letting $t = -1$ in (8.15). Since a scalar multiple of a dominant eigenvector is also a dominant eigenvector, the above calculations are producing better and better approximations to a dominant eigenvector of A.

We now show how to approximate the dominant eigenvalue once an approximation to a dominant eigenvector is known. Let λ be an eigenvalue of A and $\mathbf{x}$ a corresponding eigenvector. If $\langle\ ,\ \rangle$ denotes the Euclidean inner product, then

$$\frac{\langle \mathbf{x}, A\mathbf{x}\rangle}{\langle \mathbf{x}, \mathbf{x}\rangle} = \frac{\langle \mathbf{x}, \lambda\mathbf{x}\rangle}{\langle \mathbf{x}, \mathbf{x}\rangle} = \frac{\lambda\langle \mathbf{x}, \mathbf{x}\rangle}{\langle \mathbf{x}, \mathbf{x}\rangle} = \lambda$$

Thus, if $\tilde{\mathbf{x}}$ is an approximation to a dominant eigenvector, the dominant eigenvalue λ_1 can be approximated by

$$\lambda_1 \approx \frac{\langle \tilde{\mathbf{x}}, A\tilde{\mathbf{x}}\rangle}{\langle \tilde{\mathbf{x}}, \tilde{\mathbf{x}}\rangle} \tag{8.16}$$

The ratio in (8.16) is called the ***Rayleigh quotient***.*

Example 8

In Example 7 we obtained

$$\tilde{\mathbf{x}} = \begin{bmatrix} 509 \\ -253 \end{bmatrix}$$

as an approximation to a dominant eigenvector. Therefore

$$A\tilde{\mathbf{x}} = \begin{bmatrix} 3 & 2 \\ -1 & 0 \end{bmatrix}\begin{bmatrix} 509 \\ -253 \end{bmatrix} = \begin{bmatrix} 1021 \\ -509 \end{bmatrix}$$

* *John William Strutt Rayleigh* (1842–1919)—British physicist Rayleigh was awarded the Nobel prize in physics in 1904 for his part in the discovery of argon in 1894. His research ranged over almost the entire field of physics, including sound, wave theory, optics, color vision, electrodynamics, electromagnetism, scattering of light, viscosity, and photography.

Substituting in (8.16) we obtain

$$\lambda_1 \approx \frac{\langle \tilde{\mathbf{x}}, A\tilde{\mathbf{x}} \rangle}{\langle \tilde{\mathbf{x}}, \tilde{\mathbf{x}} \rangle} = \frac{(509)(1021) + (-253)(-509)}{(509)(509) + (-253)(-253)} \approx 2.007$$

which is a relatively good approximation to the dominant eigenvalue $\lambda_1 = 2$.

The technique illustrated in Examples 7 and 8 for approximating dominant eigenvalues and eigenvectors is often called the ***power method*** or ***iteration method***.

As illustrated in Example 7 the power method often generates vectors that have inconveniently large components. To remedy this problem it is usual to "scale down" the approximate eigenvector at each step so its components lie between $+1$ and -1. This can be achieved by multiplying the approximate eigenvector by the reciprocal of the component having largest absolute value. To illustrate, in the first step of Example 7, the approximation to the dominant eigenvector is

$$\begin{bmatrix} 5 \\ -1 \end{bmatrix}$$

The component with largest absolute value is 5; thus the scaled down eigenvector is

$$\frac{1}{5}\begin{bmatrix} 5 \\ -1 \end{bmatrix} = \begin{bmatrix} 1 \\ -.2 \end{bmatrix}$$

We now summarize the steps in the ***power method with scaling***.

Step 0. Pick an arbitrary nonzero vector $\mathbf{x}_0$.

Step 1. Compute $A\mathbf{x}_0$ and scale down to obtain the first approximation to a dominant eigenvector. Call it $\mathbf{x}_1$.

Step 2. Compute $A\mathbf{x}_1$ and scale down to obtain the second approximation, $\mathbf{x}_2$.

Step 3. Compute $A\mathbf{x}_2$ and scale down to obtain the third approximation, $\mathbf{x}_3$.

Continuing in this way, a succession, $\mathbf{x}_0, \mathbf{x}_1, \mathbf{x}_2, \ldots$, of better and better approximations to a dominant eigenvector will be obtained.

Example 9

Use the power method with scaling to approximate a dominant eigenvector and the dominant eigenvalue of the matrix A in Example 7.

Solution. We arbitrarily select

$$\mathbf{x}_0 = \begin{bmatrix} 1 \\ 1 \end{bmatrix}$$

as an initial approximation. Multiplying $\mathbf{x}_0$ by A and scaling down yields

$$A\mathbf{x}_0 = \begin{bmatrix} 3 & 2 \\ -1 & 0 \end{bmatrix}\begin{bmatrix} 1 \\ 1 \end{bmatrix} = \begin{bmatrix} 5 \\ -1 \end{bmatrix} \qquad \mathbf{x}_1 = \frac{1}{5}\begin{bmatrix} 5 \\ -1 \end{bmatrix} = \begin{bmatrix} 1 \\ -.2 \end{bmatrix}$$

Multiplying $\mathbf{x}_1$ by A and scaling down yields

$$A\mathbf{x}_1 = \begin{bmatrix} 3 & 2 \\ -1 & 0 \end{bmatrix}\begin{bmatrix} 1 \\ -.2 \end{bmatrix} = \begin{bmatrix} 2.6 \\ -1 \end{bmatrix} \qquad \mathbf{x}_2 = \frac{1}{2.6}\begin{bmatrix} 2.6 \\ -1 \end{bmatrix} = \begin{bmatrix} 1 \\ -.385 \end{bmatrix}$$

From the Rayleigh quotient the first estimate of the dominant eigenvalue is

$$\lambda_1 \approx \frac{\langle \mathbf{x}_1, A\mathbf{x}_1 \rangle}{\langle \mathbf{x}_1, \mathbf{x}_1 \rangle} = \frac{(1)(2.6) + (-.2)(-1)}{(1)(1) + (-.2)(-.2)} = 2.692$$

Multiplying $\mathbf{x}_2$ by A and scaling down yields

$$A\mathbf{x}_2 = \begin{bmatrix} 3 & 2 \\ -1 & 0 \end{bmatrix}\begin{bmatrix} 1 \\ -.385 \end{bmatrix} = \begin{bmatrix} 2.23 \\ -1 \end{bmatrix} \qquad \mathbf{x}_3 = \frac{1}{2.23}\begin{bmatrix} 2.23 \\ -1 \end{bmatrix} \approx \begin{bmatrix} 1 \\ -.448 \end{bmatrix}$$

From the Rayleigh quotient the second estimate of the dominant eigenvalue is

$$\lambda_1 \approx \frac{\langle \mathbf{x}_2, A\mathbf{x}_2 \rangle}{\langle \mathbf{x}_2, \mathbf{x}_2 \rangle} = \frac{(1)(2.23) + (-.385)(-1)}{(1)(1) + (-.385)(-.385)} = 2.278$$

Multiplying $\mathbf{x}_3$ by A and scaling down gives

$$A\mathbf{x}_3 = \begin{bmatrix} 3 & 2 \\ -1 & 0 \end{bmatrix}\begin{bmatrix} 1 \\ -.448 \end{bmatrix} = \begin{bmatrix} 2.104 \\ -1 \end{bmatrix} \qquad \mathbf{x}_4 = \frac{1}{2.104}\begin{bmatrix} 2.104 \\ -1 \end{bmatrix} = \begin{bmatrix} 1 \\ -.475 \end{bmatrix}$$

The third estimate of the dominant eigenvalue is

$$\lambda_1 \approx \frac{\langle \mathbf{x}_3, A\mathbf{x}_3 \rangle}{\langle \mathbf{x}_3, \mathbf{x}_3 \rangle} = \frac{(1)(2.104) + (-.448)(-1)}{(1)(1) + (-.448)(-.448)} = 2.125$$

Continuing in this way, we generate a succession of approximations to a dominant eigenvector and the dominant eigenvalue.

The values calculated above, together with results of further estimates are tabulated in Figure 8.3.

Figure 8.3

Step i	0	1	2	3	4	5	6	7
$x_i =$ the scaled down approximation to a dominant eigenvalue	$\begin{bmatrix} 1 \\ 1 \end{bmatrix}$	$\begin{bmatrix} 1 \\ -.2 \end{bmatrix}$	$\begin{bmatrix} 1 \\ -.385 \end{bmatrix}$	$\begin{bmatrix} 1 \\ -.448 \end{bmatrix}$	$\begin{bmatrix} 1 \\ -.475 \end{bmatrix}$	$\begin{bmatrix} 1 \\ -.488 \end{bmatrix}$	$\begin{bmatrix} 1 \\ -.494 \end{bmatrix}$	$\begin{bmatrix} 1 \\ -.497 \end{bmatrix}$
Ax_i	$\begin{bmatrix} 5 \\ -1 \end{bmatrix}$	$\begin{bmatrix} 2.6 \\ -1 \end{bmatrix}$	$\begin{bmatrix} 2.23 \\ -1 \end{bmatrix}$	$\begin{bmatrix} 2.104 \\ -1 \end{bmatrix}$	$\begin{bmatrix} 2.050 \\ -1 \end{bmatrix}$	$\begin{bmatrix} 2.024 \\ -1 \end{bmatrix}$	$\begin{bmatrix} 2.012 \\ -1 \end{bmatrix}$	—
Approximation to λ_1	—	2.692	2.278	2.125	2.060	2.029	2.014	—

There are no hard and fast rules for determining how many steps to use in the power method. We shall consider one possible procedure that is widely used.

If $\tilde{q}$ denotes an approximation to a quantity q, then the ***relative error*** in the approximation is defined to be

$$\left|\frac{q - \tilde{q}}{q}\right| \tag{8.17}$$

and the ***percentage error*** in the approximation is defined to be

$$\left|\frac{q - \tilde{q}}{q}\right| \times 100\%$$

Example 10

If the exact value of a certain eigenvalue is $\lambda = 5$, and if $\tilde{\lambda} = 5.1$ is an approximation to λ, then the relative error is

$$\left|\frac{\lambda - \tilde{\lambda}}{\lambda}\right| = \left|\frac{5 - 5.1}{5}\right| = |-.02| = .02$$

and the percentage error is

$$(.02) \times 100\% = 2\%$$

In the power method one would ideally like to decide in advance the relative error E that can be tolerated in the eigenvalue, and then stop the computations once the relative error is less than E. Thus, if $\tilde{\lambda}(i)$ denotes the approximation to the dominant eigenvalue λ_1 at the ith step, the computations would be stopped once the condition

$$\left|\frac{\lambda_1 - \tilde{\lambda}(i)}{\lambda_1}\right| < E$$

is satisfied. Unfortunately it is not possible to carry out this idea since the exact value λ_1 is unknown. To remedy this, it is usual to estimate λ_1 by $\tilde{\lambda}(i)$ and stop the computations at the ith step if

$$\left|\frac{\tilde{\lambda}(i) - \tilde{\lambda}(i - 1)}{\tilde{\lambda}(i)}\right| < E \tag{8.18}$$

The quantity on the left side of (8.18) is called the ***estimated relative error***. When multiplied by 100%, it is called the ***estimated percentage error***.

Example 11

In Example 9, how many steps should be used to assure that the estimated percentage error in the dominant eigenvalue is less than 2%?

Solution. Let $\tilde{\lambda}(i)$ denote the approximation to the dominant eigenvalue at the ith step. Thus from Figure 8.3

$$\tilde{\lambda}(1) = 2.692, \qquad \tilde{\lambda}(2) = 2.278, \qquad \tilde{\lambda}(3) = 2.125, \quad \text{etc.}$$

From (8.18) the estimated relative error after two steps is

$$\left|\frac{\tilde{\lambda}(2) - \tilde{\lambda}(1)}{\tilde{\lambda}(2)}\right| = \left|\frac{2.278 - 2.692}{2.278}\right| \approx |-.182| = .182$$

The estimated percentage error after two steps is therefore 18.2%. The estimated relative error after three steps is

$$\left|\frac{\tilde{\lambda}(3) - \tilde{\lambda}(2)}{\tilde{\lambda}(3)}\right| = \left|\frac{2.125 - 2.278}{2.125}\right| \approx |-.072| = .072$$

and the estimated percentage error is 7.2%. The remaining percentage errors are listed in the table in Figure 8.4. From this table we see that the estimated percentage is less than 2% at the end of the fifth step.

i = step number	2	3	4	5	6
$\tilde{\lambda}(i)$	2.278	2.125	2.060	2.029	2.014
Estimated relative error after i steps	.182	.072	.032	.015	.007
Estimated percentage error after i steps	18.2%	7.2%	3.2%	1.5%	.7%

Figure 8.4

OPTIONAL

We conclude this section with a proof that the power method works when A is a diagonalizable matrix with a dominant eigenvalue.

Let A be a diagonalizable $n \times n$ matrix. By Theorem 2 in Section 6.2, A has n linearly independent eigenvectors $\mathbf{v}_1, \mathbf{v}_2, \ldots, \mathbf{v}_n$. Let $\lambda_1, \lambda_2, \ldots, \lambda_n$ be the corresponding eigenvalues, and assume

$$|\lambda_1| > |\lambda_2| \geq \cdots \geq |\lambda_n| \tag{8.19}$$

By Theorem 9(*a*) in Section 4.5 the eigenvectors $\mathbf{v}_1, \mathbf{v}_2, \ldots, \mathbf{v}_n$ form a basis for R^n; thus an arbitrary vector $\mathbf{x}_0$ in R^n can be expressed in the form:

$$\mathbf{x}_0 = k_1\mathbf{v}_1 + k_2\mathbf{v}_2 + \cdots + k_n\mathbf{v}_n \tag{8.20}$$

Multiplying both sides on the left by A gives

$$\begin{aligned} A\mathbf{x}_0 &= A(k_1\mathbf{v}_1 + k_2\mathbf{v}_2 + \cdots + k_n\mathbf{v}_n) \\ &= k_1(A\mathbf{v}_1) + k_2(A\mathbf{v}_2) + \cdots + k_n(A\mathbf{v}_n) \\ &= k_1\lambda_1\mathbf{v}_1 + k_2\lambda_2\mathbf{v}_2 + \cdots + k_n\lambda_n\mathbf{v}_n \end{aligned}$$

Multiplying by A again gives

$$\begin{aligned} A^2\mathbf{x}_0 &= A(k_1\lambda_1\mathbf{v}_1 + k_2\lambda_2\mathbf{v}_2 + \cdots + k_n\lambda_n\mathbf{v}_n) \\ &= k_1\lambda_1(A\mathbf{v}_1) + k_2\lambda_2(A\mathbf{v}_2) + \cdots + k_n\lambda_n(A\mathbf{v}_n) \\ &= k_1\lambda_1^2\mathbf{v}_1 + k_2\lambda_2^2\mathbf{v}_2 + \cdots + k_n\lambda_n^2\mathbf{v}_n \end{aligned}$$

Continuing, we would obtain after p multiplications by A

$$A^p\mathbf{x}_0 = k_1\lambda_1^p\mathbf{v}_1 + k_2\lambda_2^p\mathbf{v}_2 + \cdots + k_n\lambda_n^p\mathbf{v}_n \tag{8.21}$$

Since $\lambda_1 \neq 0$ (see 8.19), (8.21) can be rewritten as

$$A^p\mathbf{x}_0 = \lambda_1{}^p\left(k_1\mathbf{v}_1 + k_2\left(\frac{\lambda_2}{\lambda_1}\right)^p\mathbf{v}_2 + \cdots + k_n\left(\frac{\lambda_n}{\lambda_1}\right)^p\mathbf{v}_n\right) \tag{8.22}$$

It follows from (8.19) that

$$\frac{\lambda_2}{\lambda_1}, \ldots, \frac{\lambda_n}{\lambda_1}$$

are all less than one in absolute value; thus $(\lambda_2/\lambda_1)^p, \ldots, (\lambda_n/\lambda_1)^p$ get steadily closer to zero as p increases, and from (8.22), the approximation

$$A^p\mathbf{x}_0 \approx \lambda_1{}^p k_1\mathbf{v}_1 \tag{8.23}$$

gets better and better.

If $k_1 \neq 0$,* then $\lambda_1^p k_1\mathbf{v}_1$ is a nonzero scalar multiple of the dominant eigenvector $\mathbf{v}_1$; thus, $\lambda_1^p k_1\mathbf{v}_1$ is also a dominant eigenvector. Therefore, by (8.23), $A^p\mathbf{x}_0$ becomes a better and better estimate of a dominant eigenvector as p is increased. ▮

EXERCISE SET 8.3

1. Find the dominant eigenvalue (if it exists).

(a) $\begin{bmatrix} -1 & 4 \\ 1 & -1 \end{bmatrix}$ (b) $\begin{bmatrix} 0 & 1 \\ 4 & 0 \end{bmatrix}$

(c) $\begin{bmatrix} 4 & 2 & 1 \\ 0 & -5 & 3 \\ 0 & 0 & 6 \end{bmatrix}$ (d) $\begin{bmatrix} 1 & -12 & 0 \\ 1 & 0 & 0 \\ 0 & 0 & 3 \end{bmatrix}$

* One cannot usually tell by inspection of the x_0 selected whether $k_1 \neq 0$. If, by accident, $k_1 = 0$, the power method still works in practical problems, since computer roundoff errors generally build up to make k_1 small but nonzero. This is one instance where errors help to obtain correct results!

2. Let

$$A = \begin{bmatrix} 3 & 4 \\ 1 & 3 \end{bmatrix}$$

(a) Use the power method with scaling to approximate a dominant eigenvector of A. Start with

$$\mathbf{x}_0 = \begin{bmatrix} 1 \\ 1 \end{bmatrix}$$

Round off all computations to three significant digits, and stop after three iterations (that is, three multiplications by A).

(b) Use the result of part (a) and the Rayleigh quotient to approximate the dominant eigenvalue of A.

(c) Find the exact values of the dominant eigenvector and eigenvalue.

(d) Find the percentage error in the approximation of the dominant eigenvaluc.

In Exercises 3–4 follow the directions given in Exercise 2.

3. $A = \begin{bmatrix} 5 & 4 \\ 3 & 4 \end{bmatrix}$ **4.** $A = \begin{bmatrix} -3 & 2 \\ 2 & 0 \end{bmatrix}$

5. Let

$$A = \begin{bmatrix} 18 & 17 \\ 2 & 3 \end{bmatrix}$$

(a) Use the power method with scaling to approximate the dominant eigenvalue and a dominant eigenvector of A. Start with

$$\mathbf{x}_0 = \begin{bmatrix} 1 \\ 1 \end{bmatrix}$$

Round off all computations to three significant digits, and stop when the estimated percentage error in the dominant eigenvalue is less than 2%.

(b) Find the exact values of the dominant eigenvalue and eigenvector.

6. Repeat the directions of Exercise 5 with

$$A = \begin{bmatrix} -5 & 5 \\ 6 & -4 \end{bmatrix}$$

7. Let

$$A = \begin{bmatrix} 2 & 1 & 0 \\ 1 & 2 & 0 \\ 0 & 0 & 10 \end{bmatrix}$$

(a) Use the power method with scaling to approximate a dominant eigenvector of A. Start with

$$\mathbf{x}_0 = \begin{bmatrix} 1 \\ 1 \\ 1 \end{bmatrix}$$

Round off all computations to three significant digits, and stop after three iterations.

(b) Use the result of part (a) and the Rayleigh quotient to approximate the dominant eigenvalue of A.

(c) Find the exact values for the dominant eigenvalue and eigenvector.
(d) Find the percentage error in the approximation of the dominant eigenvalue.

8.4 APPROXIMATING NONDOMINANT EIGENVALUES BY DEFLATION

In this section we briefly outline a method for obtaining the nondominant eigenvectors and eigenvalues of a *symmetric* matrix.

We shall need the following theorem, which we state without proof.*

Theorem 1. *Let A be a symmetric $n \times n$ matrix with eigenvalues $\lambda_1, \lambda_2, \ldots, \lambda_n$. If $\mathbf{v}_1$ is an eigenvector corresponding to λ_1, and $\|\mathbf{v}_1\| = 1$, then:*

(a) The matrix $B = A - \lambda_1 \mathbf{v}_1 \mathbf{v}_1^t$ has eigenvalues $0, \lambda_2, \ldots, \lambda_n$.
(b) If $\mathbf{v}$ is an eigenvector of B corresponding to one of the eigenvalues $\lambda_2, \ldots, \lambda_n$, then $\mathbf{v}$ is also an eigenvector of A corresponding to this eigenvalue.

(We are assuming in Theorem 1 that $\mathbf{v}_1$ is expressed as an $n \times 1$ matrix; thus $\mathbf{v}_1\mathbf{v}_1^t$ is an $n \times n$ matrix.)

Example 12

It was shown in Example 5 of Section 6.1 that

$$A = \begin{bmatrix} 3 & -2 & 0 \\ -2 & 3 & 0 \\ 0 & 0 & 5 \end{bmatrix}$$

has eigenvalues $\lambda_1 = 5, \lambda_2 = 5, \lambda_3 = 1$, and

$$\mathbf{v} = \begin{bmatrix} -1 \\ 1 \\ 0 \end{bmatrix}$$

is an eigenvector corresponding to $\lambda_1 = 5$. Normalizing $\mathbf{v}$ yields

$$\mathbf{v}_1 = \frac{1}{\sqrt{2}} \begin{bmatrix} -1 \\ 1 \\ 0 \end{bmatrix} = \begin{bmatrix} -\dfrac{1}{\sqrt{2}} \\ \dfrac{1}{\sqrt{2}} \\ 0 \end{bmatrix}$$

which is an eigenvector of norm 1 corresponding to $\lambda_1 = 5$.

* Readers interested in the proof of this theorem are referred to the references given at the end of this section.

By Theorem 1 the matrix

$$B = A - \lambda_1 \mathbf{v}_1 \mathbf{v}_1^t = \begin{bmatrix} 3 & -2 & 0 \\ -2 & 3 & 0 \\ 0 & 0 & 5 \end{bmatrix} - 5 \begin{bmatrix} -\dfrac{1}{\sqrt{2}} \\ \dfrac{1}{\sqrt{2}} \\ 0 \end{bmatrix} \begin{bmatrix} -\dfrac{1}{\sqrt{2}} & \dfrac{1}{\sqrt{2}} & 0 \end{bmatrix}$$

$$= \begin{bmatrix} 3 & -2 & 0 \\ -2 & 3 & 0 \\ 0 & 0 & 5 \end{bmatrix} - 5 \begin{bmatrix} \frac{1}{2} & -\frac{1}{2} & 0 \\ -\frac{1}{2} & \frac{1}{2} & 0 \\ 0 & 0 & 0 \end{bmatrix} = \begin{bmatrix} \frac{1}{2} & \frac{1}{2} & 0 \\ \frac{1}{2} & \frac{1}{2} & 0 \\ 0 & 0 & 5 \end{bmatrix}$$

should have eigenvalues $\lambda = 0$, 5, and 1. As a check the characteristic equation of B is

$$\det(\lambda I - B) = \det \begin{bmatrix} \lambda - \frac{1}{2} & -\frac{1}{2} & 0 \\ -\frac{1}{2} & \lambda - \frac{1}{2} & 0 \\ 0 & 0 & \lambda - 5 \end{bmatrix} = \lambda(\lambda - 5)(\lambda - 1) = 0$$

Hence the eigenvalues of B are $\lambda = 0$, $\lambda = 5$, $\lambda = 1$ as predicted by Theorem 1.

The eigenspace of B corresponding to $\lambda = 5$ is the solution space of the system

$$(5I - B)\mathbf{x} = \mathbf{0}$$

that is

$$\begin{bmatrix} \frac{9}{2} & -\frac{1}{2} & 0 \\ -\frac{1}{2} & \frac{9}{2} & 0 \\ 0 & 0 & 0 \end{bmatrix} \begin{bmatrix} x_1 \\ x_2 \\ x_3 \end{bmatrix} = \begin{bmatrix} 0 \\ 0 \\ 0 \end{bmatrix}$$

Solving this system yields $x_1 = 0$, $x_2 = 0$, $x_3 = t$. Thus the eigenvectors of B corresponding to $\lambda = 5$ are the nonzero vectors of the form

$$\mathbf{x} = \begin{bmatrix} 0 \\ 0 \\ t \end{bmatrix}$$

As predicted by part (*b*) of Theorem 1, these are also eigenvectors of A corresponding to $\lambda = 5$, since

$$A \begin{bmatrix} 0 \\ 0 \\ t \end{bmatrix} = \begin{bmatrix} 3 & -2 & 0 \\ -2 & 3 & 0 \\ 0 & 0 & 5 \end{bmatrix} \begin{bmatrix} 0 \\ 0 \\ t \end{bmatrix} = \begin{bmatrix} 0 \\ 0 \\ 5t \end{bmatrix}$$

that is

$$A \begin{bmatrix} 0 \\ 0 \\ t \end{bmatrix} = 5 \begin{bmatrix} 0 \\ 0 \\ t \end{bmatrix}$$

Similarly the eigenvectors of B corresponding to $\lambda = 1$ are also eigenvectors of A corresponding to $\lambda = 1$.

Theorem 1, to a limited extent, makes it possible to determine the non-dominant eigenvalues and eigenvectors of a *symmetric* $n \times n$ matrix A. To see how, assume the eigenvalues of A can be ordered according to the size of their absolute values as follows.

$$|\lambda_1| > |\lambda_2| > |\lambda_3| \geq \cdots \geq |\lambda_n| \tag{8.24}$$

Suppose the dominant eigenvalue and a dominant eigenvector of A have been obtained by the power method. By normalizing the dominant eigenvector, we can obtain a dominant eigenvector $\mathbf{v}_1$ having norm one. By Theorem 1 the eigenvalues of $B = A - \lambda_1 \mathbf{v}_1 \mathbf{v}_1^t$ will be $0, \lambda_2, \lambda_3, \ldots, \lambda_n$. From (8.24) these eigenvalues will ordered according to their absolute values as follows.

$$|\lambda_2| > |\lambda_3| \geq \cdots \geq |\lambda_n| \geq 0$$

Thus λ_2 is the dominant eigenvalue of B. Now, by applying the power method to B, we can approximate the eigenvalue λ_2 and a corresponding eigenvector. This technique for approximating the eigenvalue with second largest absolute value is called ***deflation***.

Unfortunately, there are practical limitations of the deflation method. Since λ_1 and $\mathbf{v}_1$ are only approximated in the power method, an error is introduced into B when deflation is used. If the deflation process is applied again, the next matrix has additional errors introduced through the approximation of λ_2 and $\mathbf{v}_2$. As the process continues, this compounding of errors eventually destroys the accuracy of the results. In practice, one should generally avoid finding more than two or three eigenvalues by deflation.

When the ratio $|\lambda_2/\lambda_1|$ is close to one, the power method has a slow rate of convergence; that is, many steps are needed to obtain a reasonable degree of accuracy. Readers interested in studying techniques for "speeding up" this rate of convergence and learning more about the numerical methods of linear algebra may wish to consult the following texts:

Analysis of Numerical Methods, E. Isaacson and H. B. Keller, John Wiley and Sons, New York, 1966.

Applied Linear Algebra, B. Noble, Prentice Hall, Inc., 1969.

Computational Methods of Linear Algebra, V. N. Faddeeva, Dover, 1959.

Additional references appear in the bibliographies of these texts.

EXERCISE SET 8.4

1. Let

$$A = \begin{bmatrix} 6 & 2 \\ 2 & 3 \end{bmatrix}$$

(a) Use the power method with scaling to approximate a dominant eigenvector. Start with

$$\mathbf{x}_0 = \begin{bmatrix} 1 \\ 1 \end{bmatrix}$$

Round off all computations to three significant digits, and stop after three iterations (i.e., three multiplications by A.)

(b) Use the result of part (a) and the Rayleigh quotient to approximate the dominant eigenvalue of A.

(c) Use deflation to approximate the remaining eigenvalue and a corresponding eigenvector; that is, apply the power method to

$$B = A - \tilde{\lambda}_1 \tilde{\mathbf{v}}_1 \tilde{\mathbf{v}}_1^t$$

where $\tilde{\mathbf{v}}_1$ and $\tilde{\lambda}_1$ are the approximations obtained in parts (a) and (b). Start with

$$\mathbf{x}_0 = \begin{bmatrix} 1 \\ 1 \end{bmatrix}$$

Round off all computations to three significant digits and stop after three iterations.

(d) Find the exact values of the eigenvalues and eigenvectors.

2. Follow the directions given in Exercise 1 with

$$A = \begin{bmatrix} 10 & 4 \\ 4 & 4 \end{bmatrix}$$

Answers to Exercises

EXERCISE SET 1.1 (page 7).

1. b, d, f

2. (a) $x = \frac{7}{6}t + \frac{1}{2},\ y = t$
   (b) $x_1 = -2s + \frac{7}{2}t + 4,\ x_2 = s,\ x_3 = t$
   (c) $x_1 = \frac{4}{3}r - \frac{7}{3}s + \frac{8}{3}t - \frac{5}{3},\ x_2 = r,\ x_3 = s,\ x_4 = t$
   (d) $v = \frac{1}{7}q - \frac{3}{2}r - \frac{1}{2}s + 2t,\ w = q,\ x = r,\ y = s,\ z = t$

3. (a) $\begin{bmatrix} 1 & -2 & 0 \\ 3 & 4 & -1 \\ 2 & -1 & 3 \end{bmatrix}$ (b) $\begin{bmatrix} 1 & 0 & 1 & 1 \\ -1 & 2 & -1 & 3 \end{bmatrix}$

   (c) $\begin{bmatrix} 1 & 0 & 1 & 0 & 0 & 1 \\ 0 & 2 & -1 & 0 & 1 & 2 \\ 0 & 0 & 2 & 1 & 0 & 3 \end{bmatrix}$ (d) $\begin{bmatrix} 1 & 0 & 1 \\ 0 & 1 & 2 \end{bmatrix}$

4. (a) $\begin{aligned} x_1 \quad\quad - x_3 &= 2 \\ 2x_1 + x_2 + x_3 &= 3 \\ - x_2 + 2x_3 &= 4 \end{aligned}$ (b) $\begin{aligned} x_1 \quad\quad &= 0 \\ x_2 &= 0 \\ x_1 - x_2 &= 1 \\ x_1 \quad\quad &= 1 \end{aligned}$

   (c) $\begin{aligned} x_1 + 2x_2 + 3x_3 + 4x_4 &= 5 \\ 5x_1 + 4x_2 + 3x_3 + 2x_4 &= 1 \end{aligned}$ (d) $\begin{aligned} x_1 &= 1 \\ x_2 &= 2 \\ x_3 &= 3 \\ x_4 &= 4 \end{aligned}$

5. $k = 6$ infinitely many solutions
   $k \neq 6$ no solutions

6. (a) The lines have no common point of intersection.
   (b) The lines intersect in exactly one point.
   (c) The three lines coincide.

EXERCISE SET 1.2 (page 16).

1. d, f **2.** b, c, f

3. (a) $x_1 = 4, x_2 = 3, x_3 = 2$
(b) $x_1 = 2 - 3t, x_2 = 4 + t, x_3 = 2 - t, x_4 = t$
(c) $x_1 = -1 - 5s - 5t, x_2 = s, x_3 = 1 - 3t, x_4 = 2 - 4t, x_5 = t$
(d) Inconsistent

4. (a) $x_1 = 4, x_2 = 3, x_3 = 2$
(b) $x_1 = 2 - 3t, x_2 = 4 + t, x_3 = 2 - t, x_4 = t$
(c) $x_1 = -1 - 5s - 5t, x_2 = s, x_3 = 1 - 3t, x_4 = 2 - 4t, x_5 = t$
(d) Inconsistent

5. (a) $x_1 = 2, x_2 = 1, x_3 = 3$
(b) $x_1 = -\frac{3}{7}t, x_2 = -\frac{4}{7}t, x_3 = t$
(c) $x_1 = 1, x_2 = 2s, x_3 = s, x_4 = -3t, x_5 = t$

7. (a) Inconsistent (b) $x_1 = -4, x_2 = 2, x_3 = 7$ (c) $x_1 = 3 + 2t, x_2 = t$

9. (a) $x_1 = 0, x_2 = -3t, x_3 = t$ (b) Inconsistent

11. (a) $x_1 = \frac{2}{3}a - \frac{1}{9}b, x_2 = -\frac{1}{3}a + \frac{2}{9}b$
(b) $x_1 = a - \frac{1}{3}c, x_2 = a - \frac{1}{2}b, x_3 = -a + \frac{1}{2}b + \frac{1}{3}c$

12. $a = 4$ infinitely many, $a = -4$ none, $a \neq \pm 4$ exactly one.

14. $\begin{bmatrix} 1 & 3 \\ 0 & 1 \end{bmatrix}$ and $\begin{bmatrix} 1 & 0 \\ 0 & 1 \end{bmatrix}$ are possible answers.

15. $\alpha = \dfrac{\pi}{2}, \beta = \pi, \gamma = 0.$

EXERCISE SET 1.3 (page 21).

1. a, c, d

2. $x_1 = 0, x_2 = 0, x_3 = 0$

3. $x_1 = -\frac{1}{4}s, x_2 = -\frac{1}{4}s - t, x_3 = s, x_4 = t$

4. $x_1 = 0, x_2 = 0, x_3 = 0, x_4 = 0$

5. $x = \dfrac{t}{8}, y = \dfrac{5t}{16}, z = t$

6. $\lambda = 4, \lambda = 2$

EXERCISE SET 1.4 (page 28).

1. (a) undefined (b) 4×2 (c) undefined
(d) undefined (e) 5×5 (f) 5×2

3. $a = 5, b = -3, c = 4, d = 1$

4. (a) $\begin{bmatrix} 12 & -3 \\ -4 & 5 \\ 4 & 1 \end{bmatrix}$ (b) $\begin{bmatrix} 7 & 6 & 5 \\ -2 & 1 & 3 \\ 7 & 3 & 7 \end{bmatrix}$ (c) $\begin{bmatrix} -5 & 4 & -1 \\ 0 & -1 & -1 \\ -1 & 1 & 1 \end{bmatrix}$

(d) $\begin{bmatrix} 9 & 8 & 19 \\ -2 & 0 & 0 \\ 32 & 9 & 25 \end{bmatrix}$ (e) $\begin{bmatrix} 14 & 36 & 25 \\ 4 & -1 & 7 \\ 12 & 26 & 21 \end{bmatrix}$ (f) $\begin{bmatrix} -28 & 7 \\ 0 & -14 \end{bmatrix}$

5. (a) Undefined (b) $\begin{bmatrix} 42 & 108 & 75 \\ 12 & -3 & 21 \\ 36 & 78 & 63 \end{bmatrix}$

(c) $\begin{bmatrix} 3 & 45 & 9 \\ 11 & -11 & 17 \\ 7 & 17 & 13 \end{bmatrix}$ (d) $\begin{bmatrix} 3 & 45 & 9 \\ 11 & -11 & 17 \\ 7 & 17 & 13 \end{bmatrix}$

(e) Undefined (f) $\begin{bmatrix} 48 & 15 & 31 \\ 0 & 2 & 6 \\ 38 & 10 & 27 \end{bmatrix}$

6. (a) $[67 \quad 41 \quad 41]$ (b) $[63 \quad 67 \quad 57]$

(c) $\begin{bmatrix} 41 \\ 21 \\ 67 \end{bmatrix}$ (d) $\begin{bmatrix} 6 \\ 6 \\ 63 \end{bmatrix}$

(e) $[24 \quad 56 \quad 97]$ (f) $\begin{bmatrix} 76 \\ 98 \\ 97 \end{bmatrix}$

7. 182

EXERCISE SET 1.5 (page 36).

3. $A^{-1} = \begin{bmatrix} 2 & -1 \\ -5 & 3 \end{bmatrix}$ $B^{-1} = \begin{bmatrix} \frac{1}{5} & \frac{3}{20} \\ -\frac{1}{5} & \frac{1}{10} \end{bmatrix}$ $C^{-1} = \begin{bmatrix} \frac{1}{2} & 0 \\ 0 & \frac{1}{3} \end{bmatrix}$

5. No

6. $\begin{bmatrix} -3 & 2 \\ \frac{5}{2} & -\frac{3}{2} \end{bmatrix}$ **7.** $\begin{bmatrix} 1 & \frac{2}{7} \\ \frac{4}{7} & \frac{1}{7} \end{bmatrix}$

8. $A^3 = \begin{bmatrix} 1 & 0 \\ 26 & 27 \end{bmatrix}$ $A^{-3} = \begin{bmatrix} 1 & 0 \\ -\frac{26}{27} & \frac{1}{27} \end{bmatrix}$ $A^2 - 2A + I = \begin{bmatrix} 0 & 0 \\ 4 & 4 \end{bmatrix}$

9. $A^{-1} = \begin{bmatrix} \frac{1}{2} & -\frac{1}{2} & \frac{1}{2} \\ \frac{1}{2} & \frac{1}{2} & -\frac{1}{2} \\ -\frac{1}{2} & \frac{1}{2} & \frac{1}{2} \end{bmatrix}$

10. $\begin{bmatrix} \cos\theta & -\sin\theta \\ \sin\theta & \cos\theta \end{bmatrix}$

11. (c) $(A + B)^2 = A^2 + AB + BA + B^2$

12. $A^{-1} = \begin{bmatrix} \frac{1}{a_{11}} & 0 & \cdots & 0 \\ 0 & \frac{1}{a_{22}} & \cdots & 0 \\ \vdots & \vdots & & \vdots \\ 0 & 0 & \cdots & \frac{1}{a_{nn}} \end{bmatrix}$

17. $0A$ and $A0$ may not have the same size.

18. $\begin{bmatrix} \pm 1 & 0 & 0 \\ 0 & \pm 1 & 0 \\ 0 & 0 & \pm 1 \end{bmatrix}$

EXERCISE SET 1.6 (page 45).

1. a, b, d, f, g

2. (a) Add -5 times the first row to the second.
   (b) Interchange the first and third rows.
   (c) Multiply the second row by $\frac{1}{8}$.

3. (a) $E_1 = \begin{bmatrix} 0 & 0 & 1 \\ 0 & 1 & 0 \\ 1 & 0 & 0 \end{bmatrix}$ (b) $E_2 = \begin{bmatrix} 0 & 0 & 1 \\ 0 & 1 & 0 \\ 1 & 0 & 0 \end{bmatrix}$

   (c) $E_3 = \begin{bmatrix} 1 & 0 & 0 \\ 0 & 1 & 0 \\ 2 & 0 & 1 \end{bmatrix}$ (d) $E_4 = \begin{bmatrix} 1 & 0 & 0 \\ 0 & 1 & 0 \\ -2 & 0 & 1 \end{bmatrix}$

4. No, since C cannot be obtained by performing a single row operation on B.

5. (a) $\begin{bmatrix} -5 & 2 \\ 3 & -1 \end{bmatrix}$ (b) $\begin{bmatrix} -5 & -3 \\ -3 & -2 \end{bmatrix}$ (c) Not invertible

6. (a) $\begin{bmatrix} \frac{3}{2} & -\frac{11}{10} & -\frac{6}{5} \\ -1 & 1 & 1 \\ -\frac{1}{2} & \frac{7}{10} & \frac{2}{5} \end{bmatrix}$ (b) Not invertible

(c) $\begin{bmatrix} \frac{1}{2} & -\frac{1}{2} & \frac{1}{2} \\ -\frac{1}{2} & \frac{1}{2} & \frac{1}{2} \\ \frac{1}{2} & \frac{1}{2} & -\frac{1}{2} \end{bmatrix}$ (d) $\begin{bmatrix} \frac{7}{2} & 0 & -3 \\ -1 & 1 & 0 \\ 0 & -1 & 1 \end{bmatrix}$

(e) $\begin{bmatrix} \frac{1}{2} & -\frac{1}{2} & \frac{1}{2} \\ 0 & 0 & 1 \\ \frac{1}{2} & \frac{1}{2} & -\frac{1}{2} \end{bmatrix}$ (f) $\begin{bmatrix} 1 & 0 & -2 \\ 3 & 1 & 2 \\ 1 & -1 & 0 \end{bmatrix}$

7. (a) $\begin{bmatrix} \frac{1}{2}\sqrt{2} & -\frac{1}{2}\sqrt{2} & 0 \\ \frac{1}{2}\sqrt{2} & \frac{1}{2}\sqrt{2} & 0 \\ 0 & 0 & 1 \end{bmatrix}$ (b) $\begin{bmatrix} 1 & 0 & 0 & 0 \\ -\frac{1}{2} & \frac{1}{2} & 0 & 0 \\ 0 & -\frac{1}{4} & \frac{1}{4} & 0 \\ 0 & 0 & -\frac{1}{8} & \frac{1}{8} \end{bmatrix}$

(c) Not invertible.

8. $A^{-1} = \begin{bmatrix} \cos\theta & -\sin\theta & 0 \\ \sin\theta & \cos\theta & 0 \\ 0 & 0 & 1 \end{bmatrix}$

9. (a) $E_1 = \begin{bmatrix} 1 & 0 \\ -3 & 1 \end{bmatrix}$, $E_2 = \begin{bmatrix} 1 & 0 \\ 0 & \frac{1}{4} \end{bmatrix}$ (b) $A^{-1} = \begin{bmatrix} 1 & 0 \\ 0 & \frac{1}{4} \end{bmatrix}\begin{bmatrix} 1 & 0 \\ -3 & 1 \end{bmatrix}$

(c) $A = \begin{bmatrix} 1 & 0 \\ 3 & 1 \end{bmatrix}\begin{bmatrix} 1 & 0 \\ 0 & 4 \end{bmatrix}$

11. $A = \begin{bmatrix} 1 & 0 & 0 \\ -2 & 1 & 0 \\ 0 & 0 & 1 \end{bmatrix}\begin{bmatrix} 1 & 0 & 0 \\ 0 & 1 & 0 \\ 0 & 1 & 1 \end{bmatrix}\begin{bmatrix} 1 & 3 & 3 & 8 \\ 0 & 1 & 7 & 8 \\ 0 & 0 & 0 & 0 \end{bmatrix}$

13. (a) $\begin{bmatrix} \frac{1}{k_1} & 0 & 0 & 0 \\ 0 & \frac{1}{k_2} & 0 & 0 \\ 0 & 0 & \frac{1}{k_3} & 0 \\ 0 & 0 & 0 & \frac{1}{k_4} \end{bmatrix}$ (b) $\begin{bmatrix} 0 & 0 & 0 & \frac{1}{k_4} \\ 0 & 0 & \frac{1}{k_3} & 0 \\ 0 & \frac{1}{k_2} & 0 & 0 \\ \frac{1}{k_1} & 0 & 0 & 0 \end{bmatrix}$

(c) $\begin{bmatrix} \frac{1}{k} & 0 & 0 & 0 \\ -\frac{1}{k^2} & \frac{1}{k} & 0 & 0 \\ \frac{1}{k^3} & -\frac{1}{k^2} & \frac{1}{k} & 0 \\ -\frac{1}{k^4} & \frac{1}{k^3} & -\frac{1}{k^2} & \frac{1}{k} \end{bmatrix}$

EXERCISE SET 1.7 (page 52).

1. $x_1 = 41, x_2 = -17$

2. $x_1 = \frac{46}{27}, x_2 = -\frac{13}{27}$

3. $x_1 = -7, x_2 = 4, x_3 = -1$

4. $x_1 = 1, x_2 = -11, x_3 = 16$

5. $x = 1, y = 5, z = -1$

6. $w = 1, x = -6, y = 10, z = -7$

7. (a) $x_1 = \frac{16}{3}, x_2 = -\frac{4}{3}, x_3 = -\frac{11}{3}$
(b) $x_1 = -\frac{5}{3}, x_2 = \frac{5}{3}, x_3 = \frac{10}{3}$
(c) $x_1 = 3, x_2 = 0, x_3 = -4$
(d) $x_1 = \frac{41}{42}, x_2 = -\frac{5}{6}, x_3 = \frac{25}{21}$

8. (a) $b_2 = 3b_1, b_3 = -2b_1$
(b) $b_3 = b_2 - b_1, b_4 = 2b_1 - b_2$

9. (a) $X = \begin{bmatrix} 0 \\ 0 \\ 0 \end{bmatrix}$ (b) $X = \begin{bmatrix} 4t \\ \frac{5t}{2} \\ t \end{bmatrix}$

EXERCISE SET 2.1 (page 60).

1. (a) 5 (b) 7 (c) 10 (d) 0 (e) 4 (f) 5

2. (a) Odd (b) Odd (c) Even (d) Even (e) Even (f) Odd

3. 5 **4.** 0 **5.** 59 **6.** $k^2 - 4k - 5$ **7.** 0 **8.** 425 **9.** 104

10. $-k^4 - k^3 + 18k^2 + 9k - 21$ **11.** (a) $\lambda = 3, \lambda = 2$ (b) $\lambda = 2, \lambda = 6$

14. 275

15. (a) 120 (b) -120

EXERCISE SET 2.2 (page 65).

1. (a) 6 (b) -16 (c) 0 (d) 0

2. 21 **3.** -5 **4.** -36 **5.** 35

6. -128 **7.** -72 **8.** $\frac{1}{6}$ **9.** 0

10. (a) 5 (b) 10 (c) 5 (d) 10

EXERCISE SET 2.3 (page 72).

1. (a) $\begin{bmatrix} 2 & -3 & 0 \\ 1 & 1 & 2 \end{bmatrix}$ (b) $\begin{bmatrix} 6 & -8 & 0 \\ 1 & 4 & 1 \\ 1 & 3 & 3 \end{bmatrix}$ (c) $\begin{bmatrix} 7 \\ 0 \\ 2 \end{bmatrix}$ (d) $\begin{bmatrix} a_{11} & a_{21} \\ a_{12} & a_{22} \\ a_{13} & a_{23} \end{bmatrix}$

4. (a) Invertible (b) Not invertible
(c) Not invertible (d) Not invertible

5. (a) 135 (b) $\frac{8}{5}$ (c) $\frac{1}{40}$ (d) -5

6. If $x = 0$, the first and third rows are proportional.
If $x = 2$, the first and second rows are proportional.

8. (a) $k = \frac{1}{2}(5 + \sqrt{17}), k = \frac{1}{2}(5 - \sqrt{17})$ (b) $k = -1$

EXERCISE SET 2.4 (page 82).

1. (a) $M_{11} = 29, M_{12} = -11, M_{13} = -19, M_{21} = 21, M_{22} = 13, M_{23} = -19$
$M_{31} = 27, M_{32} = -5, M_{33} = 19$
(b) $C_{11} = 29, C_{12} = 11, C_{13} = -19, C_{21} = -21, C_{22} = 13$
$C_{23} = 19, C_{31} = 27, C_{32} = 5, C_{33} = 19$

2. (a) $M_{13} = 36, C_{13} = 36$ (b) $M_{23} = 24, C_{23} = -24$
(c) $M_{22} = -48, C_{22} = -48$ (d) $M_{21} = -108, C_{21} = 108$

3. 152

4. (a) $\begin{bmatrix} 29 & -21 & 27 \\ 11 & 13 & 5 \\ -19 & 19 & 19 \end{bmatrix}$ (b) $(\frac{1}{152})\begin{bmatrix} 29 & -21 & 27 \\ 11 & 13 & 5 \\ -19 & 19 & 19 \end{bmatrix}$

5. 48 **6.** -66 **7.** 0 **8.** $k^3 - 8k^2 - 10k + 95$ **9.** -120

10. 0

11. $A^{-1} = \begin{bmatrix} -4 & 3 & 0 & -1 \\ 2 & -1 & 0 & 0 \\ -7 & 0 & -1 & 8 \\ 6 & 0 & 1 & -7 \end{bmatrix}$

12. $x_1 = 1, x_2 = 2$

13. $x = \frac{3}{11}, y = \frac{2}{11}, z = -\frac{1}{11}$

14. $x = \frac{26}{21}, y = \frac{25}{21}, z = \frac{5}{7}$

15. $x_1 = -\frac{30}{11}, x_2 = -\frac{38}{11}, x_3 = -\frac{40}{11}$

16. $x_1 = 3, x_2 = 5, x_3 = -1, x_4 = 8$

17. Cramer's rule does not apply.

18. $z = 2$

EXERCISE SET 3.1 (page 93).

3. (a) $\overrightarrow{P_1P_2} = (-1, 3)$ (b) $\overrightarrow{P_1P_2} = (-7, 2)$
(c) $\overrightarrow{P_1P_2} = (2, -12, -11)$ (d) $\overrightarrow{P_1P_2} = (-8, 7, 4)$

4. $\overrightarrow{PQ}$, where $Q = (9, 5, 1)$ is one possible answer.

5. $\overrightarrow{PQ}$, where $P = (0, 4, -8)$ is one possible answer.

6. (a) $(-2, 0, 4)$ (b) $(23, -15, 4)$
(c) $(-1, -5, 2)$ (d) $(-39, 69, -12)$
(e) $(-30, -7, 5)$ (f) $(0, -10, 0)$

7. $\mathbf{x} = (-\frac{1}{2}, \frac{5}{6}, 1)$

8. $c_1 = 1, c_2 = -2, c_3 = 3$

10. $c_1 = -t, c_2 = -t, c_3 = t$ (where t is arbitrary)

11. (a) $(\frac{9}{2}, -\frac{1}{2}, -\frac{1}{2})$ (b) $(\frac{23}{4}, -\frac{9}{4}, \frac{1}{4})$

12. (a) $x' = 5, y' = 8$ (b) $x = -1, y = 3$

EXERCISE SET 3.2 (page 97).

1. (a) 5 (b) $5\sqrt{2}$ (c) 3 (d) $\sqrt{3}$ (e) $\sqrt{129}$ (f) 9

2. (a) $\sqrt{13}$ (b) $2\sqrt{26}$ (c) $\sqrt{209}$ (d) $\sqrt{93}$

3. (a) $2\sqrt{3}$ (b) $\sqrt{14} + \sqrt{2}$ (c) $4\sqrt{14}$ (d) $2\sqrt{37}$
(e) $(1/\sqrt{6}, 1/\sqrt{6}, -2/\sqrt{6})$ (f) 1

4. $k = \pm\dfrac{3}{\sqrt{21}}$

7. $(1/\sqrt{3}, 1/\sqrt{3}, 1/\sqrt{3})$

8. A sphere of radius 1 centered at (x_0, y_0, z_0)

EXERCISE SET 3.3 (page 103).

1. (a) -10 (b) -3 (c) 0 (d) -20

2. (a) $-\dfrac{1}{\sqrt{5}}$ (b) $-\dfrac{3}{\sqrt{58}}$ (c) 0 (d) $-\dfrac{20}{3\sqrt{70}}$

3. (a) obtuse (b) acute (c) obtuse (d) orthogonal.

4. (a) $(\frac{12}{13}, -\frac{8}{13})$ (b) $(0, 0)$ (c) $(-\frac{80}{13}, 0, -\frac{16}{13})$ (d) $(\frac{32}{89}, \frac{12}{89}, \frac{16}{89})$

5. (a) $(\frac{14}{13}, \frac{21}{13})$ (b) $(2, 6)$ (c) $(-\frac{11}{13}, 1, \frac{55}{13})$ (d) $(-\frac{32}{89}, -\frac{12}{89}, \frac{73}{89})$

7. $\pm\left(\dfrac{2}{\sqrt{13}}, \dfrac{3}{\sqrt{13}}\right)$

8. (a) 6 (b) 36 (c) $24\sqrt{5}$ (d) $24\sqrt{5}$

10. $\cos\theta_1 = 0, \cos\theta_2 = \dfrac{3}{\sqrt{10}}, \cos\theta_3 = \dfrac{1}{\sqrt{10}}$

13. $\theta = \cos^{-1}\dfrac{2}{\sqrt{6}}$

14. $\cos\beta = \dfrac{b}{\sqrt{a^2 + b^2 + c^2}}, \cos\gamma = \dfrac{c}{\sqrt{a^2 + b^2 + c^2}}$

EXERCISE SET 3.4 (page 113).

1. (a) $(-23, 7, -1)$ (b) $(-20, -67, -9)$ (c) $(-78, -52, -26)$
(d) $(0, -56, -392)$ (e) $(24, 0, -16)$ (f) $(-12, -22, -8)$

2. (a) $(12, 30, -6)$ (b) $(-2, 0, 2)$

3. (a) $\frac{1}{2}\sqrt{374}$ (b) $9\sqrt{13}$

7. $x = (\frac{1}{2} - \frac{1}{2}t, -\frac{1}{2} + \frac{3}{2}t, t)$, where t is arbitrary

9. 227

10. (a) $\mathbf{u} = (0, 1, 0)$ and $\mathbf{v} = (1, 0, 0)$
(b) $(-1, 0, 0)$
(c) $(0, 0, -1)$

EXERCISE SET 3.5 (page 119).

1. (a) $(x - 2) + 4(y - 6) + 2(z - 1) = 0$
(b) $-(x + 1) + 7(y + 1) + 6(z - 2) = 0$
(c) $z = 0$
(d) $2x + 3y + 4z = 0$

2. (a) $x + 4y + 2z - 28 = 0$ (b) $-x + 7y + 6z - 6 = 0$
(c) $z = 0$ (d) $2x + 3y + 4z = 0$

3. (a) $(5, 0, 0)$ is a point in the plane and $\mathbf{n} = (2, -3, 7)$ is a normal vector so that $2(x - 5) - 3y + 7z = 0$ is a point normal form; other points and normals yield other correct answers.
(b) $x + 3z = 0$ is one possible answer.

4. (a) $2y - z - 1 = 0$ (b) $x + 9y - 5z - 16 = 0$

5. (a) $x = 2 + t, y = 4 + 2t, z = 6 + 5t$
(b) $x = -3 + 5t, y = 2 - 7t, z = -4 - 3t$
(c) $x = 1, y = 1, z = 5 + t$
(d) $x = t, y = t, z = t$

6. (a) $x - 2 = \dfrac{y-4}{2} = \dfrac{z-6}{5}$ (b) $\dfrac{x+3}{5} = \dfrac{y-2}{-7} = \dfrac{z+4}{-3}$

7. (a) $x = 6 + t, y = -1 + 3t, z = 5 - 9t$ or $x = 7 + t, y = 2 + 3t, z = -4 - 9t$ are possible answers
(b) $x = -t, y = -t, z = -t$ or $x = -1 - t, y = -1 - t, z = -1 - t$ are possible answers

8. (a) $x = -\frac{11}{7} + \frac{23}{7}t, y = -\frac{12}{7} - \frac{1}{7}t, z = t$
(b) $x = \frac{5}{3}t, y = t, z = 0$

9. (a) $x - 2y - 17 = 0, y + 2z - 5 = 0$ (b) $3x - 5y = 0, 2y - z = 0$

10. xy-plane: $z = 0$; xz-plane: $y = 0$; yz-plane: $x = 0$

12. $(-\frac{222}{7}, -\frac{64}{7}, \frac{78}{7})$

13. $5x - 2y + z - 30 = 0$

15. $(-17, -1, 1)$ **16.** $x - 4y + 4z + 9 = 0$

EXERCISE SET 4.1 (page 125).

1. (a) $(-3, -4, -8, 4)$ (b) $(53, 34, 49, 20)$ (c) $(-1, 2, 7, -10)$
(d) $(-99, -84, -150, 30)$ (e) $(-63, -28, -21, -69)$ (f) $(2, 6, 15, -14)$

2. $(-\frac{7}{6}, -1, -\frac{3}{2}, -\frac{1}{3})$ **3.** $c_1 = 1, c_2 = 1, c_3 = -1, c_4 = 1$

5. (a) 5 (b) $\sqrt{11}$ (c) $\sqrt{14}$ (d) $\sqrt{48}$

6. (a) $\sqrt{73}$ (b) $\sqrt{14} + 3\sqrt{7}$ (c) $4\sqrt{14}$ (d) $\sqrt{1801}$

(e) $\left(\dfrac{2}{\sqrt{6}}, 0, \dfrac{1}{\sqrt{6}}, \dfrac{1}{\sqrt{6}}\right)$ (f) 1

8. $k = \pm\dfrac{3}{\sqrt{14}}$

9. (a) -1 (b) -1 (c) 0 (d) 27

10. (a) $\left(\dfrac{2}{\sqrt{5}}, \dfrac{1}{\sqrt{5}}\right)$ and $\left(-\dfrac{2}{\sqrt{5}}, -\dfrac{1}{\sqrt{5}}\right)$

11. (a) $\sqrt{10}$ (b) $3\sqrt{3}$ (c) $\sqrt{59}$ (d) 10

EXERCISE SET 4.2 (page 130).

1. Not a vector space. Axiom 8 fails.

2. Not a vector space. Axiom 10 fails.

3. Not a vector space. Axioms 9 and 10 fail.

4. The set is a vector space under the given operations.

5. The set is a vector space under the given operations.

6. Not a vector space. Axioms 5 and 6 fail.

7. The set is a vector space under the given operations.

8. Not a vector space. Axioms 7 and 8 fail.

9. The set is a vector space under the given operations.

10. Not a vector space. Axioms 1, 4, 5, and 6 fail.

11. The set is a vector space under the given operations.

12. The set is a vector space under the given operations.

13. The set is a vector space under the given operations.

14. The set is a vector space under the given operations.

EXERCISE SET 4.3 (page 139).

1. *a, c* **2.** *b, c* **3.** *a, b, d* **4.** *b, d, e* **5.** *a, b, d*

6. (a) $(5, 9, 5) = 3\mathbf{u} - 4\mathbf{v} + \mathbf{w}$ (b) $(2, 0, 6) = 4\mathbf{u} - 2\mathbf{w}$
(c) $(0, 0, 0) = 0\mathbf{u} + 0\mathbf{v} + 0\mathbf{w}$ (d) $(2, 2, 3) = \frac{1}{2}\mathbf{u} - \frac{1}{2}\mathbf{v} + \frac{1}{2}\mathbf{w}$

7. (a) $5 + 9x + 5x^2 = 3\mathbf{p}_1 - 4\mathbf{p}_2 + \mathbf{p}_3$
(b) $2 + 6x^2 = 4\mathbf{p}_1 - 2\mathbf{p}_3$
(c) $0 = 0\mathbf{p}_1 + 0\mathbf{p}_2 + 0\mathbf{p}_3$
(d) $2 + 2x + 3x^2 = \frac{1}{2}\mathbf{p}_1 - \frac{1}{2}\mathbf{p}_2 + \frac{1}{2}\mathbf{p}_3$

8. *a, c, d*

9. (a) The vectors span (b) The vectors do not span
(c) The vectors do not span (d) The vectors span

10. *a, c*

11. The polynomials do not span P_2

12. *a, b, d*

13. $8x - 7y + z = 0$

14. $x = 2t$, $y = 7t$, $z = -t$, where $-\infty < t < +\infty$

EXERCISE SET 4.4 (page 144).

1. (a) $\mathbf{u}_2$ is a scalar multiple of $\mathbf{u}_1$.
(b) The vectors are linearly dependent by Theorem 6.
(c) $\mathbf{p}_2$ is a scalar multiple of $\mathbf{p}_1$.
(d) B is a scalar multiple of A.

2. (a) independent (b) independent (c) independent (d) dependent

3. (a) independent (b) independent (c) independent (d) independent

4. (a) independent (b) independent (c) independent (d) dependent

5. (a) dependent (b) independent (c) independent (d) dependent
(e) dependent (f) dependent

6. (a) They do not lie in a plane.
(b) They do lie in a plane.

7. (a) They do not lie on the same line.
(b) They do not lie on the same line.
(c) They do lie on the same line.

8. $\lambda = -\frac{1}{2}, \lambda = 1$

EXERCISE SET 4.5 (page 152).

1. (a) A basis for R^2 has two vectors.
(b) A basis for R^3 has three vectors.
(c) A basis for P_2 has three vectors.
(d) A basis for M_{22} has four vectors.

2. *a, b* **3.** *a, b* **4.** *c, d*

7. No basis; dimension = 0 **6.** Any two of the vectors $\mathbf{v}_1, \mathbf{v}_2, \mathbf{v}_3$

8. Basis: $(-\frac{1}{4}, -\frac{1}{4}, 1, 0), (0, -1, 0, 1)$; dimension = 2

9. No basis; dimension = 0

10. Basis: $(3, 1, 0), (-1, 0, 1)$; dimension = 2

11. No basis; dimension = 0

12. No basis; dimension = 0

13. (a) $(\frac{2}{3}, 1, 0), (-\frac{5}{3}, 0, 1)$ (b) $(1, 1, 0), (0, 0, 1)$
(c) $(2, -1, 4)$ (d) $(1, 1, 0), (0, 1, 1)$

14. (a) 3-dimensional (b) 2-dimensional (c) 1-dimensional

15. 3-dimensional

EXERCISE SET 4.6 (page 160).

1.

$\mathbf{r}_1 = (2, -1, 0, 1)$
$\mathbf{r}_2 = (3, 5, 7, -1)$
$\mathbf{r}_3 = (1, 4, 2, 7)$

$$\mathbf{c}_1 = \begin{bmatrix} 2 \\ 3 \\ 1 \end{bmatrix} \quad \mathbf{c}_2 = \begin{bmatrix} -1 \\ 5 \\ 4 \end{bmatrix} \quad \mathbf{c}_3 = \begin{bmatrix} 0 \\ 7 \\ 2 \end{bmatrix} \quad \mathbf{c}_4 = \begin{bmatrix} 1 \\ -1 \\ 7 \end{bmatrix}$$

2. (a) $(1, -3)$ (b) $\begin{bmatrix} 1 \\ 2 \end{bmatrix}$ (c) 1

3. (a) $(1, 2, 0), (0, 0, 1)$

(b) $\begin{bmatrix} 1 \\ 2 \\ 0 \end{bmatrix}, \begin{bmatrix} 0 \\ 1 \\ -1 \end{bmatrix}$

(c) 2

4. (a) $(1, 0, 1, 2), (0, 1, 1, 0), (0, 0, 0, 1)$

(b) $\begin{bmatrix} 1 \\ 0 \\ 0 \end{bmatrix}, \begin{bmatrix} 0 \\ 1 \\ 0 \end{bmatrix}, \begin{bmatrix} 0 \\ 0 \\ 1 \end{bmatrix}$

(c) 3

5. (a) $(1, 0, 5, 2, 0), (0, 1, 0, 0, 0), (0, 0, -3, 0, 1)$

(b) $\begin{bmatrix} 0 \\ 0 \\ 1 \\ 1 \\ 2 \end{bmatrix}, \begin{bmatrix} 1 \\ 0 \\ 0 \\ 1 \\ 1 \end{bmatrix}, \begin{bmatrix} 0 \\ -1 \\ 1 \\ 0 \\ 0 \end{bmatrix}$

(c) 3

6. (a) $(1, 1, -4, -3), (0, 1, -5, -2), (0, 0, 1, -\frac{1}{2})$
(b) $(1, -1, 2, 0), (0, 1, 0, 0), (0, 0, 1, -\frac{1}{6})$
(c) $(1, 1, 0, 0), (0, 1, 1, 1), (0, 0, 1, 1), (0, 0, 0, 1)$

8. 3 (b) the minimum of m and n

9. (a) $\mathbf{b} = \begin{bmatrix} 1 \\ 4 \end{bmatrix} - \begin{bmatrix} 3 \\ -6 \end{bmatrix}$

(b) $\mathbf{b}$ is not in the column space of A
(c) $\mathbf{b}$ is not in the column space of A

EXERCISE SET 4.7 (page 165).

1. (a) -12 (b) 0 (c) 0 (d) 120

2. (a) -5 (b) 0 (c) 3 (d) 52

3. (a) 16 (b) 56

4. (a) -6 (b) 0

6. (a) Not an inner product, axiom 4 fails.
(b) Not an inner product, axioms 2, 3 fail.
(c) $\langle \mathbf{u}, \mathbf{v} \rangle$ is an inner product.
(d) Not an inner product, axiom 4 fails.

16. (a) $-\frac{28}{15}$ (b) 0

17. (a) 0 (b) 1 (c) $-\frac{4}{\pi} \ln\left(\frac{1}{\sqrt{2}}\right)$

EXERCISE SET 4.8 (page 172).

1. (a) $\sqrt{21}$ (b) $\sqrt{206}$ (c) $\sqrt{2}$ (d) 0

2. (a) $\sqrt{10}$ (b) $\sqrt{85}$ (c) 1 (d) 0

3. (a) $\sqrt{6}$ (b) 5

4. (a) $\sqrt{90}$ (b) 0

5. (a) $\sqrt{45}$ (b) 0

6. (a) $\sqrt{18}$ (b) 0

7. $\sqrt{18}$ **8.** (a) $\sqrt{98}$ (b) 0

9. (a) $\frac{-1}{\sqrt{2}}$ (b) $\frac{-3}{\sqrt{73}}$ (c) 0

(d) $\frac{-20}{9\sqrt{10}}$ (e) $\frac{-1}{\sqrt{2}}$ (f) $\frac{2}{\sqrt{55}}$

10. (a) 0 (b) 0

11. (a) $\frac{19}{10\sqrt{7}}$ (b) 0

12. (a) $k = -3$ (b) $k = -2, k = -3$

14. a, b, c

15. $\pm\frac{1}{\sqrt{3249}}(-34, 44, -6, 11)$

EXERCISE SET 4.9 (page 182).

1. b **2.** b, d

3. a **4.** a

7. (a) $\left(\frac{1}{\sqrt{10}}, -\frac{3}{\sqrt{10}}\right), \left(\frac{3}{\sqrt{10}}, \frac{1}{\sqrt{10}}\right)$ (b) $(1, 0), (0, -1)$

8. (a) $\left(\frac{1}{\sqrt{3}}, \frac{1}{\sqrt{3}}, \frac{1}{\sqrt{3}}\right), \left(-\frac{1}{\sqrt{2}}, \frac{1}{\sqrt{2}}, 0\right), \left(\frac{1}{\sqrt{6}}, \frac{1}{\sqrt{6}}, -\frac{2}{\sqrt{6}}\right)$

(b) $(1, 0, 0), \left(0, \frac{7}{\sqrt{53}}, -\frac{2}{\sqrt{53}}\right), \left(0, \frac{30}{\sqrt{11925}}, \frac{105}{\sqrt{11925}}\right)$

9. $\left(0, \frac{2}{\sqrt{5}}, \frac{1}{\sqrt{5}}, 0\right), \left(\frac{5}{\sqrt{30}}, -\frac{1}{\sqrt{30}}, \frac{2}{\sqrt{30}}, 0\right),$

$\left(\frac{1}{\sqrt{10}}, \frac{1}{\sqrt{10}}, -\frac{2}{\sqrt{10}}, -\frac{2}{\sqrt{10}}\right), \left(\frac{1}{\sqrt{15}}, \frac{1}{\sqrt{15}}, -\frac{2}{\sqrt{15}}, \frac{3}{\sqrt{15}}\right)$

10. $\left(0, \frac{1}{\sqrt{5}}, \frac{2}{\sqrt{5}}\right), \left(-\frac{\sqrt{5}}{\sqrt{6}}, -\frac{2}{\sqrt{30}}, \frac{1}{\sqrt{30}}\right)$

11. $\left(\frac{1}{\sqrt{6}}, \frac{1}{\sqrt{6}}, \frac{1}{\sqrt{6}}\right), \left(\frac{1}{\sqrt{6}}, \frac{1}{\sqrt{6}}, \frac{1}{\sqrt{6}}\right), \left(\frac{2}{\sqrt{6}}, -\frac{1}{\sqrt{6}}, 0\right)$

12. $\mathbf{w}_1 = (-\frac{4}{5}, 2, \frac{3}{5}), \mathbf{w}_2 = (\frac{9}{5}, 0, \frac{12}{5})$

13. $\mathbf{w}_1 = (\frac{39}{42}, \frac{93}{42}, \frac{120}{42}), \mathbf{w}_2 = (\frac{3}{42}, -\frac{9}{42}, \frac{6}{42})$

14. $\mathbf{w}_1 = (-\frac{5}{4}, -\frac{1}{4}, \frac{5}{4}, \frac{9}{4}), \mathbf{w}_2 = (\frac{1}{4}, \frac{9}{4}, \frac{19}{4}, -\frac{9}{4})$

22. $Q(-\frac{8}{7}, -\frac{5}{7}, \frac{25}{7}); \frac{3}{7}\sqrt{35}$

23. $Q(-\frac{8}{7}, \frac{4}{7}, -\frac{16}{7})$

EXERCISE SET 4.10 (page 200).

1. (a) $(w)_S = (3, -7), [w]_S = \begin{bmatrix} 3 \\ -7 \end{bmatrix}$

(b) $(w)_S = (\frac{5}{28}, \frac{3}{14}), [w]_S = \begin{bmatrix} \frac{5}{28} \\ \frac{3}{14} \end{bmatrix}$

(c) $(w)_S = \left(a, \frac{b-a}{2}\right), [w]_S = \begin{bmatrix} a \\ \frac{b-a}{2} \end{bmatrix}$

2. (a) $(\mathbf{v})_B = (3, -2, 1)$, $[\mathbf{v}]_B = \begin{bmatrix} 3 \\ -2 \\ 1 \end{bmatrix}$

(b) $(\mathbf{v})_B = (-2, 0, 1)$, $[\mathbf{v}]_B = \begin{bmatrix} -2 \\ 0 \\ 1 \end{bmatrix}$

3. (a) $(\mathbf{p})_S = (4, -3, 1)$, $[\mathbf{p}]_S = \begin{bmatrix} 4 \\ -3 \\ 1 \end{bmatrix}$

(b) $(\mathbf{p})_B = (0, 2, -1)$, $[\mathbf{p}]_B = \begin{bmatrix} 0 \\ 2 \\ -1 \end{bmatrix}$

4. $(A)_B = (-1, 1, -1, 3)$, $[A]_B = \begin{bmatrix} -1 \\ 1 \\ -1 \\ 3 \end{bmatrix}$

5. (a) $(\mathbf{w})_S = (-2\sqrt{2}, 5\sqrt{2})$, $[\mathbf{w}]_S = \begin{bmatrix} -2\sqrt{2} \\ 5\sqrt{2} \end{bmatrix}$

(b) $(\mathbf{w})_S = (0, -2, 1)$, $[\mathbf{w}]_S = \begin{bmatrix} 0 \\ -2 \\ 1 \end{bmatrix}$

6. (a) $\mathbf{w} = (16, 10, 12)$ (b) $\mathbf{q} = 3 + 4x^2$ (c) $B = \begin{bmatrix} 15 & -1 \\ 6 & 3 \end{bmatrix}$

7. (a) $\|\mathbf{u}\| = \sqrt{2}$, $d(\mathbf{u}, \mathbf{v}) = \sqrt{13}$, $\langle \mathbf{u}, \mathbf{v} \rangle = 3$

8. (a) $\begin{bmatrix} \frac{4}{11} & \frac{3}{11} \\ -\frac{1}{11} & \frac{2}{11} \end{bmatrix}$ (b) $\begin{bmatrix} -\frac{3}{11} \\ -\frac{13}{11} \end{bmatrix}$ (d) $\begin{bmatrix} 2 & -3 \\ 1 & 4 \end{bmatrix}$

9. (a) $\begin{bmatrix} 0 & -\frac{5}{2} \\ -2 & -\frac{13}{2} \end{bmatrix}$ (b) $\begin{bmatrix} -4 \\ -7 \end{bmatrix}$ (d) $\begin{bmatrix} \frac{13}{10} & -\frac{1}{2} \\ -\frac{2}{5} & 0 \end{bmatrix}$

10. (a) $\begin{bmatrix} \frac{3}{4} & \frac{3}{4} & \frac{1}{12} \\ -\frac{3}{4} & -\frac{17}{12} & -\frac{17}{12} \\ 0 & \frac{2}{3} & \frac{2}{3} \end{bmatrix}$ (b) $\begin{bmatrix} \frac{19}{12} \\ -\frac{43}{12} \\ \frac{4}{3} \end{bmatrix}$

11. (a) $\begin{bmatrix} 3 & 2 & \frac{5}{2} \\ -2 & -3 & -\frac{1}{2} \\ 5 & 1 & 6 \end{bmatrix}$ (b) $\begin{bmatrix} -\frac{7}{2} \\ \frac{23}{2} \\ 6 \end{bmatrix}$

12. (a) $\begin{bmatrix} \frac{3}{4} & \frac{7}{2} \\ \frac{3}{2} & 1 \end{bmatrix}$ (b) $\begin{bmatrix} -\frac{11}{4} \\ \frac{1}{2} \end{bmatrix}$ (d) $\begin{bmatrix} -\frac{2}{9} & \frac{7}{9} \\ \frac{1}{3} & -\frac{1}{6} \end{bmatrix}$

13. (b) $\begin{bmatrix} \frac{1}{2} & 0 \\ -\frac{1}{6} & \frac{1}{3} \end{bmatrix}$ (c) $\begin{bmatrix} 1 \\ -2 \end{bmatrix}$ (e) $\begin{bmatrix} 2 & 0 \\ 1 & 3 \end{bmatrix}$

14. (a) $(4\sqrt{2}, -2\sqrt{2})$
(b) $(-3.5\sqrt{2}, 1.5\sqrt{2})$

15. (a) $(-1 + 3\sqrt{3}, 3 + \sqrt{3})$
(b) $(2.5 - \sqrt{3}, 2.5\sqrt{3} + 1)$

16. (a) $(.5\sqrt{2}, 1.5\sqrt{2}, 5)$
(b) $(-2.5\sqrt{2}, 3.5\sqrt{2}, -3)$

17. (a) $(-.5 - 2.5\sqrt{3}, 2, 2.5 - .5\sqrt{3})$
(b) $(.5 - 1.5\sqrt{3}, 6, -1.5 - .5\sqrt{3})$

18. (a) $(-1, 1.5\sqrt{2}, -3.5\sqrt{2})$
(b) $(1, -1.5\sqrt{2}, 4.5\sqrt{2})$

19. *a, b, d, e*

20. (a) $\begin{bmatrix} 1 & 0 \\ 0 & 1 \end{bmatrix}$ (b) $\begin{bmatrix} \frac{1}{\sqrt{2}} & \frac{1}{\sqrt{2}} \\ -\frac{1}{\sqrt{2}} & \frac{1}{\sqrt{2}} \end{bmatrix}$ (d) $\begin{bmatrix} -\frac{1}{\sqrt{2}} & 0 & \frac{1}{\sqrt{2}} \\ \frac{1}{\sqrt{6}} & -\frac{2}{\sqrt{6}} & \frac{1}{\sqrt{6}} \\ \frac{1}{\sqrt{3}} & \frac{1}{\sqrt{3}} & \frac{1}{\sqrt{3}} \end{bmatrix}$

(e) $\begin{bmatrix} \frac{1}{2} & \frac{1}{2} & \frac{1}{2} & \frac{1}{2} \\ \frac{1}{2} & \frac{5}{6} & \frac{1}{6} & \frac{1}{6} \\ \frac{1}{2} & \frac{1}{6} & \frac{1}{6} & -\frac{5}{6} \\ \frac{1}{2} & \frac{1}{6} & -\frac{5}{6} & \frac{1}{6} \end{bmatrix}$

22. (a) $\begin{bmatrix} \cos\theta & \sin\theta \\ -\sin\theta & \cos\theta \end{bmatrix}$ (b) $\begin{bmatrix} \cos\theta & \sin\theta & 0 \\ -\sin\theta & \cos\theta & 0 \\ 0 & 0 & 1 \end{bmatrix}$

23. (a) $(-2, -1)$ (b) $(-\frac{4}{5}, -\frac{22}{5})$ (c) $(-\frac{11}{5}, \frac{52}{5})$ (d) $(0, 0)$

25. *a, c*

27. (a) $(\frac{12}{5}, -\frac{9}{5}, -7)$ (b) $(2, 1, 6)$ (c) $(-\frac{42}{5}, \frac{19}{5}, -3)$
(d) $(0, 0, 0)$

29. *b*

31. (*a*) $A = \begin{bmatrix} \cos\theta & 0 & -\sin\theta \\ 0 & 1 & 0 \\ \sin\theta & 0 & \cos\theta \end{bmatrix}$ (b) $A = \begin{bmatrix} 1 & 0 & 0 \\ 0 & \cos\theta & \sin\theta \\ 0 & -\sin\theta & \cos\theta \end{bmatrix}$

32. $\begin{bmatrix} \frac{\sqrt{2}}{4} & \frac{\sqrt{6}}{4} & -\frac{\sqrt{2}}{2} \\ -\frac{\sqrt{3}}{2} & \frac{1}{2} & 0 \\ \frac{\sqrt{2}}{4} & \frac{\sqrt{6}}{4} & \frac{\sqrt{2}}{2} \end{bmatrix}$

EXERCISE SET 5.1 (page 211).

1. Linear **2.** Nonlinear **3.** Linear **4.** Linear

5. Nonlinear **6.** Linear **7.** Linear **8.** Nonlinear

9. Linear **10.** Linear **11.** Nonlinear **12.** Linear

13. Linear **14.** Nonlinear **15.** Linear **16.** Nonlinear

17. Linear **18.** Linear **19.** Linear **20.** Nonlinear

21. $F(x, y) = (-x, y)$

23. (a) $\begin{bmatrix} 1 & 3 & 4 \\ 1 & 0 & -7 \end{bmatrix}$ (b) $\begin{bmatrix} 42 \\ -55 \end{bmatrix}$ (c) $\begin{bmatrix} x + 3y + 4z \\ x - 7z \end{bmatrix}$

24. (a) $(2, 0, -1)$ (b) $T(x, y, z) = (x, 0, z)$

25. (a) $T(3, 8, 4) = (-2, 3, -1)$
(b) $T(x, y, z) = \frac{1}{3}(2x - y - z, -x + 2y - z, -x - y + 2z)$

26. (a) $T(-1, 2) = \left(-\frac{3}{\sqrt{2}}, \frac{1}{\sqrt{2}}\right)$; $T(x, y) = \left(\frac{x}{\sqrt{2}} - \frac{y}{\sqrt{2}}, \frac{x}{\sqrt{2}} + \frac{y}{\sqrt{2}}\right)$

(b) $T(-1, 2) = (1, -2)$; $T(x, y) = (-x, -y)$

(c) $T(-1, 2) = \left(-\frac{\sqrt{3}}{2} - 1, \sqrt{3} - \frac{1}{2}\right)$;

$$T(x, y) = \left(\frac{\sqrt{3}}{2}x - \frac{1}{2}y, \frac{1}{2}x + \frac{\sqrt{3}}{2}y\right)$$

(d) $T(-1, 2) = \left(-\frac{1}{2} + \sqrt{3}, \frac{\sqrt{3}}{2} + 1\right)$;

$$T(x, y) = \left(\frac{1}{2}x + \frac{\sqrt{3}}{2}y, -\frac{\sqrt{3}}{2}x + \frac{1}{2}y\right)$$

EXERCISE SET 5.2 (page 219).

1. a, c **2.** a **3.** a, b, c **4.** a

5. b **6.** a **7.** $\ker(T) = \{\mathbf{0}\}$; $R(T) = V$

8. Rank $(T) = 1$, nullity $(T) = 1$

9. Rank $(T) = 3$, nullity $(T) = 0$

10. (a) Rank $(T) = n$, nullity $(T) = 0$
(b) Rank $(T) = 0$, nullity $(T) = n$
(c) Rank $(T) = n$, nullity $(T) = 0$

11. $T(x, y, z) = (30x - 10y - 3z, -9x + 3y + z)$, $T(1, 1, 1) = (17, -5)$

12. $T(2 - 2x + 3x^2) = 8 + 8x - 7x^2$

13. (a) Nullity $(T) = 2$ (b) Nullity $(T) = 4$

(c) Nullity $(T) = 3$ (d) Nullity $(T) = 1$

14. Nullity $(T) = 0$, rank $(T) = 6$

15. (a) dimension = nullity $(T) = 3$

(b) No. In order for $A\mathbf{x} = \mathbf{b}$ to be consistent for all $\mathbf{b}$ in R^5, we must have $R(T) = R^5$. But $R(T) \neq R^5$ since rank $(T) = \dim R(T) = 4$.

16. (a) $\begin{bmatrix} 1 \\ 5 \\ 7 \end{bmatrix}, \begin{bmatrix} 0 \\ 1 \\ 1 \end{bmatrix}$ (b) $\begin{bmatrix} -\frac{14}{11} \\ \frac{19}{11} \\ 1 \end{bmatrix}$

(c) Rank $(T) = 2$, nullity $(T) = 1$

17. (a) $\begin{bmatrix} 1 \\ 2 \\ 0 \end{bmatrix}$ (b) $\begin{bmatrix} \frac{1}{2} \\ 0 \\ 1 \end{bmatrix}, \begin{bmatrix} 0 \\ 1 \\ 0 \end{bmatrix}$

(c) Rank $(T) = 1$, nullity $(T) = 2$.

18. (a) $\begin{bmatrix} 1 \\ \frac{1}{4} \end{bmatrix}, \begin{bmatrix} 0 \\ 1 \end{bmatrix}$ (b) $\begin{bmatrix} -1 \\ -1 \\ 1 \\ 0 \end{bmatrix}, \begin{bmatrix} -\frac{4}{7} \\ \frac{2}{7} \\ 0 \\ 1 \end{bmatrix}$

(c) Rank $(T) = 2$, nullity $(T) = 2$.

19. (a) $\begin{bmatrix} 1 \\ 3 \\ -1 \\ 2 \end{bmatrix}, \begin{bmatrix} 0 \\ 1 \\ -\frac{2}{7} \\ \frac{5}{14} \end{bmatrix}, \begin{bmatrix} 0 \\ 0 \\ 0 \\ 1 \end{bmatrix}$ (b) $\begin{bmatrix} -1 \\ -1 \\ 1 \\ 0 \\ 0 \end{bmatrix}, \begin{bmatrix} -1 \\ -2 \\ 0 \\ 0 \\ 1 \end{bmatrix}$

(c) Rank $(T) = 3$, nullity $(T) = 2$

22. (a) $14x - 8y - 5z = 0$ (b) $x = -t, y = -t, z = t, -\infty < t < +\infty$

26. $\ker(D)$ consists of all constant polynomials

27. $\ker(J)$ consists of all polynomials of the form kx

EXERCISE SET 5.3 (page 229).

1. (a) $\begin{bmatrix} 2 & -1 \\ 1 & 1 \end{bmatrix}$ (b) $\begin{bmatrix} 1 & 0 \\ 0 & 1 \end{bmatrix}$ (c) $\begin{bmatrix} 1 & 2 & 1 \\ 1 & 5 & 0 \\ 0 & 0 & 1 \end{bmatrix}$ (d) $\begin{bmatrix} 4 & 0 & 0 \\ 0 & 7 & 0 \\ 0 & 0 & -8 \end{bmatrix}$

2. (a) $\begin{bmatrix} 0 & 1 \\ -1 & 0 \\ 1 & 3 \\ 1 & -1 \end{bmatrix}$ (b) $\begin{bmatrix} 7 & 2 & -1 & 1 \\ 0 & 1 & 1 & 0 \\ -1 & 0 & 0 & 0 \end{bmatrix}$

(c) $\begin{bmatrix} 0 & 0 & 0 \\ 0 & 0 & 0 \\ 0 & 0 & 0 \\ 0 & 0 & 0 \\ 0 & 0 & 0 \end{bmatrix}$ (d) $\begin{bmatrix} 0 & 0 & 0 & 1 \\ 1 & 0 & 0 & 0 \\ 0 & 0 & 1 & 0 \\ 0 & 1 & 0 & 0 \\ 1 & 0 & -1 & 0 \end{bmatrix}$

3. (a) $\begin{bmatrix} 1 & 0 \\ 0 & -1 \end{bmatrix}$ (b) $\begin{bmatrix} 0 & 1 \\ 1 & 0 \end{bmatrix}$ (c) $\begin{bmatrix} -1 & 0 \\ 0 & -1 \end{bmatrix}$ (d) $\begin{bmatrix} 1 & 0 \\ 0 & 0 \end{bmatrix}$

4. (a) $(2, -1)$ (b) $(1, 2)$ (c) $(-2, -1)$ (d) $(2, 0)$

5. (a) $\begin{bmatrix} 1 & 0 & 0 \\ 0 & 1 & 0 \\ 0 & 0 & -1 \end{bmatrix}$ (b) $\begin{bmatrix} 1 & 0 & 0 \\ 0 & -1 & 0 \\ 0 & 0 & 1 \end{bmatrix}$ (c) $\begin{bmatrix} -1 & 0 & 0 \\ 0 & 1 & 0 \\ 0 & 0 & 1 \end{bmatrix}$

6. (a) $(1, 1, -1)$
(b) $(1, -1, 1)$
(c) $(-1, 1, 1)$

7. (a) $\begin{bmatrix} 0 & -1 & 0 \\ 1 & 0 & 0 \\ 0 & 0 & 1 \end{bmatrix}$ (b) $\begin{bmatrix} 1 & 0 & 0 \\ 0 & 0 & -1 \\ 0 & 1 & 0 \end{bmatrix}$ (c) $\begin{bmatrix} 0 & 0 & 1 \\ 0 & 1 & 0 \\ -1 & 0 & 0 \end{bmatrix}$

8. $\begin{bmatrix} 1 & 1 & 0 \\ 0 & -2 & -3 \end{bmatrix}$

9. (a) $\begin{bmatrix} 0 & 0 \\ -\frac{1}{2} & 1 \\ \frac{8}{3} & \frac{4}{3} \end{bmatrix}$ (b) $\begin{bmatrix} 14 \\ -8 \\ 0 \end{bmatrix}$

10. (a) $\begin{bmatrix} 1 & -\frac{3}{2} & \frac{1}{2} \\ -1 & \frac{1}{2} & \frac{1}{2} \\ 0 & \frac{1}{2} & -\frac{1}{2} \end{bmatrix}$ (b) $\begin{bmatrix} 2 \\ -2 \\ 2 \end{bmatrix}$

11. (a) $\begin{bmatrix} 0 & 0 & 0 \\ 0 & 0 & 0 \\ 1 & 1 & 4 \\ 0 & 2 & 5 \\ 1 & 3 & 1 \end{bmatrix}$ (b) $-3x^2 + 5x^3 - 2x^4$

12. (a) $[T(\mathbf{v}_1)]_B = \begin{bmatrix} 1 \\ -2 \end{bmatrix}$, $[T(\mathbf{v}_2)]_B = \begin{bmatrix} 3 \\ 5 \end{bmatrix}$

(b) $T(\mathbf{v}_1) = \begin{bmatrix} 3 \\ -5 \end{bmatrix}$, $T(\mathbf{v}_2) = \begin{bmatrix} -2 \\ 29 \end{bmatrix}$ (c) $\begin{bmatrix} \frac{19}{7} \\ -\frac{83}{7} \end{bmatrix}$

13. (a) $[T(\mathbf{v}_1)]_{B'} = \begin{bmatrix} 3 \\ 1 \\ -3 \end{bmatrix}$, $[T(\mathbf{v}_2)]_{B'} = \begin{bmatrix} -2 \\ 6 \\ 0 \end{bmatrix}$

$[T(\mathbf{v}_3)]_{B'} = \begin{bmatrix} 1 \\ 2 \\ 7 \end{bmatrix}$, $[T(\mathbf{v}_4)]_{B'} = \begin{bmatrix} 0 \\ 1 \\ 1 \end{bmatrix}$

(b) $T(\mathbf{v}_1) = \begin{bmatrix} 11 \\ 5 \\ 22 \end{bmatrix}$, $T(\mathbf{v}_2) = \begin{bmatrix} -42 \\ 32 \\ -10 \end{bmatrix}$, $T(\mathbf{v}_3) = \begin{bmatrix} -56 \\ 87 \\ 17 \end{bmatrix}$

$T(\mathbf{v}_4) = \begin{bmatrix} -13 \\ 17 \\ 2 \end{bmatrix}$ (c) $\begin{bmatrix} -31 \\ 37 \\ 12 \end{bmatrix}$

14. (a) $[T(\mathbf{v}_1)]_B = \begin{bmatrix} 1 \\ 2 \\ 6 \end{bmatrix}$, $[T(\mathbf{v}_2)]_B = \begin{bmatrix} 3 \\ 0 \\ -2 \end{bmatrix}$, $[T(\mathbf{v}_3)]_B = \begin{bmatrix} -1 \\ 5 \\ 4 \end{bmatrix}$

(b) $T(\mathbf{v}_1) = 16 + 51x + 19x^2$, $T(\mathbf{v}_2) = -6 - 5x + 5x^2$,
$T(\mathbf{v}_3) = 7 + 40x + 15x^2$

(c) $T(1 + x^2) = 22 + 56x + 14x^2$

18. (a) $\begin{bmatrix} 0 & 1 & 0 \\ 0 & 0 & 2 \\ 0 & 0 & 0 \end{bmatrix}$ (b) $\begin{bmatrix} 0 & -\frac{3}{2} & \frac{23}{6} \\ 0 & 0 & -\frac{16}{3} \\ 0 & 0 & 0 \end{bmatrix}$ (c) $-6 + 48x$

19. (a) $\begin{bmatrix} 0 & 0 & 0 \\ 0 & 0 & -1 \\ 0 & 1 & 0 \end{bmatrix}$ (b) $\begin{bmatrix} 0 & 0 & 0 \\ 0 & 1 & 0 \\ 0 & 0 & 2 \end{bmatrix}$ (c) $\begin{bmatrix} 2 & 1 & 0 \\ 0 & 2 & 2 \\ 0 & 0 & 2 \end{bmatrix}$

EXERCISE SET 5.4 (page 237).

1. $[T]_B = \begin{bmatrix} 1 & -2 \\ 0 & -1 \end{bmatrix}$ $[T]_{B'} = \begin{bmatrix} -\frac{3}{11} & -\frac{56}{11} \\ -\frac{2}{11} & \frac{3}{11} \end{bmatrix}$

2. $[T]_B = \begin{bmatrix} \frac{4}{5} & \frac{61}{10} \\ \frac{18}{5} & -\frac{19}{5} \end{bmatrix}$ $[T]_{B'} = \begin{bmatrix} -\frac{155}{10} & \frac{9}{2} \\ -\frac{375}{10} & \frac{25}{2} \end{bmatrix}$

3. $[T]_B = \begin{bmatrix} \dfrac{1}{\sqrt{2}} & -\dfrac{1}{\sqrt{2}} \\ \dfrac{1}{\sqrt{2}} & \dfrac{1}{\sqrt{2}} \end{bmatrix}$ $[T]_{B'} = \begin{bmatrix} \dfrac{13}{11\sqrt{2}} & -\dfrac{25}{11\sqrt{2}} \\ \dfrac{5}{11\sqrt{2}} & \dfrac{9}{11\sqrt{2}} \end{bmatrix}$

4. $[T]_B = \begin{bmatrix} 1 & 2 & -1 \\ 0 & -1 & 0 \\ 1 & 0 & 7 \end{bmatrix}$ $[T]_{B'} = \begin{bmatrix} 1 & 4 & 3 \\ -1 & -2 & -9 \\ 1 & 1 & 8 \end{bmatrix}$

5. $[T]_B = \begin{bmatrix} 1 & 0 & 0 \\ 0 & 1 & 0 \\ 0 & 0 & 0 \end{bmatrix}$ $[T]_{B'} = \begin{bmatrix} 1 & 0 & 0 \\ 0 & 1 & 1 \\ 0 & 0 & 0 \end{bmatrix}$

6. $[T]_B = \begin{bmatrix} 5 & 0 \\ 0 & 5 \end{bmatrix}$ $[T]_{B'} = \begin{bmatrix} 5 & 0 \\ 0 & 5 \end{bmatrix}$

7. $[T]_B = \begin{bmatrix} \frac{2}{3} & -\frac{2}{9} \\ \frac{1}{2} & \frac{4}{3} \end{bmatrix}$ $[T]_{B'} = \begin{bmatrix} 1 & 1 \\ 0 & 1 \end{bmatrix}$

EXERCISE SET 6.1 (page 244).

1. (a) $\lambda^2 - 2\lambda - 3 = 0$ (b) $\lambda^2 - 8\lambda + 16 = 0$
   (c) $\lambda^2 - 12 = 0$ (d) $\lambda^2 + 3 = 0$
   (e) $\lambda^2 = 0$ (f) $\lambda^2 - 2\lambda + 1 = 0$

2. (a) $\lambda = 3, \lambda = -1$ (b) $\lambda = 4$
   (c) $\lambda = \sqrt{12}, \lambda = -\sqrt{12}$ (d) No real eigenvalues
   (e) $\lambda = 0$ (f) $\lambda = 1$

3. (a) Basis for eigenspace corresponding to $\lambda = 3$: $\begin{bmatrix} \frac{1}{2} \\ 1 \end{bmatrix}$

   basis for eigenspace corresponding to $\lambda = -1$: $\begin{bmatrix} 0 \\ 1 \end{bmatrix}$

   (b) Basis for eigenspace corresponding to $\lambda = 4$: $\begin{bmatrix} \frac{3}{2} \\ 1 \end{bmatrix}$

   (c) Basis for eigenspace corresponding to $\lambda = \sqrt{12}$: $\begin{bmatrix} \dfrac{3}{\sqrt{12}} \\ 1 \end{bmatrix}$

   basis for eigenspace corresponding to $\lambda = -\sqrt{12}$: $\begin{bmatrix} -\dfrac{3}{\sqrt{12}} \\ 1 \end{bmatrix}$

   (d) There are no eigenspaces.

   (e) Basis for eigenspace corresponding to $\lambda = 0$: $\begin{bmatrix} 1 \\ 0 \end{bmatrix}, \begin{bmatrix} 0 \\ 1 \end{bmatrix}$

(f) Basis for eigenspace corresponding to $\lambda = 1$: $\begin{bmatrix} 1 \\ 0 \end{bmatrix}, \begin{bmatrix} 0 \\ 1 \end{bmatrix}$

5. (a) $\lambda^3 - 6\lambda^2 + 11\lambda - 6 = 0$ (b) $\lambda^3 - 2\lambda = 0$
(c) $\lambda^3 + 8\lambda^2 + \lambda + 8 = 0$ (d) $\lambda^3 - \lambda^2 - \lambda - 2 = 0$
(e) $\lambda^3 - 6\lambda^2 + 12\lambda - 8 = 0$ (f) $\lambda^3 - 2\lambda^2 - 15\lambda + 36 = 0$

6. (a) $\lambda = 1, \lambda = 2, \lambda = 3$ (b) $\lambda = 0, \lambda = \sqrt{2}, \lambda = -\sqrt{2}$
(c) $\lambda = -8$ (d) $\lambda = 2$ (e) $\lambda = 2$
(f) $\lambda = -4, \lambda = 3$

7. (a) $\lambda = 1$: basis $\begin{bmatrix} 0 \\ 1 \\ 0 \end{bmatrix}$, $\lambda = 2$: basis $\begin{bmatrix} -\frac{1}{2} \\ 1 \\ 1 \end{bmatrix}$, $\lambda = 3$: basis $\begin{bmatrix} -1 \\ 1 \\ 1 \end{bmatrix}$

(b) $\lambda = 0$: basis $\begin{bmatrix} \frac{5}{3} \\ \frac{1}{3} \\ 1 \end{bmatrix}$, $\lambda = \sqrt{2}$: basis $\begin{bmatrix} \frac{1}{7}(15 + 5\sqrt{2}) \\ \frac{1}{7}(-1 + 2\sqrt{2}) \\ 1 \end{bmatrix}$

$\lambda = -\sqrt{2}$: basis $\begin{bmatrix} \frac{1}{7}(15 - 5\sqrt{2}) \\ \frac{1}{7}(-1 - 2\sqrt{2}) \\ 1 \end{bmatrix}$

(c) $\lambda = -8$: basis $\begin{bmatrix} -\frac{1}{6} \\ -\frac{1}{6} \\ 1 \end{bmatrix}$ (d) $\lambda = 2$: basis $\begin{bmatrix} \frac{1}{3} \\ \frac{1}{3} \\ 1 \end{bmatrix}$

(e) $\lambda = 2$: basis $\begin{bmatrix} -\frac{1}{3} \\ -\frac{1}{3} \\ 1 \end{bmatrix}$ (f) $\lambda = -4$: basis $\begin{bmatrix} -2 \\ \frac{8}{3} \\ 1 \end{bmatrix}$, $\lambda = 3$: basis $\begin{bmatrix} 5 \\ -2 \\ 1 \end{bmatrix}$

8. (a) $(\lambda - 1)^2(\lambda + 2)(\lambda + 1) = 0$ (b) $(\lambda - 4)^2(\lambda^2 + 3) = 0$

9. (a) $\lambda = 1, \lambda = -2, \lambda = -1$ (b) $\lambda = 4$

10. (a) $\lambda = 1$: basis $\begin{bmatrix} 0 \\ 0 \\ 0 \\ 1 \end{bmatrix}$, $\lambda = -2$: basis $\begin{bmatrix} -1 \\ 0 \\ 1 \\ 0 \end{bmatrix}$, $\lambda = -1$: basis $\begin{bmatrix} -2 \\ 1 \\ 1 \\ 0 \end{bmatrix}$

(b) $\lambda = 4$: basis $\begin{bmatrix} \frac{3}{2} \\ 1 \\ 0 \\ 0 \end{bmatrix}$

11. (a) $\lambda = -4, \lambda = 3$
(b) Basis for eigenspace corresponding to $\lambda = -4$: $-2 + \frac{8}{3}x + x^2$
basis for eigenspace corresponding to $\lambda = 3$: $5 - 2x + x^2$

12. (a) $\lambda = 1, \lambda = -2, \lambda = -1$

(b) Basis for eigenspace corresponding to $\lambda = 1$: $\begin{bmatrix} 0 & 0 \\ 0 & 1 \end{bmatrix}$; basis for eigenspace corresponding to $\lambda = -2$: $\begin{bmatrix} -1 & 0 \\ 1 & 0 \end{bmatrix}$; basis for eigenspace corresponding to $\lambda = -1$: $\begin{bmatrix} -2 & 1 \\ 1 & 0 \end{bmatrix}$

18. $1, -1, -2^9, 2^9$.

EXERCISE SET 6.2 (page 253).

5. $P = \begin{bmatrix} \frac{4}{5} & \frac{3}{4} \\ 1 & 1 \end{bmatrix} \quad P^{-1}AP = \begin{bmatrix} 1 & 0 \\ 0 & 2 \end{bmatrix}$

6. $P = \begin{bmatrix} \frac{1}{3} & 0 \\ 1 & 1 \end{bmatrix} \quad P^{-1}AP = \begin{bmatrix} 1 & 0 \\ 0 & -1 \end{bmatrix}$

7. $P = \begin{bmatrix} 0 & 1 & 0 \\ 1 & 0 & 1 \\ -1 & 0 & 1 \end{bmatrix} \quad P^{-1}AP = \begin{bmatrix} 0 & 0 & 0 \\ 0 & 1 & 0 \\ 0 & 0 & 2 \end{bmatrix}$

8. $P = \begin{bmatrix} -2 & 0 & 1 \\ 0 & 1 & 0 \\ 1 & 0 & 0 \end{bmatrix} \quad P^{-1}AP = \begin{bmatrix} 3 & 0 & 0 \\ 0 & 3 & 0 \\ 0 & 0 & 2 \end{bmatrix}$

9. Not diagonalizable

10. $P = \begin{bmatrix} 1 & 2 & 1 \\ 1 & 3 & 3 \\ 1 & 3 & 4 \end{bmatrix} \quad P^{-1}AP = \begin{bmatrix} 1 & 0 & 0 \\ 0 & 2 & 0 \\ 0 & 0 & 3 \end{bmatrix}$

11. Not diagonalizable

12. $P = \begin{bmatrix} -\frac{1}{3} & 0 & 0 \\ 0 & 1 & 0 \\ 1 & 0 & 1 \end{bmatrix} \quad P^{-1}AP = \begin{bmatrix} 0 & 0 & 0 \\ 0 & 0 & 0 \\ 0 & 0 & 1 \end{bmatrix}$

13. Not diagonalizable

14. $P = \begin{bmatrix} 1 & 1 & 0 & 0 \\ 0 & 1 & 1 & 0 \\ 0 & 0 & 1 & 1 \\ 0 & 0 & 0 & 1 \end{bmatrix} \quad P^{-1}AP = \begin{bmatrix} -2 & 0 & 0 & 0 \\ 0 & -2 & 0 & 0 \\ 0 & 0 & 3 & 0 \\ 0 & 0 & 0 & 3 \end{bmatrix}$

15. $\begin{bmatrix} 2 \\ 1 \end{bmatrix}, \begin{bmatrix} 1 \\ -1 \end{bmatrix}$

16. $\begin{bmatrix} 1 \\ 1 \\ -1 \end{bmatrix}, \begin{bmatrix} 1 \\ 0 \\ 1 \end{bmatrix}, \begin{bmatrix} 1 \\ 1 \\ 0 \end{bmatrix}$

17. $\mathbf{p}_1 = \frac{1}{3} + x$, $\mathbf{p}_2 = x$

19. $\begin{bmatrix} 1 & 0 \\ -1023 & 1024 \end{bmatrix}$

EXERCISE SET 6.3 (page 259).

1. (a) $\lambda = 0$: 1-dimensional, $\lambda = 2$: 1-dimensional
(b) $\lambda = 1$: 1-dimensional, $\lambda = -1$: 2-dimensional
(c) $\lambda = 3$: 1-dimensional, $\lambda = 0$: 2-dimensional
(d) $\lambda = 0$: 1-dimensional, $\lambda = 6$: 2-dimensional
(e) $\lambda = 0$: 3-dimensional, $\lambda = 8$: 1-dimensional
(f) $\lambda = -2$: 3-dimensional, $\lambda = 4$: 1-dimensional

2. $P = \begin{bmatrix} \frac{1}{\sqrt{2}} & -\frac{1}{\sqrt{2}} \\ \frac{1}{\sqrt{2}} & \frac{1}{\sqrt{2}} \end{bmatrix} \quad P^{-1}AP = \begin{bmatrix} 4 & 0 \\ 0 & 2 \end{bmatrix}$

3. $P = \begin{bmatrix} \frac{\sqrt{3}}{2} & -\frac{1}{2} \\ \frac{1}{2} & \frac{\sqrt{3}}{2} \end{bmatrix} \quad P^{-1}AP = \begin{bmatrix} 8 & 0 \\ 0 & -4 \end{bmatrix}$

4. $P = \begin{bmatrix} \frac{3}{5} & -\frac{4}{5} \\ \frac{4}{5} & \frac{3}{5} \end{bmatrix} \quad P^{-1}AP = \begin{bmatrix} 25 & 0 \\ 0 & -25 \end{bmatrix}$

5. $P = \begin{bmatrix} -\frac{4}{5} & 0 & \frac{3}{5} \\ 0 & 1 & 0 \\ \frac{3}{5} & 0 & \frac{4}{5} \end{bmatrix} \quad P^{-1}AP = \begin{bmatrix} 25 & 0 & 0 \\ 0 & -3 & 0 \\ 0 & 0 & -50 \end{bmatrix}$

6. $\begin{bmatrix} \frac{1}{\sqrt{2}} & \frac{1}{\sqrt{2}} & 0 \\ \frac{1}{\sqrt{2}} & -\frac{1}{\sqrt{2}} & 0 \\ 0 & 0 & 1 \end{bmatrix}$

7. $$\begin{bmatrix} \frac{1}{\sqrt{3}} & \frac{1}{\sqrt{6}} & \frac{1}{\sqrt{2}} \\ \frac{1}{\sqrt{3}} & -\frac{2}{\sqrt{6}} & 0 \\ \frac{1}{\sqrt{3}} & \frac{1}{\sqrt{6}} & -\frac{1}{\sqrt{2}} \end{bmatrix}$$

8. $$\begin{bmatrix} 0 & 0 & \frac{1}{\sqrt{2}} & \frac{1}{\sqrt{2}} \\ 0 & 0 & \frac{1}{\sqrt{2}} & -\frac{1}{\sqrt{2}} \\ 1 & 0 & 0 & 0 \\ 0 & 1 & 0 & 0 \end{bmatrix}$$

9. $$\begin{bmatrix} \frac{1}{\sqrt{5}} & 0 & -\frac{2}{\sqrt{5}} & 0 \\ \frac{2}{\sqrt{5}} & 0 & \frac{1}{\sqrt{5}} & 0 \\ 0 & \frac{1}{\sqrt{5}} & 0 & -\frac{2}{\sqrt{5}} \\ 0 & \frac{2}{\sqrt{5}} & 0 & \frac{1}{\sqrt{5}} \end{bmatrix}$$

EXERCISE SET 7.1 (page 266).

1. (a) $y_1 = c_1e^{5x} - 2c_2e^{-x}$
$y_2 = c_1e^{5x} + c_2e^{-x}$

(b) $y_1 = 0$
$y_2 = 0$

2. (a) $y_1 = c_1e^{7x} - 3c_2e^{-x}$
$y_2 = 2c_1e^{7x} + 2c_2e^{-x}$

(b) $y_1 = -\frac{1}{40}e^{7x} + \frac{81}{40}e^{-x}$
$y_2 = -\frac{1}{20}e^{7x} - \frac{27}{20}e^{-x}$

3. (a) $y_1 = -c_2e^{2x} + c_3e^{3x}$
$y_2 = c_1e^{x} + 2c_2e^{2x} - c_3e^{3x}$
$y_3 = 2c_2e^{2x} - c_3e^{3x}$

(b) $y_1 = e^{2x} - 2e^{3x}$
$y_2 = e^{x} - 2e^{2x} + 2e^{3x}$
$y_3 = -2e^{2x} + 2e^{3x}$

4. $y_1 = (c_1 + c_2)e^{2x} + c_3e^{8x}$
$y_2 = -c_2e^{2x} + c_3e^{8x}$
$y_3 = -c_1e^{2x} + c_3e^{8x}$

5. $y = c_1e^{3x} + c_2e^{-2x}$

6. $y = c_1e^{x} + c_2e^{2x} + c_3e^{3x}$

EXERCISE SET 7.2 (page 273).

1. (a) $(1 + \pi) - 2 \sin x - \sin 2x$

(b) $(1 + \pi) - 2\left[\sin x + \dfrac{\sin 2x}{2} + \dfrac{\sin 3x}{3} + \cdots + \dfrac{\sin nx}{n}\right]$

2. (a) $\frac{4}{3}\pi^2 + 4 \cos x + \cos 2x + \frac{4}{9} \cos 3x - 4\pi \sin x - 2\pi \sin 2x - \dfrac{4\pi}{3} \sin 3x$

(b) $\frac{4}{3}\pi^2 + 4 \sum_{k=1}^{n} \dfrac{\cos kx}{k^2} - 4\pi \sum_{k=1}^{n} \dfrac{\sin kx}{k}$

3. $-\dfrac{1}{2} + \dfrac{1}{e - 1} e^x$

4. (a) $(4e - 10) + (18 - 6e)x$

5. (a) $\dfrac{3}{\pi} x$

8. $\sum_{k=1}^{\infty} \dfrac{2}{k} \sin (kx)$

EXERCISE SET 7.3 (page 283).

1. (a) $2x^2 - 3xy + 4y^2$ (b) $x^2 - xy$ (c) $5xy$ (d) $4x^2 - 2y^2$ (e) y^2

2. (a) $\begin{bmatrix} 2 & -\frac{3}{2} \\ -\frac{3}{2} & 4 \end{bmatrix}$ (b) $\begin{bmatrix} 1 & -\frac{1}{2} \\ -\frac{1}{2} & 0 \end{bmatrix}$

(c) $\begin{bmatrix} 0 & \frac{5}{2} \\ \frac{5}{2} & 0 \end{bmatrix}$ (d) $\begin{bmatrix} 4 & 0 \\ 0 & -2 \end{bmatrix}$ (e) $\begin{bmatrix} 0 & 0 \\ 0 & 1 \end{bmatrix}$

3. (a) $[x \quad y]\begin{bmatrix} 2 & -\frac{3}{2} \\ -\frac{3}{2} & 4 \end{bmatrix}\begin{bmatrix} x \\ y \end{bmatrix} + [-7 \quad 2]\begin{bmatrix} x \\ y \end{bmatrix} + 7 = 0$

(b) $[x \quad y]\begin{bmatrix} 1 & -\frac{1}{2} \\ -\frac{1}{2} & 0 \end{bmatrix}\begin{bmatrix} x \\ y \end{bmatrix} + [5 \quad 8]\begin{bmatrix} x \\ y \end{bmatrix} - 3 = 0$

(c) $[x \quad y]\begin{bmatrix} 0 & \frac{5}{2} \\ \frac{5}{2} & 0 \end{bmatrix}\begin{bmatrix} x \\ y \end{bmatrix} - 8 = 0$

(d) $[x \quad y]\begin{bmatrix} 4 & 0 \\ 0 & -2 \end{bmatrix}\begin{bmatrix} x \\ y \end{bmatrix} - 7 = 0$

(e) $[x \quad y]\begin{bmatrix} 0 & 0 \\ 0 & 1 \end{bmatrix}\begin{bmatrix} x \\ y \end{bmatrix} + [7 \quad -8]\begin{bmatrix} x \\ y \end{bmatrix} - 5 = 0$

4. (a) ellipse (b) ellipse (c) hyperbola (d) hyperbola
(e) circle (f) parabola (g) parabola (h) parabola
(i) parabola (j) circle

5. (a) $9x'^2 + 4y'^2 = 36$, ellipse
(b) $x'^2 - 16y'^2 = 16$, hyperbola
(c) $y'^2 = 8x'$, parabola
(d) $x'^2 + y'^2 = 16$, circle
(e) $18y'^2 - 12x'^2 = 419$, hyperbola
(f) $y' = -\frac{1}{7}x'^2$, parabola

6. (a) $2x'^2 - 3y'^2 = 8$, hyperbola
(b) $2\sqrt{2}x'^2 - 7x' + 9y' = 0$, parabola
(c) $7x'^2 + 3y'^2 = 9$, ellipse
(d) $4x'^2 - y'^2 = 3$, hyperbola

7. $x''^2 + 2y''^2 = 6$, ellipse

8. $13y''^2 - 4x''^2 = 81$, hyperbola

9. $2x''^2 - 3y''^2 = 24$, hyperbola

10. $6x''^2 + 11y''^2 = 66$, ellipse

11. $4y''^2 - x''^2 = 0$, hyperbola

12. $\sqrt{29}y'^2 - 3x' = 0$, parabola

13. (a) Two intersecting lines, $y = x$ and $y = -x$
(b) No graph
(c) The graph is the single point (0, 0)
(d) The graph is the line $y = x$
(e) The graph consists of two parallel lines $y' = 2$ and $y' = -2$
(f) The graph is the single point (1, 2)

EXERCISE SET 7.4 (page 289).

1. (a) $x^2 + 2y^2 - z^2 + 4xy - 5yz$
(b) $3x^2 + 7z^2 + 2xy - 3xz + 4yz$
(c) $xy + xz + yz$
(d) $x^2 + y^2 - z^2$
(e) $3z^2 + 3xz$
(f) $2z^2 + 2xz + y^2$

2. (a) $\begin{bmatrix} 1 & 2 & 0 \\ 2 & 2 & -\frac{5}{2} \\ 0 & -\frac{5}{2} & -1 \end{bmatrix}$ (b) $\begin{bmatrix} 3 & 1 & -\frac{3}{2} \\ 1 & 0 & 2 \\ -\frac{3}{2} & 2 & 7 \end{bmatrix}$

(c) $\begin{bmatrix} 0 & \frac{1}{2} & \frac{1}{2} \\ \frac{1}{2} & 0 & \frac{1}{2} \\ \frac{1}{2} & \frac{1}{2} & 0 \end{bmatrix}$ (d) $\begin{bmatrix} 1 & 0 & 0 \\ 0 & 1 & 0 \\ 0 & 0 & -1 \end{bmatrix}$

(e) $\begin{bmatrix} 0 & 0 & \frac{3}{2} \\ 0 & 0 & 0 \\ \frac{3}{2} & 0 & 3 \end{bmatrix}$ (f) $\begin{bmatrix} 0 & 0 & 1 \\ 0 & 1 & 0 \\ 1 & 0 & 2 \end{bmatrix}$

3. (a) $[x \;\; y \;\; z]\begin{bmatrix} 1 & 2 & 0 \\ 2 & 2 & -\frac{5}{2} \\ 0 & -\frac{5}{2} & -1 \end{bmatrix}\begin{bmatrix} x \\ y \\ z \end{bmatrix} + [7 \;\; 0 \;\; 2]\begin{bmatrix} x \\ y \\ z \end{bmatrix} - 3 = 0$

(b) $[x \;\; y \;\; z]\begin{bmatrix} 3 & 1 & -\frac{3}{2} \\ 1 & 0 & 2 \\ -\frac{3}{2} & 2 & 7 \end{bmatrix}\begin{bmatrix} x \\ y \\ z \end{bmatrix} + [-3 \;\; 0 \;\; 0]\begin{bmatrix} x \\ y \\ z \end{bmatrix} - 4 = 0$

(c) $[x \;\; y \;\; z]\begin{bmatrix} 0 & \frac{1}{2} & \frac{1}{2} \\ \frac{1}{2} & 0 & \frac{1}{2} \\ \frac{1}{2} & \frac{1}{2} & 0 \end{bmatrix}\begin{bmatrix} x \\ y \\ z \end{bmatrix} - 1 = 0$

(d) $[x \;\; y \;\; z]\begin{bmatrix} 1 & 0 & 0 \\ 0 & 1 & 0 \\ 0 & 0 & -1 \end{bmatrix}\begin{bmatrix} x \\ y \\ z \end{bmatrix} - 7 = 0$

(e) $[x \;\; y \;\; z]\begin{bmatrix} 0 & 0 & \frac{3}{2} \\ 0 & 0 & 0 \\ \frac{3}{2} & 0 & 3 \end{bmatrix}\begin{bmatrix} x \\ y \\ z \end{bmatrix} + [0 \;\; -14 \;\; 0]\begin{bmatrix} x \\ y \\ z \end{bmatrix} + 9 = 0$

(f) $[x \;\; y \;\; z]\begin{bmatrix} 0 & 0 & 1 \\ 0 & 1 & 0 \\ 1 & 0 & 2 \end{bmatrix}\begin{bmatrix} x \\ y \\ z \end{bmatrix} + [2 \;\; -1 \;\; 3]\begin{bmatrix} x \\ y \\ z \end{bmatrix} = 0$

4. (a) Ellipsoid. (b) Hyperboloid of one sheet. (c) Hyperboloid of two sheets.
(d) Elliptic cone. (e) Elliptic paraboloid. (f) Hyperbolic paraboloid.
(g) Sphere.

5. (a) $9x'^2 + 36y'^2 + 4z'^2 = 36$, ellipsoid
(b) $6x'^2 + 3y'^2 - 2z'^2 = 18$, hyperboloid of one sheet
(c) $3x'^2 - 3y'^2 - z'^2 = 3$, hyperboloid of two sheets
(d) $4x'^2 + 9y'^2 - z'^2 = 0$, elliptic cone
(e) $x'^2 + 16y'^2 - 16z' = 32$, hyperboloid of one sheet
(f) $7x'^2 - 3y'^2 + z' = 0$, hyperbolic paraboloid
(g) $x'^2 + y'^2 + z'^2 = 25$, sphere

6. (a) $25x'^2 - 3y'^2 - 50z'^2 - 150 = 0$, hyperboloid of two sheets
(b) $2x'^2 + 2y'^2 + 8z'^2 - 5 = 0$, ellipsoid
(c) $9x'^2 + 4y'^2 - 36z = 0$, elliptic paraboloid
(d) $x'^2 - y'^2 + z' = 0$, hyperbolic paraboloid

7. $2x''^2 - y''^2 - z''^2 = 1$, hyperboloid of two sheets

8. $x''^2 + y''^2 + 2z''^2 = 4$, ellipsoid

9. $x''^2 - y''^2 + z'' = 0$, hyperbolic paraboloid

10. $6x''^2 + 3y''^2 - 8\sqrt{2}z'' = 0$, elliptic paraboloid

EXERCISE SET 8.1 (page 296).

1. (a) $.28 \times 10^1$ (b) $.3452 \times 10^4$ (c) $.3879 \times 10^{-5}$ (d) $-.135 \times 10^0$ (e) $.17921 \times 10^2$ (f) $-.863 \times 10^{-1}$

2. (a) $.280 \times 10^1$ (b) $.345 \times 10^4$ (c) $.388 \times 10^{-5}$ (d) $-.135 \times 10^0$ (e) $.179 \times 10^2$ (f) $-.863 \times 10^{-1}$

3. (a) $.28 \times 10^1$ (b) $.35 \times 10^4$ (c) $.39 \times 10^{-5}$ (d) $-.14 \times 10^0$ (e) $.18 \times 10^2$ (f) $-.86 \times 10^{-1}$

4. $x_1 = -3, x_2 = 7$

5. $x_1 = \frac{13}{8}, x_2 = \frac{14}{8}, x_3 = \frac{21}{8}$

6. $x_1 = 1, x_2 = 2, x_3 = 3$

7. $x_1 = 0, x_2 = 0, x_3 = 1, x_4 = -1$

8. $x_1 = .997, x_2 = 1.00$

9. $x_1 = -2, x_2 = 0, x_3 = 1$

10. $x_1 = 0, x_2 = 1$ (without pivoting); $x_1 = 1, x_2 = 1$ (with pivoting)

EXERCISE SET 8.2 (page 302).

1. $x_1 \approx 2.81, x_2 \approx .94$; exact solution is $x_1 = 3, x_2 = 1$

2. $x_1 \approx .954, x_2 \approx -1.90$; exact solution is $x_1 = 1, x_2 = -2$

3. $x_1 \approx -2.99, x_2 \approx -.998$; exact solution is $x_1 = -3, x_2 = -1$

4. $x_1 \approx 0.00, x_2 \approx 1.98$; exact solution is $x_1 = 0, x_2 = 2$

5. $x_1 \approx 3.03, x_2 \approx 1.02$; exact solution $x_1 = 3, x_2 = 1$

6. $x_1 \approx 1.01, x_2 \approx -2.00$; exact solution is $x_1 = 1, x_2 = -2$

7. $x_1 \approx -3.00, x_2 \approx -1.00$; exact solution is $x_1 = -3, x_2 = -1$

8. $x_1 \approx .005; x_2 \approx 2.00$; exact solution is $x_1 = 0, x_2 = 2$

9. $x_1 \approx .492, x_2 \approx .006, x_3 \approx -.996$; exact solution is $x_1 = \frac{1}{2}, x_2 = 0, x_3 = -1$

10. $x_1 \approx 1.00, x_2 \approx .998, x_3 = 1.00$; exact solution is $x_1 = 1, x_2 = 1, x_3 = 1$

11. $x_1 \approx .499, x_2 \approx .0004, x_3 \approx -1.00$; exact solution is $x_1 = \frac{1}{2}, x_2 = 0, x_3 = -1$

12. $x_1 \approx 1.00, x_2 \approx 1.00, x_3 \approx 1.00$; exact solution is $x_1 = 1, x_2 = 1, x_3 = 1$

13. *a, d, e*

EXERCISE SET 8.3 (page 310).

1. (a) $\lambda = 3$ (b) No dominant eigenvalue (c) $\lambda = 6$ (d) $\lambda = 3$

2. (a) $\begin{bmatrix} 1.00 \\ .503 \end{bmatrix}$ (b) 5.02

(c) The dominant eigenvector is $\begin{bmatrix} 1 \\ \frac{1}{2} \end{bmatrix}$; the dominant eigenvalue is 5

(d) The percentage error is .4%

3. (a) $\begin{bmatrix} 1.00 \\ .750 \end{bmatrix}$ (b) 8.01

(c) The dominant eigenvector is $\begin{bmatrix} 1 \\ \frac{3}{4} \end{bmatrix}$; the dominant eigenvalue is 8

(d) The percentage error is .125%

4. (a) $\begin{bmatrix} 1.00 \\ -.560 \end{bmatrix}$ (b) -4.00

(c) The dominant eigenvector is $\begin{bmatrix} 1 \\ -\frac{1}{2} \end{bmatrix}$; the dominant eigenvalue is -4

(d) The percentage error is 0%

5. (a) At the end of two iterations the dominant eigenvalue and eigenvector are approximately $\lambda_1 \approx 20.1$ and $\mathbf{x} \approx \begin{bmatrix} 1 \\ .119 \end{bmatrix}$

(b) The exact values of the dominant eigenvalue and eigenvector are $\lambda_1 = 20$ and $x = \begin{bmatrix} 1 \\ \frac{2}{17} \end{bmatrix}$

6. (a) At the end of three iterations the dominant eigenvalue and eigenvector are approximately $\lambda_1 \approx -9.95$ and $\mathbf{x} \approx \begin{bmatrix} -.978 \\ 1 \end{bmatrix}$

(b) The exact values of the dominant eigenvalue and eigenvector are $\lambda_1 = -10$ and $\mathbf{x} = \begin{bmatrix} -1 \\ 1 \end{bmatrix}$

7. (a) $\begin{bmatrix} .027 \\ .027 \\ 1 \end{bmatrix}$ (b) 10.0

(c) The dominant eigenvector is $\begin{bmatrix} 0 \\ 0 \\ 1 \end{bmatrix}$; the dominant eigenvalue is 10

EXERCISE SET 8.4 (page 314).

1. (a) $\begin{bmatrix} 1 \\ .509 \end{bmatrix}$ (b) 7.00 (c) $\lambda_2 \approx 2.00$, $\mathbf{v}_2 \approx \begin{bmatrix} -.51 \\ 1 \end{bmatrix}$

(d) Exact eigenvalues 7, 2; exact eigenvectors

$$\mathbf{v}_1 = \begin{bmatrix} 1 \\ \frac{1}{2} \end{bmatrix}, \qquad \mathbf{v}_2 = \begin{bmatrix} -\frac{1}{2} \\ 1 \end{bmatrix}$$

2. (a) $\begin{bmatrix} 1 \\ .503 \end{bmatrix}$ (b) 12.0 (c) $\lambda_2 \approx 2.02$, $\mathbf{v}_2 \approx \begin{bmatrix} -.532 \\ 1 \end{bmatrix}$

(d) Exact eigenvalues 12, 2; exact eigenvectors

$$\mathbf{v}_1 = \begin{bmatrix} 1 \\ \frac{1}{2} \end{bmatrix}, \qquad \mathbf{v}_2 = \begin{bmatrix} -\frac{1}{2} \\ 1 \end{bmatrix}$$

Index